Student Supplement
for

CALCULUS
One and Several Variables
Third Edition

S. L. Salas   **Einar Hille**

by

John T. Anderson
Gordon D. Prichett
Hamilton College

JOHN WILEY & SONS, New York • Chichester • Brisbane • Toronto

$11.25E

ISBN 0 471 02882 7
Printed in the United States of America

10 9 8 7 6

Preface

These materials have been classroom tested at Hamilton College since 1972. The main body of the text is organized to parallel the Section numbers of the text by Salas/Hille. In some instances several Sections are treated together. A few Sections of the text (primarily those consisting strictly of exercises) are not treated in the Supplement. There are five occasions where comments did not properly fit under any particular Section number. Separate headings such as "Further Comments on Integration" are used. Extra exercises with solutions are provided for thirteen Sections. These Sections are marked by an asterisk in the Table of Contents.

This Student Supplement is written with the student in mind. The general style is conversational but not at the expense of the mathematics. In some Sections the primary purpose is to provide alternative explanations and perspectives on a given topic. In other Sections the discussion consists primarily of additional examples. Sometimes the commentary consists of solutions for exercises from the Salas/Hille text. Such exercises are denoted by an asterisk in the lists of Suggested Exercises appearing in the Unit Study Guides at the back of this book. An effort has been made throughout to focus the student's attention on the central ideas of the text and to suggest techniques of study and problem-solving that will help him or her master the material.

The Unit Study Guides at the end of the book are set up to facilitate self-study. This also enables these materials to be readily adapted to a self-paced (P.S.I.) course. A glance at these Unit Study Guides will show that each is made up of five parts. First there is a statement of Objectives which specifies the topics and techniques to be mastered. This is followed by the Procedures which outline the recommended means for meeting these objectives together with a list of the relevant Text Readings. The final two parts consist of Suggested Exercises from the text and a Sample Quiz. We have not provided solutions for the quizzes, for then they would have become but additional examples. In many instances the quizzes are more demanding than those we use for testing purposes. In this way a student who can complete the quiz is assuredly prepared for a regular quiz. Students using these materials would be well advised to ascertain if these Study Guides properly reflect the coverage of their course. Similarly, an instructor should note if the Objectives reflect his intended emphases and if the quizzes are close to his intended level of testing.

The thirty-six Unit Study Guides cover all nineteen Chapters of the Salas/Hille text. Each Unit covers roughly one week's work (four classes). Of course, this is but an approximation and an individual instructor may see fit to vary the time devoted to a particular Unit. The Units were set up according to material that might be covered on a 20-30 minute quiz. We have found Units 1-12 and 13-24 to fit nicely into two thirteen week semesters. Instructors interested in self paced instruction may begin by writing to the Center for Personalized Instruction, Georgetown University, Washington, D.C. 20057. Further, there are many programs throughout the country and an instructor would do well to visit at least one near him to observe such a program first-hand. An instructor wishing to use these materials for a self-paced program should carefully review and possibly revise our Study Guides to meet his purposes. Finally, anyone interested in our program may consult a report in the March 1977 Proceedings of the Rocky Mountain Mathematics Consortium or write to us directly at the Mathematics Department, Hamilton College, Clinton, New York 13323.

We are particularly grateful to William Reynolds (SUNY at Cortland) for an exhaustive review of the Preliminary Edition. We are indebted to our colleagues (David Birnbaum, Northrup Fowler, Anne Ludington, and David Smallen) and many of our students for their constructive criticisms of preliminary drafts. We wish to thank Shirley Knop for her expert proofreading of the final copy.

We owe much to Gary W. Ostedt, Editor at Wiley, and to S. L. Salas for their encouragement and close cooperation in coordinating our efforts with the revision of the Salas/Hille text.

It is a pleasure to acknowledge the tireless and expert typing of Mrs. June Darrow throughout this entire project.

Finally, we owe an immeasurable debt to our wives, Jacqueline and Jill, as well as our sons for their patience and understanding throughout the six years that have gone into this project.

John T. Anderson
Gordon D. Prichett

Hamilton College
Clinton, New York

Contents

Introduction

1

All students suffer to some degree from that old bane of the careless error. We have but two simple suggestions on this point. First, try to learn from your mistakes. For example, if you misapply the distributive law or the law of exponents as erroneously symbolized by $a(b + c) = ab + c$ and $x^{ab} = x^a x^b$, then make conscious note of your error. The correct statements here are $a(b + c) = ab + ac$ and $x^{a + b} = x^a x^b$. Second, to guard against the careless error, try to be neater, definitely try to be neater. Just think - to what extent is a sloppy paper a function of a sloppy mind?

When trying to comprehend any abstract concept, mathematical or otherwise, one must have a precise understanding of the meaning of the words involved. You will find it much easier to read the text and this Supplement if you learn, in one fashion or another, the definitions of the new words or phrases. Memorizing these few definitions will enhance your chances of understanding the material. The definitions are the building blocks - learn them.

One of the most remarkable theorems of elementary algebra is:

<u>Remainder Theorem.</u> <u>For any real number a and any polynomial p(x), if p(x) is</u>
<u>divided by x - a, the remainder is p(a).</u>

<u>Example 1.</u> What is the remainder when $x^4 + 2x^3 + x + 7$ is divided by x - 2?
<u>Solution.</u> The theorem tells us that $2^4 + 2(2)^3 + 2 + 7 = 41$ is the remainder. Otherwise, we need to execute the long division

1

$$x - 2 \overline{\smash{\big)}\ \begin{array}{l} x^3 + 4x^2 + 8x + 17 \qquad R\ 41 \\ x^4 + 2x^3 \qquad\qquad + x + 7 \end{array}}\ ,$$

which you should verify.

Example 2. What is the remainder when $p(x) = x^4 + 2x^3 + x + 7$ is divided by $x + 1$?

Solution. Since $x + 1 = x - (-1)$, the remainder is $(-1)^4 + 2(-1)^3 + (-1) + 7 = 5$.

Why would such an unexpected result be true? The primary purpose of proofs of theorems is to answer this question. Let us prove the Remainder Theorem.

Proof of the Remainder Theorem. If we divide $p(x)$ by $x - a$, we arrive at the quotient $q(x)$ and a remainder R. Thus,

$$p(x) = q(x) \cdot (x - a) + R. \tag{1}$$

(dividend is divisor times quotient plus remainder)

Substituting a for x in equation (1), we find that

$$p(a) = q(a) \cdot 0 + R = R$$

and as simply as that the remainder is $p(a)$.   Q.E.D.

A very special case of the Remainder Theorem is:

Factor Theorem. For any real number a and any polynomial $p(x)$, $x - a$ is a factor of $p(x)$ if and only if $p(a) = 0$.

The Factor Theorem is the Remainder Theorem in the special case when R = 0. Usually a theorem which is a special case of another theorem is called a corollary, but the Factor Theorem is so important it maintains its own name.

Example 3. Simplify $\dfrac{x^3 - 1}{x^2 - 1} = \dfrac{f(x)}{g(x)}$.

Solution. Since $f(1) = g(1) = 0$, $x - 1$ must be a factor of both numerator and denominator. Hence,

$$\frac{x^3 - 1}{x^2 - 1} = \frac{(x-1)(x^2 + x + 1)}{(x-1)(x + 1)} = \frac{x^2 + x + 1}{x + 1}$$

If we could simplify further, then $x + 1 = x - (-1)$ must be a factor of

$x^2 + x + 1$. But, (-1) is not a root of $x^2 + x + 1 = 0$, so we have put the expression in simplest form.

EXERCISES.

1. Find the remainders for the indicated divisions.

   (a) $(x^3 - 3x^2 + 6x - 5) \div (x - 3)$

   (b) $(x^3 + x^2 - 5x + 3) \div (x + 3)$

   (c) $(x^4 - 2x^3 - 3x^2 - 4x - 8) \div (x - 2)$

   (d) $(x^4 - 2x^3 - 3x^2 - 4x - 8) \div (x + 1)$

   (e) $(9x^3 + 6x^2 + 4x + 2) \div (3x + 1)$

2. Find some value of k for which the following statements are valid.

   (a) $x^{100} + kx + 7$ is divisible by $x + 1$

   (b) $x^3 + kx^2 + 2x - 1$ has remainder 125 when divided by $x + 2$

3. For which positive integral values of n, if $a \neq 0$, is

   (a) $x^n - a^n$ divisible by $x - a$

   (b) $x^n - a^n$ divisible by $x + a$

   (c) $x^n + a^n$ divisible by $x - a$

   (d) $x^n + a^n$ divisible by $x + a$

4. Factor completely the following polynomials.

   (a) $x^3 - 8$ (e) $x^3 + 27$

   (b) $x^4 - 1$ (f) $x^4 - 18x^2 + 81$

   (c) $x^3 + 4x^2 - 5x$ (g) $x^3 - 7x + 6$

   (d) $x^3 + 6x^2 + 12x + 8$ (h) $x^4 + 1$

SOLUTIONS.

1. (a) 13   (b) 0   (c) -28   (d) -4   (e) 1/3
2. (a) 8   (b) 69/2
3. (a) All n   (b) Even n   (c) No n   (d) Odd n
4. (a) $(x - 2)(x^2 + 2x + 4)$ (e) $(x + 3)(x^2 - 3x + 9)$
   (b) $(x - 1)(x + 1)(x^2 + 1)$ (f) $(x - 3)^2(x + 3)^2$
   (c) $x(x - 1)(x + 5)$ (g) $(x - 1)(x - 2)(x + 3)$
   (d) $(x + 2)^3$ (h) $x^4 + 1$

SECTION 1.3

In this section we consider some more problems involving lines. You should try
to solve the problems yourself before reading the solution.

Example 1. Write an equation for the line ℓ_2 that passes through the point
(-2,4) and is parallel to the line ℓ_1 which passes through the points (1,-3)
and (-2,6).
Solution. The slope of line ℓ_1 is $\frac{(-3) - 6}{1 - (-2)} = -3$. Since ℓ_1 and ℓ_2 are parallel,
the slope of ℓ_2 must also be -3. Using the point-slope form, we may now write
its equation as $y - 4 = -3(x + 2)$.

Example 2. Write an equation for the line ℓ which is perpendicular to the line
$x + 2y - 3 = 0$ and has y-intercept of 3.
Solution. We may write $x + 2y - 3 = 0$ as $y = -\frac{x}{2} + \frac{3}{2}$. The slope of this line
is - 1/2, so the slope of line ℓ is 2. Any line with slope of 2 may be written
as $y = 2x + b$ where b is the y-intercept. The equation for line ℓ is $y = 2x + 3$.

Example 3. Write an equation for the line ℓ which is parallel to the line
$y = 2x - 4$ and passes through the points (1,3) and (3,5).

Solution.   The slope of the line $y = 2x - 4$ is 2.   Using the point $(1,3)$ we may write an equation for line ℓ as $y - 3 = 2(x - 1)$ or $y = 2x + 1$.   But, notice that the point $(3,5)$ does not lie on this line: $5 \neq 2\cdot 3 + 1$.   Our algebra is correct.   The solution to this problem is not the equation of a line but a statement to the effect that there does not exist a line which meets the given conditions.   Verify that the slope of the line through $(1,3)$ and $(3,5)$ is 1, not 2.

Example 4.   Find an equation for the perpendicular bisector of the line segment joining the points $A(-4,1)$ and $B(-1,-5)$.

Solution.   The slope of the line passing through A and B is -2.   The midpoint of the line segment joining A and B has coordinates $(-\frac{5}{2}, -2)$.   An equation for the desired line is $y + 2 = \frac{1}{2}(x + \frac{5}{2})$.

SECTIONS 1.4-5

The solution of inequalities is not very different from the solution of equations. In fact, your algebraic manipulations are unaffected by the presence of "<" rather than "=" except for one very important and frequently overlooked occasion.   If an unequality is multiplied (or divided) by a negative number, then the inequality must be reversed.   That is, if $a < b$ and $c < 0$, then $ac > bc$. For example, $2 < 7$ and $-3 < 0$, but it isn't true that $-6 < -21$.   Other than this one point, you may solve inequalities as if they were equalities.   A few examples in addition to those in the text may help.

Example 1.   Solve the inequality $2x^2 + 3x + 1 > 0$

Solution. $2x^2 + 3x + 1 = (2x + 1)(x + 1) > 0$ is true if and only if $2x + 1$ and $x + 1$ have the same sign.   If both factors are positive, then $x > -\frac{1}{2}$ and $x > -1$. Consequently, $x > -\frac{1}{2}$.   On the other hand, if $2x + 1 < 0$ and $x + 1 < 0$, then $x < -\frac{1}{2}$ and $x < -1$.   So, $x < -1$.   The full solution is the union of these two solutions: $(-\infty, -1) \cup (-\frac{1}{2}, \infty)$ or $\{x : x$ is not in $[-1, -\frac{1}{2}]\}$.

<u>Example 2.</u> Solve the inequality $\frac{x - 2}{x - 5} > 3$.

<u>Solution.</u> The obvious first step is to multiply by x - 5. However, x - 5 may be positive or negative. We need to consider these two possibilities (x > 5 or x < 5) separately, for the sense (> or <) of the inequality is then affected by multiplication.

If x > 5, then x - 2 > 3(x - 5). Subject to the restriction x > 5, we find x < 13/2. The only values which satisfy both of these restrictions lie in the interval (5, 13/2). We could represent this result graphically.

x > 5

and x < 13/2

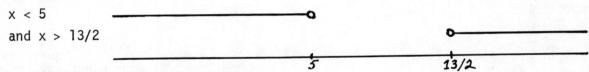

yields 5 < x < 13/2.

If x < 5, then x - 2 < 3(x - 5). Thus, if x < 5, we have x > 13/2. No value of x meets both of these restrictions.

Graphically,

x < 5

and x > 13/2

yields no common intersection.

The final solution is $(5, 13/2) \cup \emptyset = (5, 13/2)$ or $\{x : 5 < x < 13/2\}$.

<u>Example 3.</u> Solve for real x: $\frac{3}{-x^2} < 4$

<u>Solution.</u> Even though we multiply by a variable here as in the last example, we don't have two cases since $-x^2$ is never positive. Therefore, if $\frac{3}{-x^2} < 4$, $3 > 4(-x^2)$, $-x^2 < \frac{3}{4}$ and $x^2 > -\frac{3}{4}$. But the square of any real number is non-negative. The solution set here is all real numbers other than zero. Zero is excepted since $\frac{3}{-x^2} < 4$ and $x^2 > -\frac{3}{4}$ are <u>not</u> identical and zero is not a solution of the <u>original</u> problem.

<u>Example 4.</u> Solve for real x: $|3x - 2| \leq 5$.

<u>Solution.</u> Recall the definition of absolute value. We wish to determine all x such that 3x - 2 is within 5 units of the origin; the magnitude of 3x - 2 is

not more than 5.   Hence, $-5 \leq 3x - 2 \leq 5$ or $-3 \leq 3x \leq 7$.   So the solution set
is the interval [-1, 7/3].

Example 5.   Solve for real x: $|4x + 3| > 2$.
Solution.   Here we want to find all real numbers x such that $4x + 3$ is further
than 2 units from the origin;   the magnitude of $4x + 3$ is greater than 2.   Hence,
$4x + 3 < -2$ or $4x + 3 > 2$.   Thus, x is in $(-\infty, -\frac{5}{4})$ or x is in $(-\frac{1}{4}, \infty)$.

Example 6.   Solve for real x: $(x - 1)(x + 2)(x - 3)(x - 2)^2 < 0$.
Solution.   The polynomial on the left hand side has 5 linear factors.   The
inequality is true if and only if an odd number of these 5 factors are negative.
You should verify that the solution set as exhibited on the number line below is
correct.

SECTIONS 1.6-8

Much of the work in this course will be formulated directly or indirectly in
terms of functions.   It is critical that you are confident of your understanding
of the function associated terms: domain, range, one-to-one, inverse, and
composition.   These terms are clearly exampled and defined in the text.

It is possible that you may have seen functions defined in terms of relations
or sets of ordered pairs.   Descriptively, a relation is a collection of points in
the xy-plane;   the collection of first coordinates (x) constitutes the domain and
the collection of second coordinates (y) makes up the range.   We may then define
a function as follows: A relation f is a function if and only if no element in
the domain of f corresponds to (is related to) more than one element in the
range of f.

Thus each domain value specifies a single (unique) range value.   This
statement should be compared to the requirements for the more restrictive one-to-
one functions which demand in addition that each range value is specified by a
unique domain value.

The inverse f^{-1}, if it exists, simply interchanges the roles of domain and

range. Hence, it is very easy to see why a function f has an inverse if and only
if f is one-to-one. If f were many-to-one, then f^{-1} could not be a function.

This notion of interchange suggests a graphical test for one-to-oneness
and inverses. Remember that a function by definition is many-to-one, where
"many" should be read as one or more. Thus, the relation f is a function if and
only if no vertical line intersects the graph of f in more than one point. The
corresponding test for the existence of f^{-1} is made with horizontal lines.

It is preferable to discuss functions rather than relations because for each
value x in the domain of f, the range value f(x) is unique. Relations lack this
requirement of uniqueness. This uniqueness permits us to characterize range
values of a function as largest, smallest, or nicest without being ambiguous.

For emphasis, we note that a function is specified by a rule _and_ a domain
on which that rule is in force. The functions f, g, and h

$$f(x) = x^2 \text{ for } x > 0$$

$$g(x) = x^2 \text{ for } x < 1$$

$$h(x) = x^3 \text{ for } x < 1,$$

are different functions. The domain, if not stated explicitly, consists of all
values for which the given function rule makes sense. So the domain of
$F(x) = \sqrt{1 + x}$ is $x \geq -1$.

The notion of composite functions is simple as long as you remember that x
and the other symbols are just _dummy_ variables. That is, if $f(x) = x^3 + 2$ and
$g(x) = x^2 + 1$, then $f(y) = y^3 + 2$, $f(t) = t^3 + 2$, $f(g(x)) = (g(x))^3 + 2$, and
$f(g(t)) = (g(t))^3 + 2$. All express the same relationship between the domain and
range values of the function f. Namely, the domain value, whether it is x, y, t,
g(x) or g(t), is cubed and then increased by 2 to yield the proper range value
of f.

The composition of two functions is sometimes confused with the product of
two functions. But, the composition: $(f \circ g)(x) = f(g(x))$, and the product:
$(fg)(x) = f(x)g(x)$, are notationally and conceptually quite different. For
example, if we take $f(x) = x - 2$ and $g(x) = x + 5$, then

$$(f \circ g)(x) = f(g(x)) = (x + 5) - 2 = x + 3$$

whereas $$(fg)(x) = (x - 2)(x + 5) = x^2 + 3x - 10.$$

Finally, the most persistent problem which may plague your work, even after you are comfortable intuitively with ideas like inverses and composites, involves the notation itself. A function as denoted by the symbol f represents a relationship between two sets of numbers. The symbol f(x) strictly denotes elements from the second of these two sets, the range of f. The symbol f(x) may present some manipulative problems unless you carefully convince yourself at this time that each of the following statements about $f(x) = x^2 - 2$ is true.

$f(2) = 2$ $f(-3) = 7$

$f(y) = y^2 - 2$ $f(t) = t^2 - 2$

$f(y + 2) = (y + 2)^2 - 2$ $f(2x) = (2x)^2 - 2$

$\qquad = y^2 + 4y + 2$ $\qquad = 4x^2 - 2$

$f(f(x)) = (f(x))^2 - 2 = (x^2 - 2)^2 - 2 = x^4 - 4x^2 + 4 - 2 = x^4 - 4x^2 + 2$

$f(f(y)) = y^4 - 4y^2 + 2$

$f(2 + f(3)) = f(2 + 7) = f(9) = 79$

$f(\sqrt{y}) = (\sqrt{y})^2 - 2 = y - 2$

$f(x + h) = (x + h)^2 - 2$ <u>not</u> $x^2 + h - 2$

Limits and Continuity

2

Freshman Calculus is centered about the development and application of the two processes of differentiation and integration. A necessary prerequisite to these is an intuitive understanding of the concept of a limit. Further, it is necessary to develop some skills in evaluating limits and the ability to read, if not manipulate, the attending symbolism. The purpose of this introductory commentary is to lay that intuitive groundwork. The succeeding commentaries will treat the symbolic and mechanical sides of the topics of limits and continuity.

Let us begin by considering the following question. What characterizes those functions whose graphs may be drawn without lifting the pencil from the paper? Like all the other sciences, Mathematics is empirical. Even though our results are established within a strict logical framework of axioms and theorems, the formulation of these results is motivated by the examination of examples. These examples in themselves do not constitute a proof. Our empirical evidence, however, may well suggest how we may sharpen and rephrase our questions and interests, what further types of examples we might want to examine, what general results seem to hold under certain given conditions, and how we might proceed to establish these results as theorems. Throughout all our empirical searches, we are always asking what differences and similarities are exhibited by our examples.

As our first example, let us ask if we may draw the graph of the function
$$f(x) = 2x - 1 \qquad \text{for } 1 \le x \le 3$$
without lifting the pencil from the paper. More likely than not you respond

10

in the affirmative, since the graph of f is the line segment joining the points (1,1) and (3,5). (See Diagram 1.)

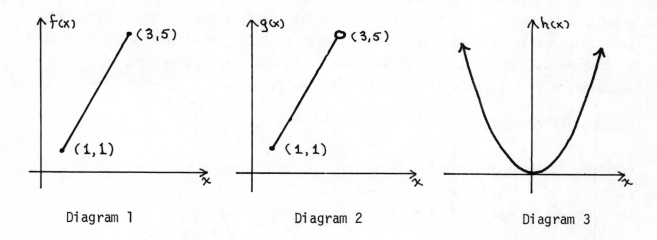

Diagram 1   Diagram 2   Diagram 3

Let us consider a very slight variation on our first example. Suppose that $g(x) = 2x - 1$ for $1 \leq x < 3$. Try to draw the graph of this function. In Diagram 2 we have a picture of this function, but have we drawn the graph of the function g in a "single stroke"? The picture implies the answer is yes, but is that open circle part of the graph? The open circle is used to indicate that the graph of g consists of the line segment joining the points (1,1) and (3,5) with the point (3,5) excluded. The open circle is, properly speaking, not part of the function but rather a tool of representation. To "draw" the graph of g, we set our pencil down at the point (1,1) and proceed along the line $y = 2x - 1$ towards the point (3,5). We will have to lift the pencil from the paper at some point (say, $(a, 2a - 1)$) <u>before</u> we get to (3,5). Thus, we will have drawn the graph of the closed line segment from (1,1) to $(a, 2a - 1)$ rather than the graph of g. The points between $(a, 2a - 1)$ and (3,5), no matter how close a is to 3, will have been omitted. Thus, we can <u>not</u> draw the graph of g without lifting the pencil from the paper, even though we give a picture (Diagram 2) that is a suggestive representation of g.

Consider the familiar parabola $h(x) = x^2$ defined for all real numbers x. We can not <u>physically</u> draw the graph of h either, even though all of us would accept Diagram 3 as representative of the principal behavior of the function h.

In each of these three examples, the picture has no apparent breaks or
"discontinuities" although some may and some may not be drawn in a single stroke.
Evidently our question needs to be rephrased since we would like to say each
graph may be drawn in a single stroke. Rather than talking about drawing the
graph, we really want to characterize those functions whose graphs are connected
or continuous. We want to exclude those instances where jumps or gaps occur, as
exemplified in Diagram 4 at the points x = A, B, C, D, E.

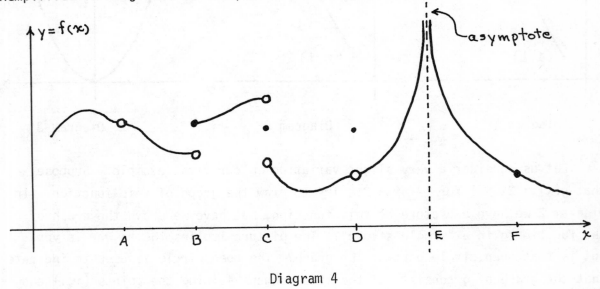

Diagram 4

Carefully examine those instances in Diagram 4 of disconnected or
discontinuous behavior. If x = A or E, there is a clear jump, gap, or break in
the graph simply because the function is not defined at that point. At
x = B, C and D the function is defined (heavy dot), but the curve does not
"pass through" the given point in the same way that it does at x = F. We are
asking, then, how we can distinguish the behavior of the function at x = F from
that at the other labelled points?

One might suggest that one can always determine from the graph when a
function behaves as it does at x = F and when it does not, but the graph of a
function is never more than a suggestive picture of predetermined characteristics
of the function. We must then learn to identify behaviors like those
represented at x = A, B, C, D, E, and F before we may draw or sketch them. To do
this we must refer to the function itself.

Since it is easy enough to spot cases like x = A or E by merely locating the values of x where the rule y = f(x) is not defined (for example, x = 0 in y = 1/x and x = -2 in y = √x), let us concern ourselves with the remaining three cases. The domain of the function in Diagram 5 is the closed interval [1,6],

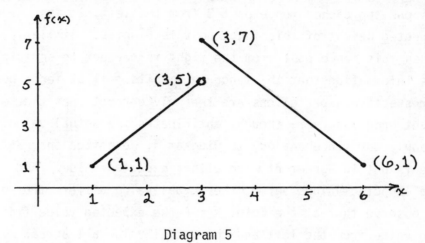

Diagram 5

since f(x) is defined for 1 ≤ x ≤ 6. As we move from left to right through x = 3 we are tempted to say that the graph jumps (up) at x = 3. (Technically, we will call x = 3 a "jump discontinuity.")

Does the graph jump in a physical sense? The answer is <u>no</u> since a jump must have a launching pad or point and a landing point. Remembering our discussion of Diagram 2, there can be no launching point here. What we are really saying, as we did for Diagram 2, is that there is no such thing as the largest real number less than 3. For the integers, it is perfectly correct to talk about consecutive numbers. The next integer after 3 is 4 and the largest integer less than 3 is 2. 3 is the only integer between 2 and 4. The rational numbers, as well as the real numbers, don't have this property of the integers.

Rather than jumping at x = 3 in a physical sense, the graph of f coasts along towards the point (3,5) from the left fully expecting to find itself at f(3) = 5. It is suddenly surprised upon arriving at x = 3 to find that f(3) = 7. This is contrary to the evidence that as x approached 3 from the left, f(x) was approaching 5. Notice that as the function coasts in from the right it expects to find f(3) = 7 and is rewarded as a good predictor since the difference between f(3) and f(x) grows ever so small as the difference x - 3 decreases to zero.

Let us look at the situation in Diagram 5 by way of a physical analogy. Suppose f(x) represents the height of a fence with fence posts at x = 1, 2, 3, 4, 5, 6. As one approaches fence post 3 from the left, it is correct to _infer_ that the _expected_ height of this fence post is 5 units. Similarly, an observer who approaches this fence post from the right is correct to anticipate on the basis of his information that the fence is 7 units tall at fence post 3. Notice that these respective expectations are the _only_ correct ones that may be drawn. Only the right hand observer, though, anticipates the _actual_ value.

If we apply our fence analogy to Diagram 4, we notice that at x = C the _actual value_ is not in agreement with either _expected value_. And, at x = D, although the expected values agree, the actual value differs from both of them. Finally, we observe that at the point x = F the expected value from the right, the expected value from the left and the actual value all agree. This last sentence translates into mathematical symbols as follows:

$$\lim_{x \uparrow F} f(x) = \lim_{x \downarrow F} f(x) = f(F).$$

Thus, for Diagram 5,

$$\lim_{x \uparrow 3} f(x) = 5 \qquad \text{and} \qquad \lim_{x \downarrow 3} f(x) = 7.$$

Formally, we will say that "the limit of f(x) from the left at x = 3 is 5" and "the limit of f(x) from the right at x = 3 is 7." In the event that the limit of f(x) from the left at some point x = s agrees with the limit of f(x) from the right at x = s, then we say that _the limit_ of f(x) at x = s exists and we can write $\lim_{x \to s} f(x)$ to denote this shared value. If observed values of f(x) both to the left and to the right but near x = s lead us to anticipate the same expected value for f(s) from either side, then $\lim_{x \to s} f(x)$ is this shared anticipated value for f(s). We may describe $\lim_{x \to s} f(x)$ as _the_ expected value of f(s).

Suppose now that $\lim_{x \to s} f(x) = f(s)$; that is, "the expected value for f(x) at x = s" equals "the actual value of f(x) at x = s." We will then say formally that f is continuous at x = s. In the same way that we have one-sided limits (left or right), we will talk about one-sided continuity. Diagram 5 indicates that f is continuous at x = 2 since $\lim_{x \uparrow 2} f(x) = \lim_{x \downarrow 2} f(x) = f(2) = 3$. Also, f is continuous from the right at x = 3 since $\lim_{x \downarrow 3} f(x) = f(3) = 7$, but f is not continuous from the left at x = 3 since $\lim_{x \uparrow 3} f(x) = 5 \neq 7 = f(3)$. Therefore, f is not continuous at x = 3.

We have rushed ahead to continuity in these preliminary remarks in part to simplify your understanding of the evaluation of limits. In most of the cases we will examine, we evaluate $\lim_{x \downarrow s} f(x)$, $\lim_{x \uparrow s} f(x)$ and $\lim_{x \to s} f(x)$ by breaking f(x) into known continuous parts and then evaluating these parts at x = s by substitution. To evaluate the limit of f(x) at x = s, we assume, when we may, that f is continuous at x = s and assign f(s) as the value of the limit. This rule of thumb is only implicit in most presentations on limits and continuity. Its apparent circularity suggests one reason why students experience unnecessary difficulties with the topics of limits and continuity. That is, to evaluate limits we assume continuity where possible, but to answer questions on continuity we need to evaluate limits. The types of limits you will probably encounter in this course are classified in Sections 2.3-4 of the Supplement with suggestions as to how they might be handled. These hints really amount to statements revealing when you may evaluate a limit by substitution (with the implicit assumption of continuity) and when you must do something else.

These preliminary comments were intended as an intuitive introduction to Chapter 2 rather than as a substitute for it. You probably hold some questioning uneasiness about the many ideas sketched above. Because the ground has now been broken for these important concepts, your concerns, which would have arisen anyway, should be more easily alleviated. Now that you have entered the tunnel of "underground Mathematics," it is time to steel your nerves and forge ahead to the real thing.

SECTIONS 2.1-2

Definition: If for each $\epsilon > 0$, there exists a $\delta > 0$ such that

$$0 < |x - c| < \delta \text{ implies } |f(x) - L| < \epsilon \tag{1}$$

then the limit of f(x) at x = c is L ($\lim_{x \to c} f(x) = L$), and conversely.
The text version (p. 46) of (1) is as follows:

$$\lim_{x \to c} f(x) = L \qquad \text{iff} \begin{cases} \text{for each } \epsilon > 0, \text{ there exists } \delta > 0 \\ \qquad \text{such that} \\ \text{if } 0 < |x - c| < \delta, \text{ then } |f(x) - L| < \epsilon \end{cases} \tag{2}$$

Statements (1) and (2) are equivalent. Many other formulations of the definition of a limit are possible. In all cases, it should first be noted that we have a definition. The <u>defined term</u> and the <u>defining phrase</u> are interchangeable. This is noted by the closing phrase "and conversely" in (1) and the "iff" (if and only if) in (2). A mathematical definition, like any definition, is actually just the presentation of a non-ambiguous but complete abbreviation for a phrase or idea.

Some points should be made concerning this definition. First, it is logically quite complex. We have an implication (if . . ., then . . .) and two universal quantifiers, "for all" (symbolically ∀) and "there exists" (symbolically ∃). It is important to note the distinction between always (for all) and some (there exists). These distinct logical structures are not inter-changeable. "∀" permits no exceptions, whereas "∃" permits all but one instance to be an exception. Certainly, you can appreciate the distinction between the two statements: all students will pass this course; and, there exists a student who will pass this course (even though at this point either statement may seem incredible). Notice the order of these quantifiers. We first permit any ε > 0 (and consequently we must consider all ε > 0) and then we assert the existence of at least one <u>corresponding</u> δ > 0. Note that each ε > 0 may have a different corresponding δ > 0.

It is always important to note the sequential structure of an implication. There is a tremendous difference between the two implications: if two lines have only one point in common, then the two lines intersect; and, if two lines intersect, then they have only one point in common. The second of these implications is false. Can you give a counterexample?

The limit definition begins

$$\text{if } 0 < |x - c| < \delta \tag{3}$$

and continues

$$\text{then } |f(x) - L| < \epsilon. \tag{4}$$

We may read this sequential relationship as: <u>provided that</u> (3) holds for some value of x, then (4) holds for that value of x. One should not interpret this implication as an undeniable truth. Rather it is part of the defining phrase or <u>condition</u> in our definition. The condition is met or is true when the defined term: $\lim\limits_{x \to c} f(x) = L$, may be used.

Another point to be made about the definition is that the value x = c is excepted. Recall that $0 < |x - c| < \delta$ may be read as

$$-\delta < x - c < \delta \text{ } \underline{but} \text{ } x \neq c \tag{5}$$

or as

$$c - \delta < x < c \text{ } \underline{or} \text{ } c < x < c + \delta. \tag{6}$$

In the jargon of the Preliminary Comments we are determining the expected or anticipated value of f(x) at x = c by examining f(x) for x near to but not equal to c.

It follows that <u>the limit of a function at a point is completely independent</u> <u>of whether or not the function is defined at that point.</u> We only become concerned with the value of f(x) at x = c when we ask whether or not the function f is continuous at x = c. Finally, note that $|f(x) - L| < \epsilon$ may be written as

$$-\epsilon < f(x) - L < \epsilon \tag{7}$$

or as

$$L - \epsilon < f(x) < L + \epsilon. \tag{8}$$

You should now slowly and carefully reread the definition of a limit as illustrated in Figure 2.2.2 of the text.

Of the many things that may be said about the limit definition, aside from its intrinsic beauty and enrapturing appeal, one other comment on the universal quantifiers is in order. Suppose that for a given specific value of $\epsilon > 0$ we are able to determine a value of $\delta > 0$ so that the implication:

$$\text{if } 0 < |x - c| < \delta, \text{ then } |f(x) - L| < \epsilon, \tag{9}$$

is true. Then, the following implications:

$$\text{if } 0 < |x - c| < \delta, \text{ then } |f(x) - L| < \epsilon' \text{ (with } 0 < \epsilon < \epsilon') \tag{10}$$

and

if $0 < |x - c| < \delta'$, then $|f(x) - L| < \epsilon$ (with $0 < \delta' < \delta$),   (11)

are also true. The point here is that there is not a one-to-one correspondence between ϵ and δ. For a given ϵ, there can be many acceptable δ's; and, a specific δ may work for more than just one given ϵ.

This idea, and others, will now be illustrated with a few examples. Consider the function $f(x) = 3x - 1$. We would like to test the statement $\lim\limits_{x \to 2} (3x-1) = 4$ by means of the limit definition. Suppose we are given $\epsilon = 3/2$. Our immediate task is to determine a value of $\delta > 0$ such that if $0 < |x - 2| < \delta$, then $|(3x - 1) - (4)| < 3/2$, where we have substituted directly in the definition. Well, if

$$0 < |x - 2| < \delta,\qquad(12)$$

then

$$|x - 2| < \delta,$$
$$-\delta < x - 2 < \delta,$$
$$2 -\delta < x < 2 + \delta,$$
$$6 - 3\delta < 3x < 6 + 3\delta,$$
$$5 - 3\delta < 3x - 1 < 5 + 3\delta,$$

and $$1 - 3\delta < (3x - 1) - (4) < 1 + 3\delta.\qquad(13)$$

At this point, we know that if (12) is true, then so is (13). However, what we want rather than (13) is

$$-3/2 < (3x - 1) - (4) < 3/2.\qquad(14)$$

If we can determine $\delta > 0$ so that (13) implies (14), then we will be done for $\epsilon = 3/2$. Now (14) follows from (13) if and only if

$$-3/2 \le 1 - 3\delta \text{ and } 1 + 3\delta \le 3/2,\qquad(15)$$

which came from writing "(13) implies (14)" as:

$$-3/2 \le 1 - 3\delta < (3x - 1) - (4) < 1 + 3\delta \le 3/2.\qquad(16)$$

We may solve the respective inequalities in (15) to find that $\delta \le 5/6$ and $\delta \le 1/6$. Therefore, any value of δ such that $0 < \delta \le 1/6$ will work when $\epsilon = 3/2$.

Have we now shown that $\lim\limits_{x \to 2}(3x - 1)= 4$? Definitely not! We have simply shown that if $\epsilon = 3/2$, then a $\delta > 0$ may be found. We have really shown that for $\epsilon \ge 3/2$, there exists a $\delta > 0$ (say, $\delta = 1/6$) such that if $0 < |x - 2| < 1/6$, then $|(3x - 1) - (4)| < \epsilon$. This is all that we have shown.

What more do we have to do to show that $\lim\limits_{x \to 2}(3x - 1)= 4$? Answer: We must still treat ϵ when $0 < \epsilon < 3/2$. Let us try another specific case, say $\epsilon = 1/2$. Inequalities (12) and (13) are unchanged, but (14) now looks like

$$-1/2 < (3x - 1) - (4) < 1/2.$$

In the same way that we found (15), we now require that

$$-1/2 \leq 1 - 3\delta \quad \underline{and} \quad 1 + 3\delta \leq 1/2 \tag{17}$$

These inequalities simplify to $\delta \leq 1/2$ and $\delta \leq -1/6$, respectively. What is our conclusion now that we cannot find a $\delta > 0$? We have shown that if we permit an error (ϵ) of up to 1/2 units in our expected or anticipated value ($L = 4$) for $f(x)$ at $x = 2$, then an open interval about $x = 2$ does <u>not</u> exist wherein $L = 4$ is a reasonable guess or anticipation. That is, $\lim\limits_{x \to 2}(3x - 1) \neq 4$.

If we still want to evaluate $\lim\limits_{x \to 2}(3x-1)$ and prove our answer, we need another guess for L. Shall we try $L = 5$? How should we proceed? Should we try $\epsilon = 3/2$? And, if everything works out all right, should we next try, say, $\epsilon = 1/2$? As long as we try specific individual values for ϵ and succeed in finding a corresponding value for δ, there will always remain smaller untested values of ϵ. What we need to do is to try all values of ϵ at once. Let us use letters instead of numbers. This approach shouldn't be all that hard, for you probably noticed that the two cases done above were almost identical. Our proof that $\lim\limits_{x \to 2}(3x - 1) = 5$ goes as follows.

If
$$0 < |x - 2| < \delta,$$
then
$$|x - 2| < \delta,$$
$$2 - \delta < x < 2 + \delta,$$
$$6 - 3\delta < 3x < 6 + 3\delta,$$
$$5 - 3\delta < 3x - 1 < 5 + 3\delta,$$
and
$$-3\delta < (3x - 1) - 5 < 3\delta,$$
which in turn implies

$$-\epsilon < (3x - 1) - 5 < \epsilon$$

iff $-\epsilon \leq -3\delta$ <u>and</u> $3\delta \leq \epsilon$. Thus, if we select δ so that $0 < \delta \leq \epsilon/3$ (say $\delta = \epsilon/3$ for neatness' sake), then the implication

if $0 < |x - 2| < \delta = \epsilon/3$, then $|(3x - 1) - (5)| < \epsilon$,

is true and $\lim\limits_{x \to 2}(3x - 1) = 5$.

You should verify that the choice: $\delta = \epsilon/10$, will also work for this example.

Notice that we obtained an explicit formula relating δ and ϵ. Such is normally the case, but the formula is not unique.

SECTIONS 2.3-4

We turn now to the practical question of how to evaluate limits. The backbone for our techniques is the collection of limit theorems appearing in Section 2.3. Let us use some of them to construct another proof of the statement:

$$\lim_{x \to 2} (3x - 1) = 5.$$

Since $\lim_{x \to 2} 3 = 3$ and $\lim_{x \to 2} x = 2$ and the limit of a product is the product of the limits provided the individual limits exist, we have that

$\lim_{x \to 2} 3x = \lim_{x \to 2} 3 \cdot \lim_{x \to 2} x = 3 \cdot 2 = 6$. Further, since the limit of a sum is the sum of the limits provided the individual limits exist and since

$\lim_{x \to 2} -1 = -1$, we have that $\lim_{x \to 2}(3x - 1) = \lim_{x \to 2} 3x + \lim_{x \to 2} -1 = 6 + (-1) = 5$.

Of course, no one writes all this or even consciously thinks all this when evaluating limits. If you were to be shown 1 + 2, you would think 3 without consciously performing an addition.

A key phrase in the above proof was "provided that." This phrase represents the hypothesis of the quoted theorem and must be satisfied before we may apply the theorem. It is incorrect to say $\lim_{x \to 0} (x^3 - \frac{2}{x}) = \lim_{x \to 0} x^3 + \lim_{x \to 0} (-\frac{2}{x})$,

since $\lim_{x \to 0} (-\frac{2}{x})$ does not exist (by Theorem 2.3.9 in the text). Further,

$\lim_{x \to 0} (x^3 - \frac{2}{x})$ does not exist, but not because $\lim_{x \to 0} (-\frac{2}{x})$ fails to exist.

Instead, $\lim_{x \to 0} (x^3 - \frac{2}{x})$ does not exist since we may write $x^3 - \frac{2}{x}$ as $\frac{x^4 - 2}{x}$ and

then apply Theorem 2.3.9 on quotients (i.e., $\lim_{x \to 0} x^4 - 2 \neq 0$ and $\lim_{x \to 0} x = 0$).

You must take particular care with limits such as this. Consider

$\lim\limits_{x \to 0} (\frac{2}{x} - \frac{2}{x})$, where clearly $\lim\limits_{x \to 0} \frac{2}{x}$ does not exist. Since $\frac{2}{x} - \frac{2}{x} = 0$ if $x \neq 0$,

we may conclude that $\lim\limits_{x \to 0} (\frac{2}{x} - \frac{2}{x}) = \lim\limits_{x \to 0} 0 = 0$ even though $\lim\limits_{x \to 0} \frac{2}{x}$ and

$\lim\limits_{x \to 0} (-\frac{2}{x})$ each fail to exist.

The basic technique for evaluating limits has just been used. Let us example it again before stating it.

$$\lim_{x \to 3} \frac{x^2 - 9}{x - 3} = \lim_{x \to 3} (x + 3) = 6.$$

We have used the fact that $\frac{x^2 - 9}{x - 3} = x + 3$ for $x \neq 3$. These expressions give rise to the same anticipated value (or limit), if it exists, at $x = 3$, since these expressions are identical for values of x near but not equal to 3. A helpful procedure is as follows:

(1) By means of the limit theorems, try to break up the given expression into simple expressions whose behavior is known or more easily determined.

(2) If step (1) leads to an impasse; namely, if you are unable to find the value of the limit or show that the limit fails to exist, then replace the given expression by one which is its equivalent for values of x near $x = c$. Now repeat step (1) for this revised but equal expression.

Graphs for the last example are in order.

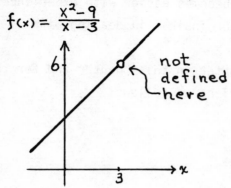

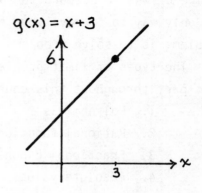

What we have done is to replace the function $f(x) = \dfrac{x^2 - 9}{x - 3}$ by the function $g(x) = x + 3$ which is identical to $f(x)$ at each $x \neq 3$. Since $g(x)$ is continuous at $x = 3$, its limiting value and actual value at $x = 3$ are in agreement. Recall the closing discussion of the Preliminary Comments.

Let us consider one more example: $\displaystyle\lim_{x \to 9} \dfrac{x - 9}{\sqrt{x} - 3}$. It is true that $\displaystyle\lim_{x \to 9}(\sqrt{x} - 3) = 0$, but it does not follow from the quotient theorem that our desired limit fails to exist, since $\displaystyle\lim_{x \to 9}(x - 9)$ is also zero. (See Theorem 2.3.9). Step 1 of our procedure an apparent failure, we now take step 2. We want to find an expression that agrees with $(x - 9)/(\sqrt{x} - 3)$ for values of x near 9. If we are lucky, the new expression will be continuous at $x = 9$ so that we may evaluate the limit by substituting 9 for x, or we may be able to show that the limit does not exist. Since $(\sqrt{x} + 3)/(\sqrt{x} + 3) = 1$, we may multiply the original expression by this quantity without changing any values. Thus,

$$\lim_{x \to 9} \frac{x - 9}{\sqrt{x} - 3} = \lim_{x \to 9} \frac{x - 9}{\sqrt{x} - 3} \cdot \frac{\sqrt{x} + 3}{\sqrt{x} + 3} = \lim_{x \to 9} \frac{(x - 9)(\sqrt{x} + 3)}{x - 9}$$

$$= \lim_{x \to 9}(\sqrt{x} + 3) = 6.$$

Notice that $(x - 9)/(\sqrt{x} - 3) = \sqrt{x} + 3$ if $x \neq 9$ and $x \geq 0$. Our principle tells us that these expressions share the same limiting value, provided one of them has a limit.

The apparent algebraic trickery of the last example is one of a few standard procedures you will develop through experience in working problems. The choice of what procedure to apply to a problem becomes easier with experience. The only way to improve your educated guessing or insight in dealing with problems is to solve problems.

The types of limit problems you will encounter in this Chapter and for the most part throughout this course may be categorized as follows:

1. Polynomials
2. Rational functions
3. Fractional exponents
4. Absolute values

5. Piecewise defined functions

6. Conglomerates

This list will eventually be lengthened and complicated when you are introduced to the trigonometric functions and logarithms. But, then you will also learn more techniques for handling limits. We will now explain case by case each of these types and a possible way to handle the associated limits.

Case 1: Polynomials. If $P(x)$ is a polynomial, then $\lim_{x \to c} P(x) = P(c)$, as proven on p. 58 of the text. That is, limits of polynomials are evaluated by direct substitution. This procedure works because all polynomial functions are continuous.

Case 2: Rational functions. A rational function $Q(x)$ is the quotient of two polynomials, say $Q(x) = \frac{N(x)}{D(x)}$. To determine $\lim_{x \to c} Q(x)$, we examine $\lim_{x \to c} N(x)$ and $\lim_{x \to c} D(x)$; that is, we evaluate $N(c)$ and $D(c)$. We then apply Theorems 2.3.7 and 2.3.9 on quotients. There are four possibilities that may result from our substitution of c for x:

$$\frac{0}{b} \quad , \quad \frac{a}{b} \quad , \quad \frac{a}{0} \quad , \quad \frac{0}{0} \qquad a, b \neq 0.$$

The first three cases are handled by the aforementioned theorems which yield $\lim_{x \to c} Q(x) = 0, \frac{a}{b}$, or "does not exist," respectively. In the event of the fourth possibility, we must reexpress the quotient to come up with one of the first three cases. For example, $\frac{x^2 - 4}{x - 2}$ becomes $x + 2$ and $\frac{x - 2}{x^3 - 8}$ becomes $1/(x^2 + 2x + 4)$. This simplification is simple because x - c is always a factor of $N(x)$ and $D(x)$, if we arrived at $\frac{0}{0}$. This rule may not carry over so easily when we talk about the quotients of functions other than polynomials.

Case 3: Fractional exponents. Suppose we are looking at some expression $f(x)$ raised to the p/q power. If q is odd, then $\lim_{x \to c} (f(x))^{\frac{p}{q}} = (\lim_{x \to c} f(x))^{p/q}$

and hence we need only evaluate $\lim\limits_{x \to c} f(x)$. If q is even, we can not handle the exponent so easily. Although $\lim\limits_{x \to -2} x$ exists, $\lim\limits_{x \to -2} x^{1/2}$ does not exist since $\sqrt{-2}$ is imaginary. If $c > 0$, then $\lim\limits_{x \to c} \sqrt{x} = (\lim\limits_{x \to c} x)^{1/2} = \sqrt{c}$. If $c < 0$, then $\lim\limits_{x \to c} \sqrt{x}$ does not exist since $\sqrt{x}$ is not defined for negative values of x. What about $\lim\limits_{x \to 0} \sqrt{x}$? One might easily think that $\lim\limits_{x \to 0} \sqrt{x} = 0$, but this is false.

To evaluate $\lim\limits_{x \to c} f(x)$, we need $f(x)$ defined for values of x near c. "Near c" means on both sides of c. Here, $\sqrt{x}$ is only defined for values on one side of zero ($x > 0$). In terms of the limit definition, there is <u>no</u> $\delta > 0$ such that if $0 < |x - 0| < \delta$, then $|\sqrt{x} - L| < \epsilon$; because, for any $\delta > 0$ there are x between $-\delta$ and zero where $\sqrt{x}$ is not defined. Thus, $|\sqrt{x} - L| < \epsilon$ cannot be evaluated or tested and the implication is false.

Consequently, Case 3 is just as easy and straightforward as the previous two, as long as we check the domain of the function in question whenever q is even. For example, the domain of $\sqrt{x^2 - 4}$ is $x \geq 2$ or $x \leq -2$. Thus, $\lim\limits_{x \to c} \sqrt{x^2 - 4} = \sqrt{c^2 - 4}$ iff $c > 2$ or $c < -2$ (and fails to exist otherwise).

<u>Case 4</u>: It is not always the case that $\lim\limits_{x \to c} |f(x)| = |\lim\limits_{x \to c} f(x)|$. For example, $\lim\limits_{x \to 0} \left| \dfrac{|x|}{x} \right| = 1$, whereas $\left| \lim\limits_{x \to 0} \dfrac{|x|}{x} \right|$ does not exist. Just compare the graphs of $\left| \dfrac{|x|}{x} \right|$ and $\dfrac{|x|}{x}$ to see this. If $\lim\limits_{x \to c} f(x) = L$, then we may say that $\lim\limits_{x \to c} |f(x)| = |L|$. This is problem 4 on page 67 of the text. But, as just shown (by example), it is <u>not</u> always the case that if $\lim\limits_{x \to c} f(x)$ fails to exist then $\lim\limits_{x \to c} |f(x)|$ also fails to exist. In this instance and any others where you are unsure of the correct approach, the safest and best way to handle the absolute value symbol is to eliminate it using the definition.

A single non-trivial example should suffice.

$$|(x - 1)(x - 2)(x - 3)| = (x - 1)(x - 2)(x - 3)$$

whenever $(x - 1)(x - 2)(x - 3)$ is non-negative, just by the definition of absolute value. Further, $|(x - 1)(x - 2)(x - 3)| = -(x - 1)(x - 2)(x - 3)$ whenever $(x - 1)(x - 2)(x - 3)$ is negative. Thus, we may eliminate the absolute value symbol here as follows:

$$|(x - 1)(x - 2)(x - 3)| = \begin{cases} (x - 1)(x - 2)(x - 3) & \text{when } x \geq 3 \text{ or } 1 \leq x \leq 2 \\ -(x - 1)(x - 2)(x - 3) & \text{when } x < 1 \text{ or } 2 < x < 3. \end{cases} \tag{1}$$

Case 5: Piecewise defined functions. Equation (1) is an example of a "piecewise defined function." These are functions given by formally different algebraic rules over different parts of their domains. Here again the limit of $f(x)$ at $x = c$ is found by examination of $f(x)$ for values of x <u>near</u> c and consequently on both sides of c.

Let us completely illustrate this point by evaluating

$$\lim_{x \to 5} f(x), \lim_{x \to 3} f(x) \text{ and } \lim_{x \to -1} f(x)$$

for

$$f(x) = \begin{cases} x^2 & \text{iff } -2 \leq x < -1 \\ x - 1 & \text{iff } -1 \leq x \leq 3 \\ \sqrt{x^2 - 5} & \text{iff } 3 < x \end{cases} \tag{2}$$

To evaluate $\lim_{x \to 5} f(x)$, our notion of near but on both sides of 5 shows that all we need to know about the function f is that $f(x) = \sqrt{x^2 - 5}$ when $x > 3$. On the other hand, only the first line of the definition of $f(x)$ is immaterial when we evaluate $\lim_{x \to 3} f(x)$. Similarly, our evaluation of $\lim_{x \to -1} f(x)$ requires that we concern ourselves with the first two lines of the definition of f. You should check that the values of these three limits are $2\sqrt{5}$, 2, and "does not exist," respectively. If you quickly see that $\lim_{x \to -2} f(x)$ does not exist, you are more than ready for our last case.

Case 6: Conglomerates. A conglomerate limit problem is a combination of some or all of the above cases spiced with occasional instructor deviousness. To handle these problems, you take a smattering of the above techniques and add a pinch of ingenuity. Would you believe that

$$\lim_{x \to 2} \left(\frac{\frac{x^2 - 1}{x - 1} - \sqrt{11 - |x|}}{3 + |x - 1|} \right)^{3/4}$$

does not exist? Would you further believe that harder questions may be posed without all this mess?

A footnote. Two important theorems have not been mentioned. There are some problems on p. 67 of the text which bring the Pinching Theorem into play. As for the Uniqueness Theorem on p. 56 of the text, just imagine the ambiguity and resulting uselessness of the theory of limits if we were unable to prove this very critical theorem.

These two theoretical results are very closely related to what is known as the completeness axiom for the real number system. One formulation of this axiom states that every infinite sequence of real numbers which is increasing and bounded above converges to (has as a limit) a unique real number. For example, 1/2, 3/4, 7/8, 15/16, etc. is clearly increasing and is bounded above by any number greater than or equal to 1. The sequence converges to 1 (has 1 as its limit). The set of rational numbers, however, does not satisfy the completeness axiom. For example, the sequence, 1, 1.4, 1.41, 1.414, etc. is both increasing and bounded above. It converges, but the limit: $\sqrt{2}$, is not a rational number.

EXERCISES

Evaluate $\lim_{x \to 2} f(x)$ for each of the following.

1. $f(x) = \begin{cases} x^2, & x < 2 \\ 2x, & x > 2. \end{cases}$

2. $f(x) = \begin{cases} x^3, & x < 1 \\ x + 3, & x > 1. \end{cases}$

3. $f(x) = \begin{cases} 5 + x, & x < 3 \\ 4, & x > 3. \end{cases}$

4. $f(x) = \begin{cases} x^2, & x < 1 \\ x + 2, & x > 3. \end{cases}$

5. $f(x) = \begin{cases} x^2 & , x < 2 \\ \dfrac{x^2 - 4}{x - 2} & , x > 2. \end{cases}$

6. $f(x) = \begin{cases} x^2 & , x \le 2 \\ 3x & , x > 2. \end{cases}$

7. $f(x) = (x - 2)^{1/3}.$

8. $f(x) = (x - 2)^{1/4}.$

SOLUTIONS

1. 4   2. 5   3. 7   4. does not exist   5. 4

6. does not exist   7. 0   8. does not exist

SECTIONS 2.5-6

Now that you have mastered the notion of a limit, the topic of one-sided limits should be easy. You already have the theoretical background and practical tools under your belt. Thus, a brief statement on why and when we encounter one-sided limits should be sufficient.

Recall the discussion of $\lim\limits_{x \to 0} \sqrt{x}$ on p.24 of the Supplement. In convincing you that this limit does not exist, we argued intuitively that the anticipated value of a function f at a point x = c is inferred from values of f(x) for x <u>near</u> and on <u>both</u> sides of c. Since $\sqrt{x}$ is not defined for x < 0, no inference (and thus no limit) is possible. The limit of f(x) at x = c from the right or $\lim\limits_{x \downarrow c} f(x)$ is simply an inference based on values of f(x) for x near but only to the right of (or greater than) c. Thus, $\lim\limits_{x \downarrow 0} \sqrt{x} = 0.$

The limit of f at x = c from the left is the analogous limit of values to the left of (or less than) c. We shall write $\lim\limits_{x \uparrow c} f(x)$. Thus $\lim\limits_{x \uparrow 0} \sqrt{x}$ does not exist.

The relationship between a "full" or two-sided limit (or more simply just limit) and one-sided limits is immediate. That is, $\lim\limits_{x \to c} f(x)$ exists if and only if $\lim\limits_{x \uparrow c} f(x)$ and $\lim\limits_{x \downarrow c} f(x)$ both exist and have the same value. This common value is $\lim\limits_{x \to c} f(x)$. Thus, $\lim\limits_{x \to 0} \sqrt{x}$ fails to exist because $\lim\limits_{x \downarrow 0} \sqrt{x} = 0$ but

$\lim\limits_{x \uparrow 0} \sqrt{x} \neq 0$.

Of what use are one-sided limits? They arise in a very natural way. The two most common ways are piecewise defined functions and absolute values, which when interpreted may give rise to piecewise defined functions. To evaluate $\lim\limits_{x \to c} f(x)$ for the function given in (2) on p. 25 of the Supplement, you only need one-sided limits if c = -2, -1, or 3. However, it is not always obvious when a given limit needs to be handled using one-sided limits. For example, $\lim\limits_{x \to 0} \sqrt{x}$ requires that we utilize the concept of a one-sided limit, but this is not immediately apparent.

The idea of a continuous function (at a point or on an interval) is relatively simple. It provides most of the theoretical punch behind many of the mechanical procedures and theorems to be developed in this course. Given that you can determine the existence (and value, when it exists) of $\lim\limits_{x \to c} f(x)$, then all that is left to ask is whether or not this value is f(c). If so, f is continuous at x = c.

Now perhaps you can see why we introduced the subject of continuity in the Preliminary Comments. For, how did you evaluate $\lim\limits_{x \to c} f(x)$? Well, you <u>assumed</u> f was continuous and evaluated f(c). The two statements:

"evaluate $\lim\limits_{x \to c} f(x)$ by substituting c for x"

and

"f is continuous at x = c"

are equivalent. That is, you may evaluate a limit by substitution iff the function is continuous. You may assume a function to be continuous, however, if and only if it is.

What do you do in practice? Those instances where the assumption of continuity is not valid and other techniques are needed to evaluate limits (and eventually answer questions on continuity as well) were essentially given before when we classified and described different types of limits. For example, limits of rational functions may be evaluated by substitution only if the denominator is nonzero.

Given an arbitrary function, how do you determine whether or not it is continuous at a specific point, say x = c? The test is simple. All three of the quantities: $f(c)$, $\lim\limits_{x \downarrow c} f(x)$ and $\lim\limits_{x \uparrow c} f(x)$, must exist and be equal in value. This is what the definition of continuity asserts by the statement

$$\lim_{x \to c} f(x) = f(c).$$

If you look at Diagram 4 of the Preliminary Comments, you will find five different examples of ways in which this condition may not be met. There are other cases, just as easy to ascertain.

Another question you might be asked is to determine the points of discontinuity, if any, for a given function. How, then, do you locate <u>possible</u> points of discontinuity? We have already mentioned the case where zeroes appear in the denominator of a rational function and the case involving endpoints of the domain. Other obvious cases (especially in the event of a piecewise defined function) are points where the function is not defined ("holes" in the domain) and points where $\lim\limits_{x \downarrow c} f(x) \neq \lim\limits_{x \uparrow c} f(x)$.

The notion of one-sided continuity is tested for in the obvious way. Like one-sided limits, its usefulness lies in our ability to then be more explicit as to why a function is not continuous at a point. For example, $f(x) = \sqrt{x}$ is discontinuous at x = 0 because it is continuous from the right but not from the left.

Two important theorems are given without proof in these Sections.

<u>Intermediate-Value Theorem</u>: If f is continuous on [a,b] and C is a number between f(a) and f(b), then there is at least one number c between a and b such that f(c) = C.

<u>Maximum-Minimum Theorem</u>: If f is continuous on [a,b] then, somewhere on [a,b], f takes on a maximum value M and a minimum value m.

A few other critical theorems in the Calculus will be presented in later Chapters, so we will take this opportunity to discuss the structure and meaning

of a theorem. A theorem is an implication that consists of two parts: hypothesis
(if clause) and conclusion (then clause). The oft repeated error in using
theorems is to ignore the hypothesis and take the conclusion of the theorem as an
absolute truth. One should rather first determine if the hypothesis of the
theorem is satisfied. And, then, the conclusion may be taken as true. The truth
of the hypothesis of a theorem is <u>sufficient</u> to guarantee the truth of the
conclusion. However, it is not <u>necessary</u> that the hypothesis of a theorem be true
in order for the conclusion to be true. In such cases, the truth of the
conclusion may not be inferred by the theorem but must then follow by some other
argument. The graphs in Diagram 1 illustrate cases when the conclusion of the
Intermediate-Value Theorem is true, but not by virtue of the theorem, since one
or more parts of the hypothesis are not true.

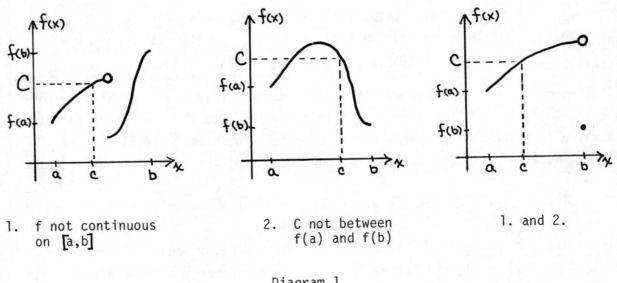

1. f not continuous
   on [a,b]

2. C not between
   f(a) and f(b)

1. and 2.

Diagram 1

Associated with every implication (H implies C) is an equivalent statement:
its contrapositive (not C implies not H). The contrapositives of our two
theorems are as follows (rephrased slightly for improved readability):

<u>Contrapositive of Intermediate-Value Theorem</u>: Suppose the function f is defined on the interval [a,b] and C is some fixed real number. If there is no real number c between a and b such that f(c) = C, then f is not continuous on [a,b] or C is not between f(a) and f(b).

<u>Contrapositive of Maximum-Minimum Theorem</u>: If the function f does not take on a maximum value somewhere on [a,b] or if f does not take on a minimum value somewhere on [a,b], then f is not continuous on [a,b].

We noted earlier that the hypothesis is sufficient to guarantee the truth of the conclusion. On the other hand, if the conclusion is false, it is necessary that some part of the hypothesis is false. Such was just illustrated by the two given contrapositives.

You probably saw a specific application of the Intermediate Value Theorem in one of your high school algebra courses when you were studying techniques for finding roots of polynomial equations. Just consider what the Intermediate Value Theorem says in the event that f is a polynomial, C = 0 and f(a) and f(b) are of opposite signs. The veracity of this technique was probably argued intuitively as follows. If the curve in question passes from one side of the x-axis to the other, then it must cross the x-axis at least once. Underlying whatever persuasion was given is the fact that polynomials are continuous.

This is probably your first contact with the Maximum-Minimum Theorem in any form, but again the essential element of the proof is the continuity of f on a closed interval. Proofs for these two theorems appear in Appendix B of the text. You should look at these proofs at least once, if for no other reason than to make sure they are there. The given proofs are existential as opposed to constructive in nature. That is, the theorems specify that f(c) = C has a solution and that extrema (maxima and minima) exist, but the proofs don't provide a method for actually finding these quantities. A constructive proof shows existence by giving a method for finding that object which the theorem says exists. Numerous procedures for finding "roots" of equations are given in a Junior Level course entitled "Numerical Analysis." During the intervening two years, see if you can solve the equation $x + \cos x = 0$ and generalize your method.

The determination of extrema is postponed only until Chapter 4.

We conclude with two examples on continuity. The second example may help you handle the examples and problems which use the greatest integer function.

Example 1. Determine where the following piecewise defined function is discontinuous and fully justify your conclusions.

$$f(x) = \begin{cases} x^3 & , & x \leq -1 \\ x^2 - 2 & , & -1 < x < 0 \\ 3 - x & , & 0 \leq x < 2 \\ \dfrac{4x - 1}{x - 1} & , & 2 \leq x < 4 \\ \dfrac{15}{7 - x} & , & 4 < x < 7 \\ 5x + 2 & , & 7 \leq x \end{cases}$$

Solution. Each of the six pieces of this function are continuous, respectively, on the open intervals $(-\infty,-1)$, $(-1,0)$, $(0,2)$, $(2,4)$, $(4,7)$, $(7,\infty)$. We need only examine the behavior of f at x = -1, 0, 2, 4, and 7.

f is continuous at x = -1 since $\lim\limits_{x \uparrow -1} f(x) = \lim\limits_{x \uparrow -1} x^3 = -1$,

$$\lim\limits_{x \downarrow -1} f(x) = \lim\limits_{x \downarrow -1}(x^2 - 2) = -1,$$

and $$f(-1) = (-1)^3 = -1.$$

f is discontinuous at x = 0 as f(0) = 3 - (0) = 3 but

$$\lim\limits_{x \uparrow 0} f(x) = \lim\limits_{x \uparrow 0}(x^2 - 2) = -2.$$

$\lim\limits_{x \downarrow 2} f(x) = 7 \neq 1 = \lim\limits_{x \uparrow 2} f(x)$, so f is discontinuous at x = 2.

f(4) is not defined. Since $\lim\limits_{x \to 4} f(x)$ exists, the resulting discontinuity is removable. (If we define f(4) = 5, f is now continuous at x = 4.)

The discontinuity at x = 7 is essential (not removable) since $\lim\limits_{x \uparrow 7} f(x)$ does not exist.

Example 2. Consider the function $f(x) = [x^2 + 1]$ with domain of $[-1,2)$. Where is f discontinuous?

Solution. We could begin by writing f as:

$$f(x) = \begin{cases} 2 & \text{for } x = -1 \\ 1 & -1 < x < 1 \\ 2 & 1 \le x < \sqrt{2} \\ 3 & \sqrt{2} \le x < \sqrt{3} \\ 4 & \sqrt{3} \le x < 2 \end{cases}$$

On the following graph of f, the graph of $x^2 + 1$ is indicated by the dotted curve. What do you notice?

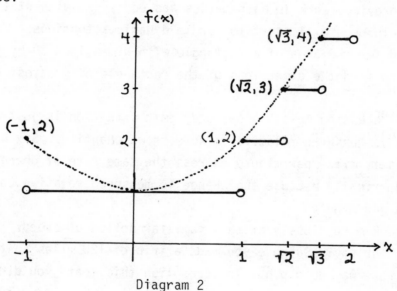

Diagram 2

f is discontinuous at $x = 1$ since $\lim\limits_{x \uparrow 1} f(x) = 1$ but $\lim\limits_{x \downarrow 1} f(x) = f(1) = 2$. One could support the discontinuity of f at $x = -1$ in a number of ways. $\lim\limits_{x \downarrow -1} f(x) = 1 \neq f(-1)$ or one could say that $\lim\limits_{x \to -1} f(x)$ does not exist. The other discontinuities of f occur for $x = \sqrt{2}$ and $\sqrt{3}$. Can you give supporting evidence?

Differentiation

3

Much of your previous work in Mathematics was concerned with static situations. With the Calculus you will start to handle dynamic situations. Whereas before you were asked for the area of a rectangle with dimensions 30 by 40, you will soon be able to find the dimensions of the rectangle of greatest area which has perimeter 140.

The first question to ask about a dynamic situation is just what is changing. Change is always measured relative to some known quantity. Six hundred miles per hour and ten miles per minute express the same rate of change of an object but differ numerically because the change is measured relative to two different units: hours and minutes.

Let us look more closely at this familiar notion of speed. What does 60 miles per hour say? Suppose you made a trip of 120 miles in two hours. Your _average_ rate was 60 m.p.h. To accomplish this feat, you didn't have to travel at the _instantaneous_ rate of 60 m.p.h. for each instant of the trip. As an aside, can you prove that during this trip you had to be going 60 m.p.h. for at least one instant? Try using the Intermediate Value Theorem. If you can solve this problem, then you can understand why the time of entry is marked on your toll-road ticket and you have a future as a toll booth attendant. Many other careers await the trained mathematician.

Instantaneous velocity is defined in a very natural way as the limiting value of average velocity over shorter and shorter time intervals. The average velocity between time t_0 and t_1 is $\dfrac{d(t_1) - d(t_0)}{t_1 - t_0}$, where $d(t)$ is distance in terms of time t.   This ratio is the slope of the straight line in Diagram 1.

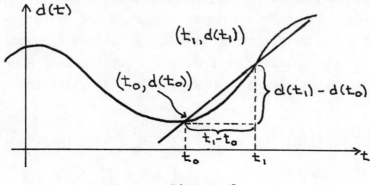

Diagram 1

Changing t_1, we get a different line.

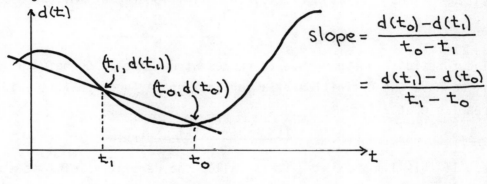

$$slope = \frac{d(t_0) - d(t_1)}{t_0 - t_1}$$

$$= \frac{d(t_1) - d(t_0)}{t_1 - t_0}$$

Diagram 2

What happens as t_1 gets closer to t_0?

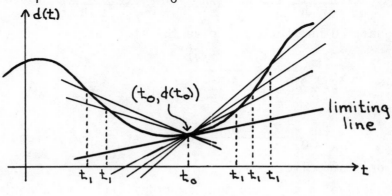

Diagram 3

It appears that we have a tangent line to the curve at the point $(t_0, d(t_0))$.
The slope of each "secant line" in the last diagram is the average velocity
during the interval from t_0 to t_1 (or t_1 to t_0). As this difference $t_1 - t_0$
(or $t_0 - t_1$) shrinks closer to zero, we presumably get a better approximation of
the instantaneous velocity at time t_0. Graphically, the instantaneous velocity
at t_0:

$$v(t_0) = \lim_{t \to t_0} \frac{d(t) - d(t_0)}{t - t_0} , \qquad (1)$$

is the slope of the "limiting line" in Diagram 3.

Let us denote the difference between t_0 and t by h; that is, let $t_0 + h = t$.
Remember that since t may be less than or greater than t_0, h can take on negative
or positive values. We may reformulate definition (1) as follows:

$$v(t_0) = \lim_{h \to 0} \frac{d(t_0 + h) - d(t_0)}{h} \qquad (2)$$

The function d doesn't have to represent distance. Changing a few symbols
in (2), we have the definition of a derivative: A function f is said to be
differentiable at x iff

$$\lim_{h \to 0} \frac{f(x + h) - f(x)}{h} = f'(x)$$

exists. If this limit exists, it is called the derivative of f at x and may be
interpreted as the instantaneous rate of change of $f(x)$ relative to x.

It should be noted that this is a "full limit." We need both

$$\lim_{h \downarrow 0} \frac{f(x + h) - f(x)}{h} \quad \text{and} \quad \lim_{h \uparrow 0} \frac{f(x + h) - f(x)}{h}$$

to exist and be equal. Other texts may give the limit as

$$\lim_{x_2 \to x_1} \frac{f(x_2) - f(x_1)}{x_2 - x_1}$$

or as

$$\lim_{\Delta x \to 0} \frac{f(x + \Delta x) - f(x)}{\Delta x} \qquad (\Delta x \text{ is read as "delta x"})$$

The difference is only symbolic.

Your first objective in applying the limit definition of a derivative is to
rewrite the numerator so you may divide out the "h" of the denominator. You

probably then can evaluate the limit by substitution without getting a zero in the denominator. One rule covers all cases for this initial simplification of the numerator: perform the indicated operations. For example, in $(x + h)^2 - x^2$ first "expand" and in $\frac{1}{x + h} - \frac{1}{x}$ first "add" the fractions. Then, just take the limit, for that is all the derivative is.

Finally, the proof of the theorem on p. 93 that differentiability implies continuity is standard, but elegant. It deserves a close reading. The converse of this theorem is not true. The best way to remember which way this implication goes is to keep the example $f(x) = |x|$ in mind. This function is continuous at $x = 0$ but has no derivative at $x = 0$. Figuratively, to be "smooth" is to be "unbroken," but not conversely.

SECTION 3.2

"Messy" functions, and other ones too, may be formed by adding, subtracting, multiplying and dividing other functions. Fortunately, we have rules for differentiating such combinations. If f and g are differentiable at x,

$$(f + g)'(x) = f'(x) + g'(x)$$
$$(f - g)'(x) = f'(x) - g'(x)$$
$$(fg)'(x) = f'(x)g(x) + f(x)g'(x)$$
$$(f/g)'(x) = \frac{g(x)f'(x) - f(x)g'(x)}{[g(x)]^2} \text{ with } g(x) \neq 0.$$

Coupling these rules with the fact that $f'(x) = nx^{n-1}$ when $f(x) = x^n$ for n an integer, you have all you need to differentiate most of the functions with which you are familiar.

You will be delighted to learn that the derivative of x^n relative to x is nx^{n-1} even when n is not an integer. For example, the derivative of $x^{1/2}$ with respect to x is $\frac{1}{2} x^{-1/2}$. This surprising result is presented later on.

Some helpful mechanical hints are in order. First, even though you could differentiate $4x^3$ using the product rule:

(*)
$$(4x^3)' = (4)'(x^3) + (4)(x^3)'$$
$$= (0)(x^3) + (4)(3x^2)$$
$$= 12x^2,$$

it is more efficient and less open to error to apply the rule
$$[c \cdot f(x)]' = c \cdot f'(x)$$
for c a constant to obtain
$$(4x^3)' = 4(x^3)' = 4(3x^2) = 12x^2.$$

You will soon omit the intermediate steps of this last line and revel in your prowess at noting that $(6x^2)' = 12x$, $(-3x^5)' = -15x^4$, and so on.

There is always more than one way to do a problem, even an easy one. Suppose you wish to differentiate the product $(4x + 3)(2 - x^2)$. You could write

(*)
$$[(4x + 3)(2 - x^2)]' = (4x + 3)'(2 - x^2) + (4x + 3)(2 - x^2)'$$
$$= 4(2 - x^2) + (4x + 3)(-2x)$$
$$= -12x^2 - 6x + 8.$$

Or, you could write $(4x + 3)(2 - x^2)$ as $-4x^3 - 3x^2 + 8x + 6$ and then differentiate term by term to obtain the same answer.

Suppose you wanted to differentiate $\frac{4x^2 - 3}{x}$ relative to x. In this instance, there are at least three routes to the correct answer, and of course many more yet that lead to something grossly incorrect. We could treat this problem with the quotient rule.

(*)
$$\left(\frac{4x^2 - 3}{x}\right)' = \frac{(x)(4x^2 - 3)' - (4x^2 - 3)(x)'}{x^2}$$
$$= \frac{(x)(8x) - (4x^2 - 3)(1)}{x^2} = \text{etc.}$$

Or, we could handle this problem with the product rule by first writing $(4x^2 - 3)/x$ as $(4x^2 - 3)(x^{-1})$. Then,

(*)
$$[(4x^2 - 3)(x^{-1})]' = (4x^2 - 3)'(x^{-1}) + (4x^2 - 3)(x^{-1})'$$
$$= (8x)(x^{-1}) + (4x^2 - 3)(-x^{-2}) = \text{etc.}$$

Finally, we could call into play the rigged nature of this problem and write

$(4x^2 - 3)/x$ as $4x - 3x^{-1}$. Consequently,

(*) $(\frac{4x^2 - 3}{x})' = (4x - 3x^{-1})' = 4 + 3x^{-2}$

which you should verify as the simplified form of the result of the two previous approaches.

Look back now at the five equations marked by an asterisk on the left. Notice that in each instance we differentiated only to the extent of applying the form of the rule we were using. We didn't try to do too much in one step. Such is perhaps the best advice we can offer for mastering this process or anything else. Notice that we can reread our work to check whether or not we correctly applied the rules, as well as checking on the algebra involved. The case of the quotient rule, where one may easily forget to square the denominator or misplace or even omit the "minus sign," substantiates this suggestion about carefulness.

SECTION 3.3

In the last section we wrote f'(x) to denote the derivative of f(x) relative to or <u>with respect to</u> x. The prime was used to denote the derivative with respect to the variable we were using. The notation $\frac{d}{dx}$ is introduced in this section. It is read the same way: the derivative with respect to x of ..., but notice that the "x" or variable is made explicit. Both notations are useful. We prefer to use the prime notation for functional type statements like f'(x) = 2x and (f + g)'(x) = f'(x) + g'(x). We prefer to use the $\frac{d}{dx}$ for algebraic manipulations, since it is more difficult to overlook or misinterpret a "$\frac{d}{dx}$" than it is a " ' " in involved calculations. Just compare the two statements:

$$\frac{d}{dx}\left(\frac{4x^2 - 3}{x}\right) = \frac{x \frac{d}{dx}(4x^2 - 3) - (4x^2 - 3)\frac{d}{dx}(x)}{x^2}$$

and

$$\left(\frac{4x^2 - 3}{x}\right)' = \frac{x(4x^2 - 3)' - (4x^2 - 3)x'}{x^2} .$$

The symbol dx is given a meaning of its own in some texts. This is prompted

by the historical development of the notation dx. We will turn to this point
in Section 4.9 of the Supplement.

SECTION 3.4

There is a simple two step procedure for handling "rate of change" problems.
Suppose you want to calculate the rate of change of quantity A relative to (or
with respect to) quantity B. First, you find an equation involving A and B.
Second, you differentiate this equation with respect to B and solve for $\frac{dA}{dB}$.
This simple procedure is amply illustrated by the text.

As a related question, what do we obtain when we differentiate the
expression LW with respect to L? By the product rule, we have that

$$\frac{d(LW)}{dL} = L \frac{dW}{dL} + W \frac{dL}{dL} .$$

If LW represents the area A of a rectangle of width W and length L, we may
interpret this result as $\frac{dA}{dL}$, the rate of change of the area of the rectangle
with respect to its length. Can we simplify the equation

$$\frac{dA}{dL} = L \frac{dW}{dL} + W \frac{dL}{dL} ?$$

Common sense alone suggests that $\frac{dL}{dL} = 1$. That is, the rate of change of something relative to itself is identically one. In general, then, to find $\frac{dA}{dL}$ we
need values for L, W and $\frac{dW}{dL}$, since

$$\frac{dA}{dL} = L \frac{dW}{dL} + W .$$

In the event that the length and width of the rectangle are <u>independent</u>;
that is, changes in values of one variable do not affect changes in values of the
other, we algebraically interpret such independence by $\frac{dW}{dL} = 0$. Consequently,

$$\frac{dA}{dL} = L \frac{dW}{dL} + W = W.$$

Literally, if the length of a rectangle is increased by a unit measure, then the area is changing relative to the length at the rate of W square units.

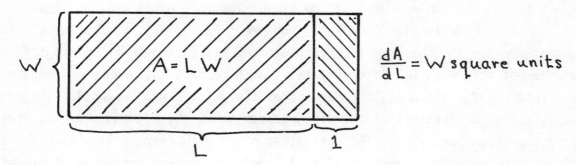

Remember that for this special case L and W are independent.

     After pondering the following semantic puzzle, you should set about the
task of becoming a whiz at differentiation.

PUZZLE: We have seen that the limit of a sum is the sum of the limits provided
the individual limits exist. The derivative is a limit and the derivative of a
sum is the sum of the derivatives provided the individual derivatives exist.
Why isn't it the case that the derivative of a product is the product of the
derivatives:

$$(fg)'(x) = f'(x) \cdot g'(x),$$

since the limit of a product is the product of the limits provided the
individual limits exist?

!əlzzud əht fo tnemetats eht ni alumrof deyalpsid eht yb deledom

yticorta eht gnitucexe rof esucxe on evah won uoY :laroM .tluser tnereffid

a ot sdael noitavired eht ;yaw taht tuo krow t'nseod tI :R̲E̲W̲S̲N̲A̲

SECTION 3.5

It should help to keep the following two brief notions in mind. First, if A
changes m times as fast as B and B changes·n times as fast as C then A changes

mn times as fast as C. Second, a derivative is taken relative to or with respect to some variable. We look now at an algebraic embodiment of these ideas.

What is the derivative of x^6? If you are thinking $6x^5$, you could be either right or wrong. The question is incomplete and misleading in its present form. You were not told what to differentiate x^6 with respect to. Pause and ponder that statement for a moment, for here comes the juicy part. We would like to calculate four derivatives of x^6: with respect to x, x^2, x^3, and x^6. The first and last problems are easy. The derivative of x^6 with respect to x is $6x^5$. The derivative of x^6 with respect to x^6 is 1. Only the second assertion might raise some eyebrows. The statement: Arnold can run S times as fast as himself, only makes sense if S is 1. An algebraic argument is just as simple.

What of the other two problems? How do we find the derivative of x^6 with respect to x^2? Let us begin by writing x^6 as $(x^2)^3$. We should think of x^2 as the name of a variable, so we will simplify matters by writing $t = x^2$. Now we are asking for the derivative of $x^6 = t^3$ with respect to $x^2 = t$. This is easy. The result is $3t^2$. Thus, the derivative of x^6 with respect to x^2 is $3(x^2)^2$. Perhaps you can mimic this argument to convince yourself that the derivative of x^6 with respect to x^3 is $2(x^3)^1 = 2x^3$. Our four differentiations of x^6 are summarized below.

$$\text{Table 1: Certain derivatives of } x^6$$

$$\frac{d}{dx}[x^6] = 6x^5 \qquad\qquad \frac{d}{d(x^3)}[x^6] = 2x^3$$

$$\frac{d}{d(x^2)}[x^6] = 3x^4 \qquad\qquad \frac{d}{d(x^6)}[x^6] = 1$$

Clearly, these derivatives of x^6 are not equal. Table 1 should provide ample and final persuasion for the necessity and meaning of the phrase "with respect to" whenever we differentiate.

Now we close in on the chain rule.

Table 2: Some useful derivatives

$$\frac{d}{dx} [x] = 1 \qquad\qquad\qquad\qquad \frac{d}{dx} [x^3] = 3x^2$$

$$\frac{d}{dx} [x^2] = 2x \qquad\qquad\qquad\qquad \frac{d}{dx} [x^6] = 6x^5$$

What happens when we multiply corresponding entries of Tables 1 and 2? Answer this question by filling in the following table.

Table 3: Four chain rule derivations of $\frac{d}{dx}(x^6)$

$$\frac{d}{dx} [x^6] \cdot \frac{d}{dx} [x] = \qquad\qquad\qquad \frac{d}{d(x^3)} [x^6] \cdot \frac{d}{dx} [x^3] =$$

$$\frac{d}{d(x^2)} [x^6] \cdot \frac{d}{dx} [x^2] = \qquad\qquad\qquad \frac{d}{d(x^6)} [x^6] \cdot \frac{d}{dx} [x^6] =$$

Your answer in each case should be $6x^5$. All four of these calculations have something very special in common. In each case, we have found the derivative of x^6 with respect to some variable (x, x^2, x^3 or x^6) and then multiplied by the derivative with respect to x of that same intermediate variable. Symbolically, each entry in Table 3 fits the form

$$\frac{d}{dt} [x^6] \cdot \frac{dt}{dx} = \frac{d}{dx} [x^6].$$

If we let $y = x^6$, the form is even neater:

$$\frac{dy}{dt} \cdot \frac{dt}{dx} = \frac{dy}{dx} .$$

What does $f'(x^2 + 1)$ mean? Well, $f'(x^2 + 1)$ denotes the derivative with respect to $x^2 + 1$ of the function $f(x^2 + 1)$. Suppose $f(t) = t^3 + 2$. Since $f(x^2 + 1) = (x^2 + 1)^3 + 2$, we have that $f'(x^2 + 1) = 3(x^2 + 1)^2$. Compare this result with $f'(t) = 3t^2$ or $f'(y) = 3y^2$ or $f'(2z) = 3(2z)^2$. Notice that with the " ' " notation it is necessary to look to the argument of f to determine what it is we are differentiating "with respect to." The form $f'(x^2 + 1)$ is neater than

$\dfrac{d}{d(x^2 + 1)}$ $[f(x^2 + 1)]$ and just as explicit. But, how would you write the

derivative with respect to x of $f(x^2 + 1)$? You could write: $f'(x^2 + 1) \cdot (x^2 + 1)'$ but here the "dee" notation is more explicit: $\dfrac{d}{dx} [f(x^2 + 1)]$. Perhaps now you can see better why we prefer the "dee" notation except for simple "functional" statements. To complete this example, we observe that

$$\frac{d}{dx} [f(x^2 + 1)] = f'(x^2 + 1) \cdot \frac{d}{dx} (x^2 + 1)$$

$$= \frac{d}{d(x^2 + 1)} [(x^2 + 1)^3 + 2] \cdot \frac{d}{dx} [x^2 + 1]$$

$$= \frac{d}{d(x^2 + 1)} [(x^2 + 1)^3 + 2] \cdot (2x)$$

$$= 3(x^2 + 1)^2(2x)$$

If we let $y = f(x^2 + 1)$ and $t = x^2 + 1$, we have just executed the chain rule as symbolized by

$$\frac{dy}{dx} = \frac{dy}{dt} \cdot \frac{dt}{dx} = \frac{d}{dt} [t^3 + 2] \cdot \frac{d}{dx} [x^2 + 1]$$

$$= 3t^2 \cdot 2x = 3(x^2 + 1)^2(2x).$$

The formal statement of the chain rule is usually couched in terms of the composition of two functions.

Chain Rule: If g is differentiable at x and f is differentiable at g(x), then
$$(f \circ g)'(x) = \frac{d}{dx}[f(g(x))] = f'(g(x)) \cdot g'(x). \tag{1}$$

For our previous example, we select $f(x) = x^3 + 2$ and $g(x) = x^2 + 1$. Since
$$f'(g(x)) = \frac{d}{d(g(x))}[(g(x))^3 + 2] = 3(g(x))^2 = 3(x^2 + 1)^2$$
and
$$g'(x) = \frac{d}{dx} [x^2 + 1] = 2x,$$
it should be no surprise that equation (1) yields
$$(f \circ g)'(x) = \frac{d}{dx}[(x^2 + 1)^3 + 2] = 3(x^2 + 1)^2(2x).$$
As a second example using these same functions, we observe that
$$(g \circ f)'(x) = g'(f(x)) \cdot f'(x).$$

Consequently, since

$$g'(f(x)) = \frac{d}{d(f(x))}[(f(x))^2 + 1] = 2f(x)$$

and

$$f'(x) = 3x^2,$$

we have that

$$(g \circ f)'(x) = \frac{d}{dx}[(x^3 + 2)^2 + 1] = 2(x^3 + 2)(3x^2).$$

Note that this does not equal $(f \circ g)'(x)$ as calculated above.

<u>Example 1</u>.   Calculate $\frac{dy}{dx}$ for $y = (x^2 + 1)^{31} + 4x$.

<u>Solution</u>.

$$\frac{dy}{dx} = 31(x^2 + 1)^{30} \frac{d}{dx}(x^2 + 1) + 4 \tag{2}$$

$$= 31(x^2 + 1)^{30}(2x) + 4.$$

It is incorrect to say that we have differentiated y with respect to $x^2 + 1$ in equation (2), since what we have is $\frac{dy}{dx}$, not $\dfrac{dy}{d(x^2 + 1)}$. We just haven't yet evaluated the $\frac{d}{dx}(x^2 + 1)$ in the first term.

It is incorrect to write either

$$\frac{dy}{dx} = 31(x^2 + 1)^{30} + 4 \tag{3}$$

or

$$\frac{dy}{dx} = [31(x^2 + 1)^{30} + 4](2x) \tag{4}$$

as the solution to this problem. Equations (3) and (4) typify the usual misapplications of the chain rule. These errors could be more easily avoided here if equation (2) were preceded, <u>at least in thought</u>, by the sum rule for derivatives

$$\frac{dy}{dx} = \frac{d}{dx}[(x^2 + 1)^{31}] + \frac{d}{dx}[4x].$$

The person who makes the type of error exampled in equation (3) has failed to apply the chain rule at all. He is leaving the $\frac{dt}{dx}$ out of $\frac{d}{dx}[t^n] = nt^{n-1} \cdot \frac{dt}{dx}$. The person who makes the type of error exampled in equation (4) has remembered (?) but misapplied the chain rule. His error is more apparent when we rewrite equation (4) as $31(x^2 + 1)^{30}(2x) + 4(\underline{2x})$.

Any error you might make involving the chain rule is some combination of these two types of errors. You will more easily avoid such a comedy (of errors) by remembering to <u>apply the chain rule to those terms, and only those terms, in which it is called for</u>. For example, if

$$y = (x^2 + 1)^{31} + (x^3 - x)^{-2} + x^2,$$

then

$$\frac{dy}{dx} = 31(x^2 + 1)^{30}(2x) - 2(x^3 - x)^{-3}(3x^2 - 1) + 2x.$$

<u>Example 2</u>. Calculate $\frac{dz}{dx}$ for $z = [(x^2 + 1)^{31} + 4x]^7$

<u>Solution</u>. $\frac{dz}{dx} = 7[(x^2 + 1)^{31} + 4x]^6 \cdot \frac{d}{dx}\{(x^2 + 1)^{31} + 4x\}$

$$= 7[(x^2 + 1)^{31} + 4x]^6 \cdot \{31(x^2 + 1)^{30}(2x) + 4\}$$

Notice how this example builds upon example 1. Namely, $z = y^7$ and we may write the first line of our solution as

$$\frac{dz}{dx} = \frac{dz}{dy} \cdot \frac{dy}{dx} = 7y^6 \cdot \frac{dy}{dx} .$$

<u>Example 3</u>. Calculate $\frac{dt}{dx}$ for $t = \{x^3 + [(x^2 + 1)^{31} + 4x]^7\}^5$.

<u>Solution</u>. $\frac{dt}{dx} = 5\{x^3 + [(x^2 + 1)^{31} + 4x]^7\}^4 \cdot \frac{d}{dx}\{x^3 + [(x^2 + 1)^{31} + 4x]^7\}$

$$= 5\{x^3 + [(x^2 + 1)^{31} + 4x]^7\}^4 \cdot \left(3x^2 + 7[(x^2 + 1)^{31} + 4x]^6 \cdot \right).$$

$$\frac{d}{dx}[(x^2 + 1)^{31} + 4x]$$

Notice how this example keeps a good thing going. If we let $u = x^3$, then $t = (u + z)^5$ and we can detail the first few steps of the solution as follows:

$$\frac{dt}{dx} = \frac{d}{dx}[(u + z)^5] = 5(u + z)^4 \cdot \frac{d}{dx}(u + z)$$

$$= 5(u + z)^4 \cdot \left(\frac{du}{dx} + \frac{dz}{dx}\right)$$

$$= 5(u + z)^4 \cdot \left(\frac{du}{dx} + \frac{dz}{dy} \cdot \frac{dy}{dx}\right)$$

$$= 5(u + z)^4(3x^2 + 7y^6 \frac{dy}{dx}) = \text{etc.}$$

<u>Example 4</u>. Calculate $\frac{dy}{dx}$ for $y = (f \circ g \circ h)(x) = f(g(h(x)))$, where $h(x) = x^2 + 1$, $g(t) = t^3 - 2t$, and $f(z) = z^5$.

<u>Solution</u>. If we let $z = g(h(x))$ and $t = h(x)$, then a repeated application of the

chain rule gives $\frac{dy}{dx} = f'(g(h(x))) \cdot g'(h(x)) \cdot h'(x)$,
or more briefly,

$$\frac{dy}{dx} = \frac{dy}{dz} \cdot \frac{dz}{dt} \cdot \frac{dt}{dx} . \qquad (5)$$

Equation (5) yields

$$\frac{dy}{dx} = 5z^4(3t^2 - 2)(2x), \qquad (6)$$

which in terms of x is

$$\frac{dy}{dx} = 5\{(x^2 + 1)^3 - 2(x^2 + 1)\}^4 \cdot [3(x^2 + 1)^2 - 2] \cdot (2x) \qquad (7)$$

Of course, the other way to find $\frac{dy}{dx}$ here is to first express y in terms of x.

$$y = z^5 = (t^3 - 2t)^5 = [(x^2 + 1)^3 - 2(x^2 + 1)]^5$$

Consequently,

$$\frac{dy}{dx} = \underbrace{5[(x^2 + 1)^3 - 2(x^2 + 1)]^4}_{\frac{dy}{dz}} \cdot \underbrace{\{3(x^2 + 1)^2(2x) - 2(2x)\}}_{\frac{dz}{dx} = \frac{dz}{dt} \cdot \frac{dt}{dx}},$$

which you should compare for accuracy and simplicity with the earlier derivation.
The difference in approach is simply a question of whether you solve in terms of
x before or after you differentiate. It is our feeling that the first approach
is an easier application of the chain rule.

Finally, suppose we wanted to evaluate $\frac{dy}{dx}$ when x = 1. Note that equation (7)
is not necessary, for we could learn from the equations given in the statement of
the problem that t = 2, z = 4 and y = 32 when x = 1. Then we may turn to equation
(6) to compute $\frac{dy}{dx} = 5(4)^4(3 \cdot 2^2 - 2)(2)$ when x = 1.

Example 5. Given that $y = t^2 + 1$, $t = u + \sqrt{u}$ and $u = x^3 - 2x$, find $\frac{dy}{dx}$ when x = 2.
Solution. $\frac{dy}{dt} = 2t, \frac{dt}{du} = 1 + \frac{1}{2\sqrt{u}}$ and $\frac{du}{dx} = 3x^2 - 2$.

When x = 2, $u = (2)^3 - 2(2) = 4$ and $t = 4 + \sqrt{4} = 6$ so that

$$\frac{dy}{dt} = 12, \frac{dt}{du} = \frac{5}{4} \text{ and } \frac{du}{dx} = 10;$$

whence

$$\frac{dy}{dx} = \frac{dy}{dt} \cdot \frac{dt}{du} \cdot \frac{du}{dx} = (12)(\frac{5}{4})(10) = 150.$$

SECTION 3.6

We will have occasion to use the second derivative in Chapter 4 and other higher order derivatives in Chapter 13, so this topic is not frivolous. The second derivative of an expression is simply the first derivative of the first derivative. Each time we differentiate with respect to the same variable. Thus

$$\frac{d^3}{dx^3} \text{ (some expression)} = \frac{d}{dx}(\frac{d}{dx}(\frac{d}{dx} \text{ (some expression)})).$$

Since we have used the "dee" notation in chain rule examples of derivatives taken with respect to quantities not represented by a single symbol, we should caution you that $\frac{d^3}{dx^3}$ can only mean the third derivative with respect to x. The third derivative with respect to x^2 is written as $\frac{d^3}{d(x^2)^3}$. The first derivative with respect to x^3 is written as $\frac{d}{d(x^3)}$. Sometimes $\frac{d^3}{dx^3}$ is confused with $(\frac{d}{dx})^3$. Notice though, that $\frac{d^3}{dx^3}(x^3 + 1) = 6$ whereas $[\frac{d}{dx}(x^3 + 1)]^3 = [3x^2]^3 = 27x^6$. The notation $\frac{d^k}{d(A)^n}$ only makes sense if k = n and, further, we will only see it when A is a single symbol as in $\frac{d^4}{dy^4}$ or $\frac{d^7}{dt^7}$.

SECTION 3.7

The business of differentiating inverse functions is presented primarily to extend the rule $\frac{d}{dx}(x^n) = nx^{n-1}$ to include rational values for n, not just integral values. The graph on p. 123 of the text illustrates the main features of inverse functions. Literally, the statement

$$(f^{-1})'(f(x)) = 1/f'(x) \quad \text{or} \quad \frac{dx}{dy} = \frac{1}{dy/dx}$$

may be modeled by saying that if A changes 3 times as fast as B, then B changes one-third as fast as A.

Some qualifications accompany the rule

$$\frac{d}{dx}(x^{p/q}) = \frac{p}{q} x^{\frac{p}{q} - 1} , \tag{1}$$

where p and q are integers, $q \neq 0$ and $\frac{p}{q} < 1$ is in reduced form. If q is even, then equation (1) holds iff $x > 0$. We exclude $x < 0$, as $x^{p/q}$ is not defined for $x < 0$. We also exclude $x = 0$ since the derivative is a full or two-sided limit, and the left hand limit doesn't exist here (Why?). If q is odd, we only exclude $x = 0$ since $x^{p/q - 1}$ is not defined at $x = 0$ when $p/q < 1$.

The graphs below illustrate what happens when $0 < \frac{p}{q} < 1$. What happens when $\frac{p}{q} > 1$?

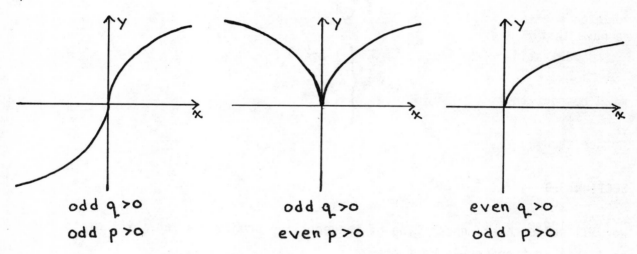

odd q >0   odd q >0   even q >0
odd p >0   even p >0   odd p >0

In the first graph, we have a <u>vertical tangent</u> which some might say has "infinite" slope. For such a case, we might write $\lim\limits_{h \to 0} \frac{(0 + h)^{p/q} - 0}{h} = +\infty$ to indicate more precisely just why the limit doesn't exist. The second graph has no tangent line at $x = 0$, vertical or otherwise, but rather a <u>cusp</u>. We would write

$$\lim\limits_{h \uparrow 0} \frac{(0 + h)^{p/q} - 0}{h} = -\infty \quad \text{and} \quad \lim\limits_{h \downarrow 0} \frac{(0 + h)^{p/q} - 0}{h} = +\infty$$

to be more explicit as to just why the limit (derivative) doesn't exist.

We can handle more complex problems by remembering that the derivative exists and is given by our mechanical rules wherever the mechanically derived rule makes sense. Of course, if endpoints of intervals occur in the domain of f, these endpoints are omitted from the domain of f'. For example, if $f(x) = (x^2 - 1)^{3/2}$, then $f'(x) = 3x(x^2 - 1)^{1/2}$ and f' exists wherever this formula is meaningful and excludes endpoints from the domain of f. As the domain of f is $(-\infty, -1] \cup [1, \infty)$, the domain of f' is $(-\infty, -1) \cup (1, \infty)$. The following is a more complicated example illustrating these ideas. Notice that f has no derivative at x = 8 even though it is continuous there.

For

$$f(x) = \begin{cases} x^{1/3} & , x < 8 \\ \\ \dfrac{x}{4} & , x \geq 8 \end{cases}$$

we have that

$$f'(x) = \begin{cases} \dfrac{1}{3} x^{-2/3} & , x < 8 \text{ but } x \neq 0 \\ \\ \dfrac{1}{4} & , x > 8 \end{cases}$$

SECTION 3.9

We will solve the toughest type of tangent line problem we can think of to show that this section should be a breeze.

<u>Example</u>. Find the equations of all the straight lines, if any, which pass through the point (-2,2) and are tangent to the curve $y = f(x) = x^3 - x$.

If the point (-2,2) were on the curve, we could evaluate

$$y - f(-2) = f'(-2)(x - (-2))$$

and be done. But such is not the case as $f(-2) = 6 \neq 2$. However, it doesn't hurt to be on the watch for something special that could greatly simplify the problem at hand. Since (-2,2) is not the point of tangency, let us say that (a,b) is the point of tangency. Since (a,b) is on the curve, we know that

$$b = f(a) = a^3 - a. \tag{1}$$

The slope of our tangent line can be represented as

$$\frac{b - (2)}{a - (-2)} , \qquad (2)$$

since (a,b) and (-2,2) are distinct points on the line. Because f'(a) is another expression for the same slope, we may equate this value with that in (2):

$$f'(a) = 3a^2 - 1 = \frac{b - 2}{a + 2} . \qquad (3)$$

We have represented all that we can about this problem with the given information: (-2,2) lies on the tangent line, (a,b) lies on the tangent line and the curve; and the slope of the tangent line is given by the "two-point formula" and by the derivative evaluated at the point of tangency. Fortunately, this is enough. The quantities a and b are "unknowns" but equations (1) and (3) relate them. Your immediate temptation to solve simultaneously is brilliant. Substituting equation (1) in equation (3), we find that

$$3a^2 - 1 = \frac{(a^3 - a) - 2}{a + 2} ,$$

$$3a^3 - a + 6a^2 - 2 = a^3 - a - 2,$$

$$2a^3 + 6a^2 = 0$$

$$a^2(a + 3) = 0$$

Thus, a = 0 or a = -3. As f(0) = 0, f'(0) = -1, f(-3) = -24 and f'(-3) = 26, the equations of the two lines that fit the conditions are

$$y - 0 = -1(x - 0) \text{ and } y + 24 = 26(x + 3).$$

SECTION 3.10

The differentiation of an equation like

$$x^3 + xy^2 - y^4 + 7 = 0 \qquad (1)$$

involving two variables is called implicit since equation (1) _may_ implicitly define one or more functions of one of the variables in terms of the other variable.

In general, it is not possible to solve an equation like (1) for y in terms of x. Even when we can, we might not obtain a function. Consider the circle $x^2 + y^2 = 4$. We may obtain $y = \pm\sqrt{4 - x^2}$, but this expression is not a function as some values of x give rise to more than one value for y. Here we have two implicitly defined functions: the upper and lower semicircles, which may though be explicitly expressed:

$$y = \sqrt{4 - x^2} \quad \text{for } -2 \le x \le 2$$

and
$$y = -\sqrt{4 - x^2} \quad \text{for } -2 < x < 2. \tag{2}$$

These are not the only possibilities. (Remember, a function is specified both by a rule and a domain. For example, $y = x^2$ for $x > 0$ and $y = x^2$ for $1 < x < 2$ are two different functions.) One of the many changes we could make in equation (2) is

$$y = \sqrt{4 - x^2} \quad \text{for } -2 < x < 2$$

and
$$y = -\sqrt{4 - x^2} \quad \text{for } -2 \le x \le 2. \tag{3}$$

Any collection of points or any curve in the plane may be partitioned into one or more functions. It is common practice to make the minimum number of partitions. Even though it is correct to say that $x^2 + y^2 = 4$ implicitly defines the following three functions:

$$y = \sqrt{4 - x^2}, \ x \text{ in } [-2,1];$$

$$y = -\sqrt{4 - x^2}, \ x \text{ in } [0,2);$$

$$y = \begin{cases} \sqrt{4 - x^2}, \ x \text{ in } (1,2] \\ -\sqrt{4 - x^2}, \ x \text{ in } (-2,0) \end{cases}$$

the equations (2) or (3) would be the most likely choice.

The point of implicit differentiation is that we can calculate the derivative without first explicitly setting up the functions implicitly defined by an equation like (1).

The implicit differentiation of $x^2 + y^2 = 4$ yields $2x + 2y\frac{dy}{dx} = 0$ or $\frac{dy}{dx} = -x/y$ for $y \ne 0$. Looking at equation (2), if $y = \sqrt{4 - x^2}$, then

$$\frac{dy}{dx} = -x/\sqrt{4 - x^2} = -x/y$$

for $|x| < 2$. (Notice that $y = 0$ iff $|x| = 2$.) Next, if $y = -\sqrt{4 - x^2}$, then $\frac{dy}{dx} = +x/\sqrt{4 - x^2} = -x/y$ for $|x| < 2$, which again agrees with the implicitly derived result.

The implicit differentiation of $x^3 + y^3 = 7$ yields $3x^2 + 3y^2\frac{dy}{dx} = 0$ or $\frac{dy}{dx} = -x^2/y^2$ for $y \neq 0$. Here, only one function: $y = (7 - x^3)^{1/3}$, is defined implicitly. Differentiating this function, we find that

$$\frac{dy}{dx} = \frac{1}{3}(7 - x^3)^{-2/3}(-3x^2) = -x^2/y^2 \text{ for } y \neq 0.$$

These two brief examples suggest, as is the case, that implicit differentiation is mechanically correct. Finally, from $x^2 + y^2 = -1$, we can calculate implicitly that $\frac{dy}{dx} = -x/y$ for $y \neq 0$. However, no functions are defined by $x^2 + y^2 = -1$ (why?). Our derivation, though, of $\frac{dy}{dx}$ was correct. The derivative fails to exist because there is no function. The meaning of our calculations, as exampled here, is that if there had been a function and if it were differentiable, then $\frac{dy}{dx} = -x/y$ would be the expression for that derivative.

We will now examine some more complex implicit differentiations. All the usual rules of differentiation apply.

<u>Example 1.</u> Calculate $\frac{dy}{dx}$ and $\frac{d^2y}{dx^2}$ for equation (1).

<u>Solution.</u>
$$3x^2 + \underbrace{(y^2 + x(2y\frac{dy}{dx}))}_{\frac{d}{dx}(xy^2) \text{ by product rule.}} - 4y^3\frac{dy}{dx} + 0 = 0 \tag{4}$$

so that
$$\frac{dy}{dx} = \frac{3x^2 + y^2}{4y^3 - 2xy}. \tag{5}$$

There are two ways we could proceed in order to find $\frac{d^2y}{dx^2}$. First, we could differentiate equation (5) with respect to x.

$$\frac{d^2y}{dx^2} = \frac{(4y^3 - 2xy)(6x + 2y\frac{dy}{dx}) - (3x^2 + y^2)(12y^2\frac{dy}{dx} - 2y - 2x\frac{dy}{dx})}{(4y^3 - 2xy)^2}, \tag{6}$$

which may be written in terms of x and y by substitution of (5) in (6).

Second, we could differentiate equation (4) with respect to x.

$$6x + 2y\frac{dy}{dx} + \left(2y\frac{dy}{dx} + 2x\left(\frac{dy}{dx}\right)^2 + 2xy\frac{d^2y}{dx^2}\right) - 12y^2\left(\frac{dy}{dx}\right)^2 - 4y^3\frac{d^2y}{dx^2} = 0$$

$$\underbrace{\phantom{2y\frac{dy}{dx} + 2x\left(\frac{dy}{dx}\right)^2 + 2xy\frac{d^2y}{dx^2}}}$$

$$\frac{d}{dx}\left[x\left(2y\frac{dy}{dx}\right)\right] \text{ by product rule twice}$$

You should verify that we obtain the same result for $\frac{d^2y}{dx^2}$ by both approaches.

This second approach is preferable when the derivatives need only be evaluated for a few specific values of x and y.

Frequently, students calculate $\frac{d^2y}{dx^2}$ by squaring $\frac{dy}{dx}$. The following example should convince you that this approach is wrong.

Example 2.   Calculate $\frac{d^2y}{dx^2}$ for xy = 1.

Solution.   Differentiating xy = 1 implicitly twice, we obtain

$$x\frac{dy}{dx} + y = 0$$

and then

$$\frac{dy}{dx} + x\frac{d^2y}{dx^2} + \frac{dy}{dx} = 0$$

so that

$$\frac{d^2y}{dx^2} = -\frac{2}{x}\frac{dy}{dx} = -\frac{2}{x}\left(-\frac{y}{x}\right) = \frac{2y}{x^2} = \frac{2}{x^3}.$$

But,

$$\left(\frac{dy}{dx}\right)^2 = \left(\frac{-y}{x}\right)^2 = \left(-\frac{1}{x^2}\right)^2 = \frac{1}{x^4} \neq \frac{d^2y}{dx^2}.$$

Finally, compare the following example to Example 1.

Example 3.   Calculate $\frac{dx}{dy}$ for equation (1).

Solution. $$3x^2\frac{dx}{dy} + y^2\frac{dx}{dy} + 2xy - 4y^3 = 0$$

so that $$\frac{dx}{dy} = \frac{4y^3 - 2xy}{3x^2 + y^2}.$$

The Mean-Value Theorem and Applications

4

The structure of theorems was discussed in Section 2.5. The points of that discussion are illustrated below in terms of Rolles Theorem and The Mean Value Theorem.

Rolles Theorem:

    Hypothesis: If f is differentiable on (a,b) and continuous on [a,b] and if
               f(a) = f(b) = 0, then

    Conclusion: there is at least one real number c in (a,b) at which f'(c) = 0.

The Mean Value Theorem:

    Hypothesis: If f is differentiable on (a,b) and continuous on [a,b], then

    Conclusion: there is at least one real number c in (a,b) at which

$$f'(c) = \frac{f(b) - f(a)}{b - a} \ .$$

To apply either of these theorems, we need first to verify that the respective hypotheses are true. It is possible for the conclusion of a theorem to be true even though the hypothesis fails to hold. The graphs on the next page example this possibility for The Mean Value Theorem. Notice that in each instance the conclusion of The Mean Value Theorem is true, but one or both parts of the hypothesis of The Mean Value Theorem are not satisfied. Thus, the theorem does not apply in these cases, even though its conclusion is true. We would have to appeal to something other than the theorem for a proof. It would be enough just to find (if we could) a value for c in (a,b) such that

$$f'(c) = [f(b) - f(a)]/(b - a).$$

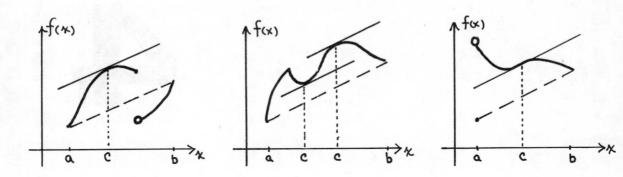

There are two obviously related but different questions that may be put to
you. We will example them for Rolles Theorem.

1.   Does Rolles Theorem apply to the function f on [a,b]?

2.   Does the conclusion of Rolles Theorem hold for f on [a,b]?

Example 1.   Suppose $f(x) = 4 - 3x - x^2$, a = -4, b = 1.

Note that f(-4) = f(1) = 0. Also, f is differentiable on (-4,1) and
continuous on [-4,1] since it is a polynomial. The hypothesis of Rolles Theorem
is satisfied and, consequently, the theorem may be applied to the function f.

We have answered question 2 as well. As the hypothesis of Rolles Theorem is
satisfied for f on [-4,1], the conclusion of Rolles Theorem must hold for f. We
may prefer, however, to take a constructive rather than existential view and
simply determine if there is a real number c between -4 and 1 where f'(c) = 0.
One would proceed as follows.

$$f(x) = 4 - 3x - x^2$$
$$f'(x) = -3 - 2x$$
$$0 = -3 - 2x$$
$$-3/2 = x$$

Since f'(-3/2) = 0 and -3/2 is between -4 and 1, the answer to question 2 is yes.
The answer to question 1 could still be no. Only if these calculations had
yielded no values for c could we have resolved question 1 using this constructive
approach. What would have been the answer then and why?

Suppose the following question had been asked.

3. <u>Find</u> those points in (a,b), if any, where the conclusion of Rolles Theorem holds.

If you take the first approach, you will either determine that values of c <u>exist</u> or you will determine that the hypothesis of Rolles Theorem is not true. In either event, you will still need to execute the algebra to determine the specific values of c, if any. It is preferable then to take a mechanical or constructive approach when you are asked to determine <u>specific</u> values as opposed to resolving just <u>existence</u>. If you find no values for c in (a,b), then it might be advisable to try the first approach. If the hypothesis is satisfied, then values of c exist but you missed them. In any event, since question 3 is essentially algebraic, you can only be sure of your answer if your algebra is complete and accurate.

<u>Example 2</u>. $f(x) = x + \dfrac{1}{x} + \dfrac{3}{2}$, a = -1, b = 3.

The answer to question 1 is <u>no</u>. There are three equally good reasons you could give to support this answer: f is not continuous at x = 0; f is not differentiable at x = 0; $f(b) = f(3) \neq 0$.

As the answer to question 1 is no, you can say nothing about the conclusion of Rolles Theorem. There might yet be values of c in (a,b) where f'(c) = 0. Your work to resolve questions 2 and 3 is the same.

$$f(x) = x + \frac{1}{x} + \frac{3}{2}$$

$$f'(x) = 1 - \frac{1}{x^2}$$

$$0 = 1 - \frac{1}{x^2}$$

$$1 = x^2$$

$$x = 1 \text{ or } -1.$$

The conclusion of Rolles Theorem holds, since f'(1) = 0 and 1 is in the interval (-1,3). The value c = -1 does not satisfy the conclusion of Rolles Theorem (why?).

The hypothesis of Rolles Theorem is weaker than that of The Mean Value Theorem, for the additional conditions: f(a) = f(b) = 0, are hypothesized. When we substitute these conditions into the equation in the conclusion of The Mean Value Theorem: $f'(c) = \dfrac{f(b) - f(a)}{b - a}$, we get $f'(c) = 0$. This is why Rolles Theorem is a special case of The Mean Value Theorem. In fact, The Mean Value Theorem could be retitled the Generalized Rolles Theorem, since it requires less and says more.

      You should recall from the discussion in Section 2.5 that the contrapositive of each of these theorems is a true implication.

Contrapositive of The Mean Value Theorem:
      If for no real number c between a and b is $f'(c) = \dfrac{f(b) - f(a)}{b - a}$, then f is
      not differentiable on (a,b) or f is not continuous on [a,b].
Can you write the contrapositive of Rolles Theorem?

SECTION 4.2

Increasing and decreasing are function properties <u>on an interval, not at a point</u>. (See Definition on p. 140 of the text.) Frequently, students <u>incorrectly</u> assert that $f(x) = x^2$ is decreasing on $(-\infty, 0)$ where $f'(x) < 0$, increasing on $(0, \infty)$ where $f'(x) > 0$, and constant at $x = 0$ where $f'(x) = 0$.

      The converse of Theorem 4.2.2 is not true. For example, f is increasing on the interval I if $f'(x)$ is positive for all x in I. However, we may not conclude that $f'(x)$ is positive for all x in I if f is increasing on I. Note that $f(x) = x^3$ is increasing on $(-1,2)$ but $f'(0)$ is not positive. Further, the function graphed below is increasing on the given interval I but does not have a positive derivative at all points in I.

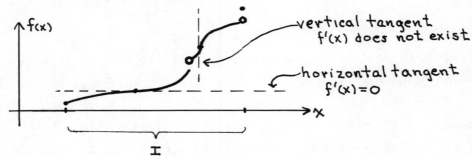

To establish the increasing nature of the last two functions, it is not enough to just examine the sign of the derivative. We need to appeal to the definition of increasing and to the theorems which extend the term increasing for continuous functions from open to closed intervals.

A simple two step procedure for determining increasing and decreasing will always apply. First, differentiate the given function and make the necessary algebraic manipulations to determine those open <u>intervals</u> on which $f'(x)$ is positive, negative or zero. f is then increasing, decreasing or constant on those respective <u>intervals</u>. Next, apply the definitions and remaining theorems to determine in which intervals the remaining points: endpoints of intervals, points of discontinuity and points where $f'(x)$ is not defined, <u>might</u> be included.

Surely, it is trivial to determine if a function g is positive for some specific value of c, but how do we determine all values of c for which $g(c)$ is positive? Ponder for a moment the following two questions. If $g(a)$ and $g(b)$ are positive, under what conditions is $g(x)$ positive for all x between a and b? If $g(a)$ is positive and $g(b)$ is negative, what must happen between a and b? The simple, but non-constructive, answer to the first question is that $g(x)$ must never be zero or negative for any x between a and b. This answer leads us to the second question, which is completely resolved by the Intermediate Value Theorem. Namely, g is either not continuous on $[a,b]$ or $g(c) = 0$ for some c in (a,b). This result implies the following procedure:

1. Make a complete list of all points where g is either discontinuous or zero.
2. At some convenient point between each pair of successive points just listed determine if g is positive or negative. It follows that g is correspondingly positive or negative on the entire interval bounded by the two successive points.

<u>Example 1</u>. Find the intervals on which f is increasing and those on which f is decreasing, if

$$f(x) = \begin{cases} x & \text{for } x < 1 \\ x^3 - 9x^2 + 24x - 15 & \text{for } x \geq 1. \end{cases}$$

<u>Solution</u>. First we calculate f'.

$$f'(x) = \begin{cases} 1 & \text{for } x < 1 \\ 3x^2 - 18x + 24 & \text{for } x > 1. \end{cases}$$

Notice that $f'(x)$ is not defined for $x = 1$. Hence, f is continuous but not differentiable at $x = 1$ (Why?). Also, $f'(x) = 0$ when $x = 2$ or 4. Thus, the list to be drawn up under step 1 of our procedure consists of the points $x = 1$, 2 and 4. This gives us four intervals: $(-\infty,1)$, $(1,2)$, $(2,4)$ and $(4,\infty)$. Since $f'(0) = 1$, $f'(x) > 0$ for all x in $(-\infty,1)$. Since $f'(\frac{3}{2}) = \frac{15}{4}$, $f'(x) > 0$ for all x in $(1,2)$. We treat the other two intervals similarly. The results by step 2 of our procedure could be represented on a number line.

Sign of $f'(x)$

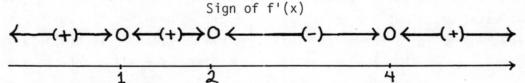

Notice that the signs <u>don't</u> alternate for this example.

Thus, f is decreasing on $(2,4)$ and increasing on $(-\infty,1)$, on $(1,2)$, and on $(4,\infty)$. It remains to consider the points $x = 1$, 2 and 4. Since f is continuous, f is decreasing on $[2,4]$ and increasing on $(-\infty,1]$, on $[1,2]$ and on $[4,\infty)$. Notice how we have used Theorem 4.2.3 of the text.

There remains the question of whether or not we can say f is increasing on $(-\infty,2]$ because f is increasing on the successive intervals $(-\infty,1]$ and $[1,2]$. For this example, the answer is yes because f is continuous at $x = 1$. The general question of just when you may say f is increasing (decreasing) on (a,b) if it is increasing (decreasing) on (a,c) and on (c,b), for $a < c < b$, is addressed in Examples 3 through 6.

<u>Example 2</u>. $f(x) = x^2$.

We have $f'(x) > 0$ for $x > 0$ and $f'(x) < 0$ for $x < 0$. Thus, f is increasing on $(0,\infty)$ and decreasing on $(-\infty,0)$. Since f is continuous for all x, f is increasing on $[0,\infty)$ and decreasing on $(-\infty,0]$. The inclusion of $x = 0$ in both intervals does not say f is both increasing and decreasing at x = 0. In fact f is neither _at_ x = 0 since these properties apply _on intervals_, not at individual points.

Example 3. f(x) = x - [x] for 1 < x < 3.

Recall that [x] denotes the greatest integer function. Since $\frac{d}{dx}([x]) = 0$ if x is not an integer and does not exist if x is an integer, we find that

$$f'(x) = \begin{cases} 1 & \text{for } 1 < x < 2 \\ \text{undefined} & \text{for } x = 2 \\ 1 & \text{for } 2 < x < 3. \end{cases}$$

So, f is increasing on (1,2) and on (2,3). Notice that f is not continuous on (1,3). In itself, this fact does not prohibit f from being increasing on (1,3). It just suggests that you give the matter some thought. Well, since

$$f(\tfrac{3}{2}) = \tfrac{1}{2} > 0 = f(2),$$

it is impossible for f to be increasing on (1,3). Since f is continuous at x = 2 from the _right_, we conclude that f is increasing on (1,2) and on [2,3). Note that 2 is contained only in the second interval.

Example 4. g(x) = x + [x] for 1 < x < 3.

We have

$$g'(x) = \begin{cases} 1 & \text{for } 1 < x < 2 \\ \text{undefined} & \text{for } x = 2 \\ 1 & \text{for } 2 < x < 3, \end{cases}$$

so g is increasing on (1,2) and on (2,3). Like the function f of the last example, g is only continuous from the right at x = 2. Thus, g is increasing on (1,2) and on [2,3). Is g increasing on (1,3)? The answer is yes, since g(x) < g(2) < g(t) for x < 2 < t. Notice that it would be incorrect to say g is increasing on [1,3] because g is not defined at x = 1 or 3.

Example 5. $f(x) = \dfrac{1}{x}$.

As $f'(x) = -\dfrac{1}{x^2} < 0$ for $x \neq 0$, f is decreasing on $(-\infty,0)$ and on $(0,\infty)$. But,

f is not decreasing on $(-\infty,\infty)$ since $f(0)$ is not defined.

Example 6. $f(x) = \begin{cases} 1/x & \text{for } x \neq 0 \\ 0 & \text{for } x = 0. \end{cases}$

Here, we have an example of a function where a point in its domain may <u>not</u> be included in any interval where the function is increasing, decreasing or constant. This simple extension of example 6 is also decreasing on $(-\infty,0)$ and on $(0,\infty)$. We may not say f is decreasing on $(-\infty,\infty)$ nor may the point $x = 0$ be included in $(-\infty,0)$ or $(0,\infty)$ since $f(x) < 0 < f(t)$ if $x < 0 < t$.

This section of the text closes with a critical theorem (4.2.4) preparing us for the companion process to differentiation; namely, integration. It will be the focal point of our attention beginning with Chapter 5. In the last Chapter we developed some expertise at answering the question: Given $f(x)$, what is $f'(x)$? The present theorem is the beginning of the assault on the question: Given $f'(x)$, what is $f(x)$? The theorem states that $f'(x) = g'(x)$ for all x in the open interval I if and only if f and g differ by a constant on I. One might be tempted to abbreviate this result as follows: f - g = a constant iff $f' = g'$. Something important has been omitted (What?). As a hint, the functions of examples 3 and 4 don't differ by a constant.

SECTIONS 4.3-4

Local extrema occur in open intervals within the domain of a function. A function is therefore defined on both sides of the number $x = c$ where a local extremum occurs. If $f'(c)$ exists, it must be zero. If $f'(c)$ does not exist, then f may or may not be continuous at $x = c$. Each of these possibilities identifies $x = c$ as a critical point or value where a local extremum may occur. The following table lists which tests may be applied in each possible situation.

Condition	$f'(c) = 0$	f'(c) does not exist	
		f continuous at x = c	f not continuous at x = c
Test	Definition 1st derivative test 2nd derivative test	Definition 1st derivative test	Definition

TABLE 1: Conditions and Tests for Local Extrema

The second derivative test does not apply if f'(c) does not exist simply because f"(c) does not exist. The first derivative test does not apply if f is not continuous since continuity is part of the hypothesis of this test. For example, consider the function in Diagram 1.

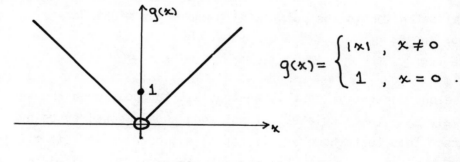

$$g(x) = \begin{cases} |x| & , x \neq 0 \\ 1 & , x = 0 \end{cases}$$

Diagram 1

Although $g'(x) = -1 < 0$ for $x < 0$ and $g'(x) = 1 > 0$ for $x > 0$, g does not have a local minimum at x = 0. In fact, g has a local maximum at x = 0. (Why? Use Definition 4.3.1 of the text.)

In determining points where f'(x) does not exist, you should recall that the domain of f' is always a subset of the domain of f. You can then inspect the function f' on the domain of f to see which points are exceptions. If $f'(x) = 1/x^2$, then f'(0) does not exist. If $f'(x) = \sqrt{x}$, then f'(x) is not defined for $x \leq 0$. If the domain of f is [a,b], then f' does not exist at the endpoints a and b. In general, it is relatively easy to determine the domain of

f' except perhaps in the case of piecewise defined functions. You should draw
a graph to convince yourself that if

$$f(x) = \begin{cases} x^2 & \text{for} & x \leq 1 \\ 2x & & 1 < x < 2 \\ x^2 - 2x + 4 & & 2 \leq x \leq 3 \\ 3x - 2 & & 3 < x \end{cases}$$

then

$$f'(x) = \begin{cases} 2x & \text{for} & x < 1 \\ (\text{d.n.e.}) & & x = 1 \\ 2 & & 1 < x \leq 2 \\ 2x - 2 & & 2 < x < 3 \\ (\text{d.n.e.}) & & x = 3 \\ 3 & & 3 < x \end{cases}$$

Pay special attention to the point $x = 1$, where you might have expected $f'(1) = 2$,
as $\lim_{x \to 1} f'(x) = 2$.

The second derivative test is the easiest to apply when the second derivative
is relatively easy to calculate. If not, one should consider using one of the
other tests. This test, however, is not always conclusive. If $f''(c) = 0$, we
must rely on some other test. For example, there is a local minimum at $x = 0$ for
$f(x) = x^4$, but $f''(0) = 0$ rather than being positive. Again, since
$f''(0) = f'(0) = 0$ for $f(x) = x^5$, we must apply other tests before we may see that
$f(x) = x^5$ has neither a local minimum nor a local maximum at $x = 0$.

There is a natural tendency to align a negative second derivative with a
minimum and a positive second derivative with a maximum. To avoid this confusion,
simply bear in mind the example $f(x) = x^2$ which has a local minimum at $x = 0$
where $f''(x)$ is positive. Looking at $f''(x)$ as the first derivative of $f'(x)$, see
if you can explain why a positive value of f'' is linked with a minimum and a
negative value with a maximum.

Answer: If $f'' > 0$, then f' is increasing. So, if $f'(c) = 0$ and $f''(c) > 0$, then f' must be changing from negative to positive at $x = c$. This means that f is changing at $x = c$ from a decreasing to an increasing function. Therefore, f has a minimum at $x = c$. A similar argument can be used when $f'' < 0$.

The first derivative test requires that we examine the sign of $f'(x)$ for values of x to the left of and to the right of but near $x = c$. Rather than writing "$f'(x) < 0$ for x in $(c - \delta, c)$" or "$f'(x) < 0$ for x sufficiently close to c on the left," we would suggest you just write "$f'(c^-) < 0$." Similarly, we will use "c^+" to denote the interval $(c, c + \delta)$ of values near and to the right of $x = c$. This notation permits us to express quite briefly our supportive reasons in the analysis of a function.

Example 1. Find the local extrema of $f(x) = x^3 - 3x$.
Solution. $f(x) = x^3 - 3x$, all real x (is the domain)
        $f'(x) = 3(x - 1)(x + 1)$, all real x
Critical points: $f'(x) = 0$: $x = 1, -1$
                 $f'(x)$ does not exist (d.n.e.): none
f has a local minimum at $(1,-2)$ since $f'(1^-) < 0$, $f'(1) = 0$, and $f'(1^+) > 0$.
f has a local maximum at $(-1,2)$ since $f'(-1^-) > 0$, $f'(-1) = 0$, and $f'(-1^+) < 0$.

Observations. Notice first that by immediately listing the domains of f and f', we make it easier to check at a later time for points where f' does not exist. Next, we have calculated $f(c)$ when discussing the critical point c as we will probably need this value (for graphs or word problems), and this is a good way to organize all we know about f at c.

Testing by the definition on p. 146 of the text is a last resort. The graphs given in Diagram 2 typify what might happen at a critical point $x = c$ where f' does not exist and f is not continuous (so neither the first nor the second derivative tests may be used). Of course, many more graphs than the six given in Diagram 2 are possible. A little common sense should serve to resolve any problems you may encounter like these.

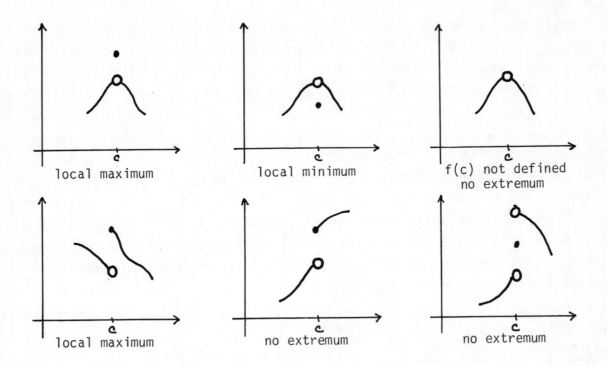

Diagram 2: Examples of critical values needing
special attention

If you know where to look, candidates for endpoint extrema are easy to find! Suppose f is continuous on [a,b], then f has an endpoint <u>min</u>imum at x = a if f'(a$^+$) is <u>pos</u>itive, whereas f has an endpoint <u>max</u>imum at x = b if f'(b$^-$) is <u>pos</u>itive. If the function f is <u>not</u> continuous at the endpoint, we must change our approach. Let f be continuous on [a,b), and have a jump discontinuity at x = b. If $\lim\limits_{x \uparrow b}$ f(x) < f(b), f has an endpoint maximum at x = b, whereas f has

an endpoint minimum at x = b if $\lim\limits_{x \uparrow b}$ f(x) > f(b).

<u>Example 2</u>. Does f(x) = x^2 on [-1,2] have an endpoint minimum at x = -1?
<u>Solution</u>. f(x) = x^2, [-1,2]
        f'(x) = 2x, (-1,2)
Since f is continuous from the right at x = -1 and f'(-1$^+$) < 0, f has an endpoint maximum at the point (-1,1). The answer therefore is no!

<u>Example 3</u>. Does f have an endpoint maximum at x = 1 where

$$f(x) = \begin{cases} x + 2, & x < 1 \\ 22/7, & x = 1. \end{cases}$$

<u>Solution</u>. Since $\lim_{x \uparrow 1} f(x) = 3 \neq f(1)$, f is discontinuous at the endpoint 1.
As $\lim_{x \uparrow 1} f(x) < f(1)$, f has an endpoint maximum at x = 1.

An absolute maximum, <u>if it exists</u>, is the largest of all the endpoint and local maxima of the function f. We are only guaranteed the existence of absolute extrema if f is continuous on a closed interval. (Maximum-Minimum Theorem). Consequently, some thought is required if f is discontinuous at some point and/or does not have a closed interval for a domain. Diagram 3 exemplifies two such instances which require special care.

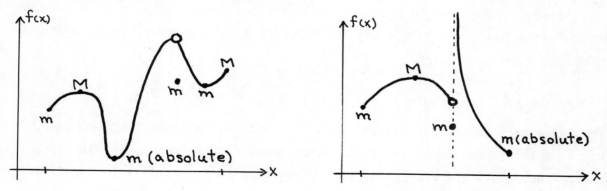

Diagram 3: Two functions lacking an absolute maximum.
M denotes a local or endpoint maximum, m a minimum.

If the domain of f is not closed, we should examine the behavior of f near the "endpoint(s)" of the domain. On the domain (a,b) if either $\lim_{x \downarrow a} f(x) = +\infty$ or $\lim_{x \uparrow b} f(x) = +\infty$, f has no absolute maximum on (a,b). In fact, if $\lim_{x \downarrow a} f(x) = M$ and f(x) < M for all x in (a,b), f has no absolute maximum. For a domain like [a,∞) we would examine $\lim_{x \to +\infty} f(x)$ for similar results. For example, $f(x) = x^2$ has no absolute maximum on (-∞,5]. Similarly f(x) = 1/x has no absolute extrema on (3,∞). Naturally, analogous remarks hold for the determination of an absolute minimum.

If we are solely interested in finding absolute extrema our task is far simpler. We do not need the first or second derivative tests. We do not need to first individually classify local and endpoint extrema. We need only evaluate the function at each critical point and evaluate a one-sided limit at any missing endpoint. This last step is necessary since the behavior of the function near a missing endpoint may rule out the existence of an absolute extremum.

Example 4. Examine the function $f(x) = x^4 - 2x^2 + 7$ on $[-1,3)$ for absolute extrema.

Solution. First we calculate the derivative

$$f'(x) = 4x^3 - 4x = 4x(x - 1)(x + 1).$$

Next, $f'(x) = 0$ for $x = 1$ and 0; note that $x = -1$ is not in the domain of f'. However, we do consider $x = -1$ as a critical point since it is an endpoint. On the other hand, $x = 3$ is a missing endpoint. Consequently, we compute the following four quantities:

$$f(-1) = 6, \qquad f(0) = 7, \qquad f(1) = 6, \qquad \lim_{x \uparrow 3} f(x) = 70.$$

The largest of these four values is 70. It "occurs" at a missing endpoint, so f has no absolute maximum. The smallest of these four values is 6. Since at least one of the points (here both) at which the value of 6 occurs is in the domain of f, the function has an absolute minimum of 6.

The topic of maxima and minima can seem confusing. This confusion usually arises because there are a variety of tests one can use and it is not always clear how to approach a problem. Most problems, though, are quite straightforward. For the rest, common sense is usually enough. In all cases, if you consciously write complete arguments and fully justify your work, you will catch your own mistakes and know when you've done enough work to resolve the problem. A careful rereading now of the Summary on p. 155 of the text should help you pull it all together.

SECTION 4.5

"Word problems" are really quite easy. Even though they are rigged, they should
appeal to you as they are less artificial than some other topics you may study.
The following procedure should serve as a helpful guide.
  (1) Read the problem carefully.
  (2) Write down what you are asked to find.
  (3) Reread the problem.
  (4) Draw and label whatever diagram may model the problem.
  (5) List whatever information is given.
  (6) Using this information and the diagram, formulate an equation containing
       the variable to be maximized or minimized.
  (7) Determine the domain of the function just formulated.
  (8) Find the required extremum.
  (9) Answer the original question with a complete sentence.
 (10) Breathe deeply.
The major points of this procedure will now be discussed and exampled. Step 10
is left for you as an exercise.

     The first step in the successful solution of a maximum-minimum word problem
is to <u>carefully</u> read the statement of the problem at least twice. The purpose
of these initial readings is to get an overview of the problem, not to solve it.
You may start to see a geometric or algebraic model for the situation being
described, but <u>before</u> you begin recording equations and facts you should determine
what the problem is asking.

     The second step in attacking any word problem therefore is to establish
exactly what the problem is asking you to find. Suppose you are asked to find
the largest area that can be enclosed by a rectangle whose perimeter is 40 inches.
After carefully reading the question, you should ascertain that "area" is the
quantity to be maximized. You may find it helpful to actually note that as the
first line of your written solution with a comment like "maximize the area of a
rectangle." To blindly attempt to write down equations with no expressed goal in
mind actually results in frustration as a function of confusion.

     The remaining steps of the procedure will now be illustrated using the above problem about a rectangle. It is easy to see the diagram or model involved here and with it you should list all the relevant information.

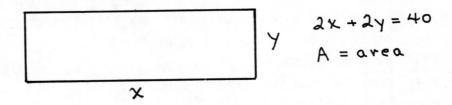

One should now formulate an equation containing the variable to be maximized, A; the equation A = xy is readily apparent.

We don't know which variable — x or y — to differentiate A "with respect to" in order to search out the maximal A. It doesn't matter which variable we choose, but it is usually best to first write the variable A as a function of just one of the other variables of the problem. This is where the remaining information in the problem comes into play. Although some problems are stocked with unnecessary information, it will generally be your experience in this course that all the given information is needed for the solution of the problem. Under this assumption, we should be able to use

$$A = xy \tag{1}$$

and
$$2x + 2y = 40 \tag{2}$$

to write A as a function of <u>one</u> variable. Solving equation (2) for y and substituting this result in equation (1), we arrive at

$$A(x) = x(20 - x) \tag{3}$$

Our next step is to determine the domain for the functional rule (3) that is dictated by the restrictions of the problem. Here, $0 < x < 20$. Why? Applying now the techniques of Sections 4.3-4, one discovers that "the maximum area that can be enclosed by a rectangle whose perimeter is 40 inches is 100 square inches." (You should verify this result.)

Consider the following variations on this problem.
1. If the perimeter of a rectangle is 40 inches and its length is 4 inches, what is the maximum possible area of the rectangle?
2. If the perimeter of a rectangle is 40 inches and its length is 100

inches, what is the maximum possible area of the rectangle?

　　3.　What is the maximum possible area of a rectangle?

Each of these legitimate extremum word problems has a solution since a complete response may be given to the question posed.

<u>Solution to problem 1</u>.　The diagram and information for this problem lead us to
$$A(x) = x(20 - x) \text{ for } \underline{x = 4}.$$
That is, there is only <u>one</u> rectangle which meets the conditions of the problem. The area of this rectangle, 64 square inches, is the desired maximum.

<u>Solution to problem 2</u>.　The diagram and information for this problem lead us to
$$A(x) = x(20 - x) \text{ for } 0 < x < 20$$
and
$$x = 100.$$
There are <u>no</u> rectangles which meet the conditions of the problem, so there is <u>no</u> rectangle of maximum area.

<u>Solution to problem 3</u>.　The diagram and information for this problem lead us to
$$A(x) = xy \text{ for } 0 < x, 0 < y.$$
As x and y can both be arbitrarily large, there is clearly <u>no</u> rectangle of maximum area.

　　These three problems represent the sort of problems that could and do arise when one enters the real world to design models and apply the techniques of this and other courses.　About the trickiest thing we might do is to give you a problem requesting a minimum when there is none.　Of course, if you correctly analyze the function (in the spirit of Sections 4.3-4), such "trickery" will be clear to you and the solution is simply the supported statement that there is no minimum.　Remember, <u>the solution to a problem is not just and always a number; it is the complete and justified response to the question posed</u>.

　　Once you have found a functional representation for the variable to be maximized or minimized, your next step is to specify the domain of this function. Invariably, the domain permitted by the problem is not as generous as that permitted by the algebraic sense of the function rule.　For example,

A(x) = x(20 - x) is algebraically sensible for all real x. But, since we are representing the area of a rectangle, we must have 0 < x < 20 to insure that A is positive. The determination of the domain is usually very simple, but unfortunately overlooked by too many too often. You should take special care in determining whether or not the problem permits the domain to be a closed interval. This decision, together with the question of the continuity of your function, determine the legitimacy of applying the Maximum-Minimum Theorem.

Before you read the solution to each of the following examples, see if you can establish exactly what you are asked to find and set up the function needed to find it. Comments that are not part of the solutions are enclosed in parentheses.

Example 1. Determine the coordinates of the point on the curve y = f(x) = 3x + 2 which is closest to the point (-1,4).

Solution. We wish to minimize the distance between the point (-1,4) and the curve y = 3x + 2. An arbitrary point on this curve is (x, 3x + 2). Its distance from the point (-1,4) is given by

$$d(x) = \sqrt{(x + 1)^2 + (3x + 2 - 4)^2} \ , \quad \text{all real x.}$$

Thus,

$$d'(x) = \frac{5(2x - 1)}{d(x)} \quad , \quad \text{all real x.}$$

Hence, d'(1/2) = 0. As d(x) > 0 for all real x it follows that d'(x) < 0 for x < 1/2 and d'(x) > 0 for x > 1/2. So, d is decreasing on (-∞,1/2] and increasing on [1/2,∞). Consequently, d(x) ≥ d(1/2) for all real x and the absolute minimum of d(x) occurs when x = 1/2. The closest point on the curve y = 3x + 2 to the point (-1,4) is (1/2,7/2).

Example 2. What are the dimensions of the rectangle of largest area that may be inscribed in a circle of radius one unit?

Solution. We need to maximize the area of a rectangle. (None of the generality of the problem is lost by placing this geometry problem in an analytic plane with center of the circle at the origin and the rectangle symmetrically located with respect to the axes.)

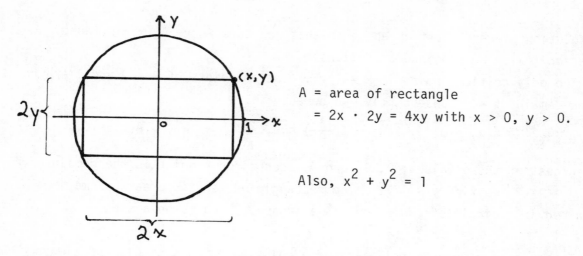

A = area of rectangle
= 2x · 2y = 4xy with x > 0, y > 0.

Also, $x^2 + y^2 = 1$

If we can find the vertex (x,y) in the first quadrant of the rectangle of largest area we will have essentially solved the problem. By substitution one sees that

$$A(x) = 4x\sqrt{1 - x^2} \quad \text{for} \quad 0 < x < 1.$$

(Notice that we used $y = +\sqrt{1 - x^2}$ rather than $y = -\sqrt{1 - x^2}$ since y > 0.)

$$A'(x) = \frac{4(1 - 2x^2)}{\sqrt{1 - x^2}} \quad \text{for} \quad 0 < x < 1. \tag{4}$$

Now, A'(x) = 0 for $x = \pm 1/\sqrt{2}$. Since $x = -1/\sqrt{2}$ is not in the domain of A, we have $x = 1/\sqrt{2}$ as the only critical point. Since $A'(\frac{1^-}{\sqrt{2}}) > 0$ and $A'(\frac{1^+}{\sqrt{2}}) < 0$, there is a local maximum at $x = 1/\sqrt{2}$. (We need to examine the behavior of A near the endpoints 0 and 1 (Why?)). As $\lim_{x \downarrow 0} A(x) = 0 = \lim_{x \uparrow 1} A(x)$, the local maximum is the absolute maximum. From the equation of the circle, $y = 1/\sqrt{2}$ when $x = 1/\sqrt{2}$. Consequently, the rectangle of largest area that may be inscribed in a circle of radius one is a square, $\sqrt{2}$ units on a side.

Alternate Solution. This argument mimics the work of Example 4 from Sections 4.3-4 of the Supplement. Namely, we evaluate the function at each critical point and calculate one-sided limits at missing endpoints and points of discontinuity. The largest value in this list of values is the absolute maximum if it occurs for a point in the domain of the function. Otherwise, there is no absolute maximum.

Our solution proceeds just as above to that juncture where A'(x) is found (equation (4)). We continue as follows.

The only critical point is $x = 1/\sqrt{2}$. The endpoints of $x = 0$ and 1 are not in the domain of A. Consequently, we compute:

$$\lim_{x \downarrow 0} A(x) = 0, \qquad A(\frac{1}{\sqrt{2}}) = 2, \qquad \lim_{x \uparrow 1} A(x) = 0.$$

The largest of these values: 0, 2, and 0, occurs for a point in the domain of A. When $x = 1/\sqrt{2}$, we find $y = 1/\sqrt{2}$. The rectangle of largest area that may be inscribed in a circle of radius one is a square, $\sqrt{2}$ units on a side.

The technique used in this alternate solution is very simple and always applies. We and the text have not opted for exclusive use of this technique since a variety of arguments better serves to strengthen your understanding of previous topics like increasing, decreasing, local extrema, et al.

Example 3. A rectangular poster is to have side margins of 2 inches, top and bottom margins of 4 inches, and contain 50 square inches of printed matter. What are the dimensions of the poster of least area which meets these specifications? Solution. We need to <u>minimize the area</u>.

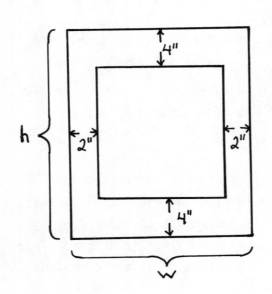

A = area of poster = hw

Also, $50 = (h - 8)(w - 4)$

Thus, $w = 4 + \dfrac{50}{h - 8}$ and

$A(h) = 4h + \dfrac{50h}{h - 8}$, $8 < h < \infty$

$A'(h) = 4 - \dfrac{400}{(h - 8)^2}$, $8 < h < \infty$

$A''(h) = + \dfrac{800}{(h - 8)^3}$, $8 < h < \infty$

Now, A'(h) = 0 if h = 18 or -2. The only critical point is h = 18. Also,
A"(h) > 0 for all h > 8. Consequently, A'(h) < 0 on (8,18) and A'(h) > 0 on
(18,∞); that is, A is decreasing on (8,18) and increasing on (18,∞). The
absolute minimum occurs for h = 18. When h = 18, w = 9. The poster of least
area has height 18 inches and width 9 inches. (Notice that h and w stand in the
same proportion as the margins. You might find it interesting to test this
coincidence (?) by changing the relative sizes of the margins.)

Example 4. A lighthouse is located 3 miles offshore to the North of a point A
on a straight beach. Five miles down the beach to East of point A is a general
store. The lighthouse keeper can row at the rate of 4 miles per hour and walk
at the rate of 6 miles per hour. How fast can the keeper get to the store from
the lighthouse?
Solution. We need to minimize the time.

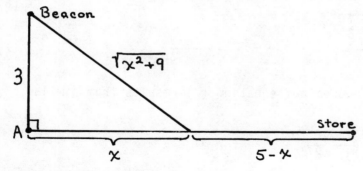

Suppose the keeper lands
his boat x miles to the
East of point A. Certainly,
0 ≤ x ≤ 5 or else his trip
is unnecessarily long.

The total time of the trip is given by

$$T(x) = \frac{\sqrt{x^2 + 9}}{4} + \frac{5 - x}{6} \quad \text{for} \quad 0 \le x \le 5. \tag{5}$$

Then

$$T'(x) = \frac{x}{4\sqrt{x^2 + 9}} - \frac{1}{6} \quad \text{for} \quad 0 < x < 5.$$

Solving T'(x) = 0, we find that $6x = 4\sqrt{x^2 + 9}$ or $x = \pm 6\sqrt{5}/5$. Since T is
continuous on the closed interval [0,5], T has an absolute minimum. It may occur
at but three places: x = 0, 5, or $6\sqrt{5}/5$. Substitution in equation (5) yields
T(0) ≈ 1.58 hours, T(5) ≈ 1.46 hours, and T($6\sqrt{5}/5$) ≈ 1.39 hours. Consequently,

the quickest the keeper can get from the beacon to the store is approximately
1.39 hours, as accomplished by landing $6\sqrt{5}/5$ miles to the East of point A.

Example 5. Two hallways are 8 feet and 27 feet wide respectively and they meet
at a right angle. What is the length of the _longest_ ladder that may be carried
horizontally down one hallway into the other?

Solution. (It makes no difference which hallway we start in. More importantly,
at the point of tightest fit, the longest ladder will just touch the inside
corner. Thus, we can view the problem as though we were _minimizing the length_
of a ladder that is in the following position.)

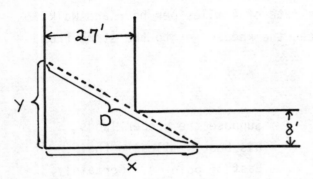

D = length of ladder

$$D^2 = x^2 + y^2 \qquad (6)$$

$$\frac{y}{x} = \frac{8}{x - 27} \qquad (7)$$

(Equation (6) is the Pythagorean Theorem and equation (7) results from similar
triangles. How?) Thus,

$$D(x) = \sqrt{x^2 + (\frac{8x}{x - 27})^2} \qquad \text{for } 27 < x < \infty$$

and

$$D'(x) = \frac{x}{D(x)} (1 - (\frac{12}{x - 27})^3) \quad \text{for } 27 < x < \infty.$$

(You should definitely verify this calculation of D'.)

Hence D'(39) = 0. Since D'(x) < 0 for 27 < x < 39 and D'(x) > 0 for x > 39, D is
decreasing on (27,39] and increasing on [39,∞). Thus, $D(x) \geq D(39)$ for all x in
the domain of D and the absolute minimum occurs when x = 39. Consequently, the
longest ladder that may negotiate this corner is $13\sqrt{13}$ feet long.

Alternate Solution. As above we find that the only critical point is x = 39.
Since the smallest value $(13\sqrt{13})$ in the following list:

$$\lim_{x \downarrow 27} D(x) = +\infty, \qquad D(39) = 13\sqrt{13}, \qquad \lim_{x \to +\infty} D(x) = +\infty$$

occurs at a point (x = 39) in the domain of D, we know that D has an absolute minimum. Consequently, the longest ladder that may negotiate this corner is $13\sqrt{13}$ feet long.

You should now put yourself to work solving problems. You can only learn to solve word problems through experience. You should try some exercises before reading the following.

ADDITIONAL COMMENTS. One of the reasons we claim that the word problems in this course are essentially easy is that there are relatively few mathematical models at our disposal. The apparently large number of different problems is due to two factors. An example will suffice. There are three formulas normally associated with a rectangle: LW = A, 2L + 2W = P and $L^2 + W^2 = D^2$, involving the five variables: length, width, area, perimeter and diagonal. Not all of these variables can be independent at the same time, nor will a given word problem necessarily require all three equations. We might need to maximize the area for given perimeter, maximize perimeter for given area, minimize diagonal for given area, and so on. (The first source of "different" problems is then the variety of ways we combine these formulas to isolate one of the variables.) Next, a rectangle could be the model for a poster, a window, a field, a rug, a desk top, and so on. If in each case we need to maximize area under some conditions like fixed perimeter, then the difference in the problems is strictly non-mathematical.

The variety of problems is limited by the number of reasonable (not complex) geometric models and by the number of the descriptive equations for each model. The student who has difficulty in setting up the function to be analyzed either cannot solve simultaneous equations or has not put down all the relevant equations. For emphasis, we note that the most commonly overlooked equations are the Pythagorean Theorem and the relationships available from similar and congruent figures, especially triangles. Invariably, this is why students have difficulty with one of the classic problems, which was Example 5 in the Supplement.

The necessity of determining whether or not the endpoints are part of your function's domain has already been noted. Briefly, if the endpoint is in the domain, it could, as an endpoint extremum, be the absolute extremum you are looking for. On the other hand, if the endpoint is not in the domain, the behavior of the function near the endpoint could rule out the existence of the absolute extremum you are looking for. Examples 2 and 3 on p. 66 of the Supplement amply illustrate these points.

It is sometimes useful once you have completed a word problem to ponder whether or not your result is (intuitively) reasonable. Recall Example 2 and the problem of maximizing the area of a rectangle whose perimeter is fixed. The complete symmetry of these problems does not distinguish between the length and the width of the rectangle. As there is no apparent reason to make one larger than the other, something special like an extremum (but not so clearly the maximum) occurs when the rectangle is a square. Remember, though, that it was our intuition, if not a semantic game, that led us to falsely conjecture that the derivative of a product is the product of the derivatives. To test your intuition, think about Example 2 in terms of a semicircle instead of a circle and then derive the result. Is it a surprise?

Finally, one way to check your analysis of a function for an absolute extremum is to see if you can respond to the question had it asked you to, say, find the maximum rather than the minimum. What are the correct responses if the exercises you have worked had been so altered?

SECTIONS 4.6-7

It helps to remember that $f'' = (f')'$; that is, $f''(x)$ is the first derivative of $f'(x)$. This implies $f'(x)$ is increasing whenever $f''(x)$ is positive. If $f''(x)$ is positive, f is concave upward. If $f''(x)$ is negative, f is concave downward. Since concavity is defined in terms of increasing and decreasing f', it is an interval property. We have a point of inflection at $x = c$, if $f''(c^+) \cdot f''(c^-) < 0$. This inequality is a brief way of saying that $f''(c^+)$ and $f''(c^-)$ are of opposite signs.

For each of the examples below, we will determine the maxima and minima (local, endpoint, and absolute), concavity, and points of inflection. These examples represent in form and substance what you should be able to do. Comments which are explanatory but not part of the solution proper have been enclosed in parentheses. This is not the only style of problem you may encounter. You might also be given a function and its first two derivatives along with some assertions concerning that function's extrema et al. You would be required then to support or deny those assertions with the same brief, but complete, reasons as done in the examples. Some suggestions on graphing techniques will be given after these examples and then a graph will be executed for each of the examples. It would be advisable for you to sketch the graph for each example prior to looking at the one provided.

<u>Example 1.</u> Analyze $f(x) = \frac{x^3}{3} + x^2 - 8x + 4$ on $(-\infty,\infty)$.

$$f(x) = \frac{x^3}{3} + x^2 - 8x + 4 \quad , \quad \text{all real } x$$

$$f'(x) = (x + 4)(x - 2) \quad , \quad \text{all real } x$$

$$f''(x) = 2(x + 1) \quad , \quad \text{all real } x$$

(The derivatives were calculated and factored on the side, but listed closely together for easy reference. <u>Your</u> calculations should be accessible to check for errors. Recall that d.n.e. is an abbreviation for "does not exist.")

possible
 extrema:
$\begin{cases} f'(x) = 0: \quad x = -4, 2 \\[2em] f' \text{ d.n.e.} \quad \begin{aligned} &\text{endpoints:} \quad \text{none} \\ &\text{others} \quad : \quad \text{none} \end{aligned} \end{cases}$

possible points
 of inflection:
$\begin{cases} f''(x) = 0: \quad x = -1 \\[1.5em] f'' \text{ d.n.e.:} \quad \text{none} \end{cases}$

$(-4, 30\frac{2}{3})$: $f'(-4) = 0$, $f''(-4) < 0$;   a local <u>max</u>imum.

$(2, -5\frac{1}{3})$: $f'(2) = 0$, $f'(2^-) < 0$, $f'(2^+) > 0$;   a local <u>min</u>imum.

$(-1, 12\frac{2}{3})$: $f''(-1) = 0$, $f''(-1^-) < 0$, $f''(-1^+) > 0$;   a point of inflection.

(Second derivative test would have been briefer at $x = 2$. We list points as $(x, f(x))$ for later purposes of graphing and/or determining absolute extrema.)

Concavity: $f''(x) > 0$ for $x > -1$ so f is concave <u>up</u> on $[-1,\infty)$.

   $f''(x) < 0$ for $x < -1$ so f is concave <u>down</u> on $(-\infty,-1]$.

(The intervals of concavity were "closed up" here since the test for concavity simply involves the increase and decrease of f'. Remember (Theorem 4.2.3) that if a function (here, f') is continuous on $[a,b]$ and its derivative (here, f'') is positive on (a,b), then that function is increasing on $[a,b]$.)

Absolute extrema: As $\lim\limits_{x \to +\infty} f(x) = +\infty$ and $\lim\limits_{x \to -\infty} f(x) = -\infty$ (notice that the $\frac{x^3}{3}$ term "dominates for large x"), there is no absolute maximum and no absolute minimum.

<u>Example 2</u>. Analyze $f(x) = \frac{x^3}{3} + x^2 - 8x + 4$ on $[-3,6]$.

   $f(x) = \frac{x^3}{3} + x^2 - 8x + 4$, $[-3,6]$

   $f'(x) = (x + 4)(x - 2)$, $(-3,6)$

   $f''(x) = 2(x + 1)$, $(-3,6)$.

possible
 extrema
 $\begin{cases} f'(x) = 0: \quad x = 2 \qquad\qquad (x = -4 \text{ is not in the domain of } f) \\[2em] f' \text{ d.n.e.} \begin{cases} \text{endpoints: } x = -3,6 \\ \text{others} \quad : \text{ none} \end{cases} \end{cases}$

possible points
 of inflection
 $\begin{cases} f'' = 0: \quad x = -1 \\[1.5em] f'' \text{ d.n.e.: none} \qquad\qquad (\text{compare domains of } f' \text{ and } f'') \end{cases}$

$(2,-5\frac{1}{3})$: $f'(2) = 0$, $f''(2) > 0$; a local <u>min</u>imum.

$(-3,28)$: f continuous from right at $x = -3$ and $f'(-3^+) < 0$; an endpoint <u>max</u>imum.

$(6,64)$: f continuous from left at $x = 6$ and $f'(6^-) > 0$; an endpoint <u>max</u>imum.

$(-1,12\frac{2}{3})$: $f''(-1) = 0$, $f''(-1^-) < 0$, $f''(-1^+) > 0$; a point of inflection.

Concavity: $f''(x) > 0$ for x in $(-1,6)$ so f is concave <u>up</u> on $[-1,6)$.

   $f''(x) < 0$ for x in $(-3,-1)$ so f is concave <u>down</u> on $(-3,-1]$.

Absolute extrema: (Since f is continuous on a closed interval, the Maximum-Minimum Theorem tells us they exist as the largest and smallest of all local and endpoint extrema.

Absolute maximum is 64 and occurs at x = 6.

Absolute minimum is $-5\frac{1}{3}$ and occurs at x = 2.

Example 3.   Analyze $f(x) = x^4 - 4x^3 + 1$ on [-1,5).

$$f(x) = x^4 - 4x^3 + 1, \quad [-1,5)$$

$$f'(x) = 4x^2(x - 3) \quad , \quad (-1,5)$$

$$f''(x) = 12x(x - 2) \quad , \quad (-1,5)$$

possible $\begin{cases} f'(x) = 0: \quad x = 0,3 \\ \\ f' \text{ d.n.e.} \begin{cases} \text{endpoints:} \quad x = -1 \\ \text{others} \quad : \quad \text{none} \end{cases} \end{cases}$
extrema:

possible points $\begin{cases} f''(x) = 0: \quad x = 0,2 \\ \\ f'' \text{ d.n.e.:} \quad \text{none} \end{cases}$
of inflection:

(0,1) : f'(0) = 0, (f''(0) = 0, so 2nd derivative test is inconclusive)

  $f'(0^-) < 0$ and $f'(0^+) < 0$; so not an extremum.

  $f''(0) = 0, f''(0^-) > 0, f''(0^+) < 0$; a point of inflection.

(3,-26): f'(3) = 0, f''(3) > 0; a local minimum.

(-1,6) : f is continuous from the right at x = -1 and $f'(-1^+) < 0$; an endpoint
  maximum.

(2,-15): $f''(2) = 0, f''(2^-) < 0, f''(2^+) > 0$; a point of inflection.

Concavity: f''(x) > 0 for x in (2,5) and in (-1,0) so f is concave up on [2,5)
  and on (-1,0].

  f''(x) < 0 on (0,2) so f is concave down on [0,2].

Absolute extrema: $\lim\limits_{x \uparrow 5} f(x) = 126^-$. ($126^-$ means f(x) approaches 126 from

  "below.") Combining this with the determined extrema, f has no

  absolute maximum and the absolute minimum of -26 occurs at x = 3.

Example 4:   Analyze $f(x) = \dfrac{2x^2}{1 - x}$.

$$f(x) = \frac{2x^2}{1 - x} \qquad , \quad x \neq 1$$

$$f'(x) = \frac{2x(2 - x)}{(1 - x)^2} \quad , \quad x \neq 1$$

$$f''(x) = \frac{4}{(1 - x)^3} \qquad , \quad x \neq 1$$

possible
 extrema:
$$\begin{cases} f'(x) = 0: \quad x = 0,2 \\ \\ f' \text{ d.n.e.:} \quad \text{none} \end{cases}$$

possible points
 of inflection:
$$\begin{cases} f''(x) = 0: \quad \text{none} \\ \\ \\ f'' \text{ d.n.e.:} \quad \text{none} \end{cases}$$

(0,0) : f'(0) = 0, f"(0) > 0; a local <u>min</u>imum.

(2,-8): f'(2) = 0, f"(2) < 0; a local <u>max</u>imum.

Concavity: f"(x) > 0 for x < 1 so f is concave <u>up</u> on $(-\infty,1)$. (Notice that the
           interval is left open. Why?)

      f"(x) < 0 for x > 1 so f is concave <u>down</u> on $(1,\infty)$.

Absolute extrema: None, since $\lim\limits_{x \downarrow 1} f(x) = -\infty$ and $\lim\limits_{x \uparrow 1} f(x) = +\infty$.

         (Alternatively, one could point out that $\lim\limits_{x \to +\infty} f(x) = -\infty$ and
      $\lim\limits_{x \to -\infty} f(x) = +\infty$.)

(Other information: x = 1 and y = -2x - 2 are asymptotes. Only the first one
should have been apparent to you.)

<u>Example 5</u>. Analyze $f(x) = 1 + (x - 2)^{5/3}$.

$$f(x) = 1 + (x - 2)^{5/3} \quad , \text{ all real } x$$

$$f'(x) = \frac{5}{3}(x - 2)^{2/3} \qquad , \text{ all real } x$$

$$f''(x) = \frac{10}{9}(x - 2)^{-\frac{1}{3}} \qquad , \quad x \neq 2$$

possible $\cdot$ extrema $\begin{cases} f'(x) = 0 : x = 2 \\ \\ f' \text{ d.n.e. : none} \end{cases}$

possible points $\cdot$ of inflection $\begin{cases} f''(x) = 0 : \text{none} \\ \\ f'' \text{ d.n.e. : } x = 2 \end{cases}$

$(2,1)$: $f'(2) = 0$, $f'(2^-) > 0$ and $f'(2^+) > 0$; not an extremum.

$\qquad\qquad$ $f''(2)$ d.n.e., $f''(2^-) < 0$, $f''(2^+) > 0$; a point of inflection.

Concavity: $f''(x) > 0$ for $x > 2$ so f is concave up on $[2,\infty)$.

$\qquad\qquad$ $f''(x) < 0$ for $x < 2$ so f is concave down on $(-\infty,2]$.

Absolute extrema: None, since $\lim\limits_{x \to +\infty} f(x) = +\infty$ and $\lim\limits_{x \to -\infty} f(x) = -\infty$.

$\qquad$ (Though apparently messy, these "infinite limits" are easy. The second just notes that if x is arbitrarily large and negative then so is x - 2, so is $(x - 2)^{5/3}$, and so is $f(x)$.)

Example 6. Analyze $f(x) = 6x^{5/3} - 15x^{2/3}$.

$\qquad$ $f(x) = 6x^{5/3} - 15x^{2/3}$, all real x

$\qquad$ $f'(x) = 10x^{-\frac{1}{3}}(x - 1)$, $x \neq 0$

$\qquad$ $f''(x) = \dfrac{10}{3} x^{-\frac{4}{3}}(2x + 1)$, $x \neq 0$

possible $\cdot$ extrema $\begin{cases} f'(x) = 0 : x = 1 \\ \\ f' \text{ d.n.e. : } x = 0 \end{cases}$

possible points $\cdot$ of inflection $\begin{cases} f''(x) = 0 : x = -\dfrac{1}{2} \\ \\ f'' \text{ d.n.e. : } x = 0 \end{cases}$

$(1,-9)$: $f'(1) = 0$, $f''(1) > 0$, a local minimum.

$(0,0)$: f is continuous at $x = 0$, $f'(0^-) > 0$, $f'(0^+) < 0$; a local maximum.

$\qquad\qquad$ Also, $f''(0)$ d.n.e., $f''(0^-) > 0$ and $f''(0^+) > 0$; not a point of inflection.

$\left(-\dfrac{1}{2}, \dfrac{-18}{\sqrt[3]{4}}\right)$: $f''(-\dfrac{1}{2}) = 0$, $f''(-\dfrac{1}{2}^-) < 0$, $f''(-\dfrac{1}{2}^+) > 0$; a point of inflection.

Concavity: $f''(x) > 0$ for x in $(-\dfrac{1}{2},0)$ and x in $(0,\infty)$ so f is concave up on

$[-\frac{1}{2},0)$ and on $(0,\infty)$. (x = 0 not included since f'(0) d.n.e.)

$f''(x) < 0$ for $x < -\frac{1}{2}$ so f is concave <u>down</u> on $(-\infty,-\frac{1}{2}]$.

Absolute extrema: Since $f(x) \to +\infty$ as $x \to +\infty$ and $f(x) \to -\infty$ as $x \to -\infty$, there
          are no absolute extrema.

(Other information: As $x \uparrow 0$, $f'(x) \to +\infty$ and as $x \downarrow 0$, $f'(x) \to -\infty$. Consequently,
f approaches the vertical at x = 0. It has a cusp there rather than a vertical
tangent.)

   It is a relatively simple matter to sketch the graph of a function once it
has been analyzed. You should first plot each point of interest and record its
coordinates. You may find it helpful to indicate at each such point the result
of your analysis. Some examples are given below.

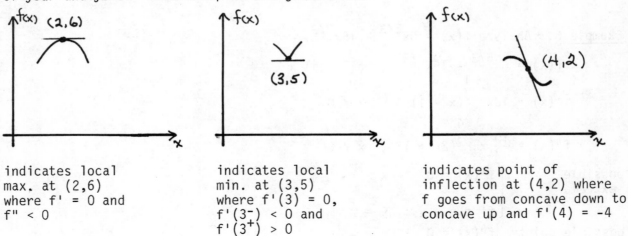

indicates local   indicates local   indicates point of
max. at (2,6)   min. at (3,5)   inflection at (4,2) where
where f' = 0 and   where f'(3) = 0,   f goes from concave down to
f" < 0 $f'(3^-) < 0$ and   concave up and f'(4) = -4
                           $f'(3^+) > 0$

Then it is simply a matter of connecting the "dots" (if f is continuous). A
more accurate graph may be sketched by not only plotting more points, but also
short line segments indicating the slope of the curve at some points. The
purpose of the graph you sketch is to represent the general shape of the curve and
the important characteristics which you discovered in your analysis. It is not
our goal to derive a highly accurate graph which may be used for interpolation.
In fact, as it is sometimes difficult to sketch the graph in the small space
allotted, the labelling of plotted points becomes necessary.

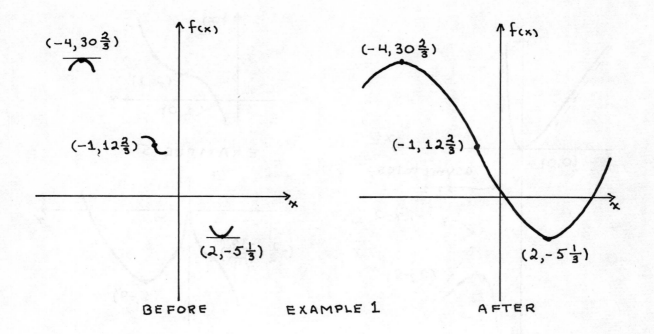

$(-4, 30\frac{2}{3})$

$(-1, 12\frac{2}{3})$

$(2, -5\frac{1}{3})$

BEFORE   EXAMPLE 1   AFTER

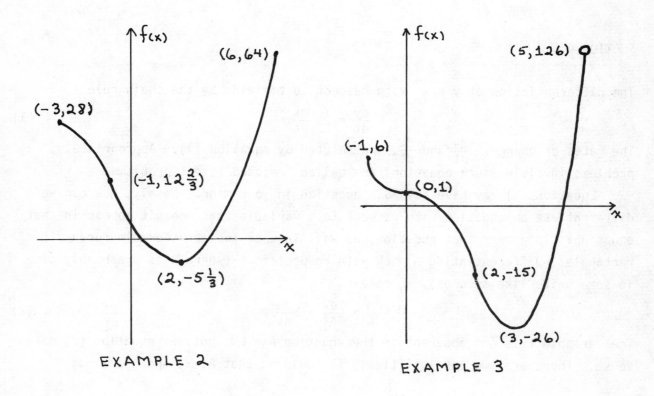

$(6, 64)$

$(-3, 28)$

$(-1, 12\frac{2}{3})$

$(2, -5\frac{1}{3})$

EXAMPLE 2

$(5, 126)$

$(-1, 6)$

$(0, 1)$

$(2, -15)$

$(3, -26)$

EXAMPLE 3

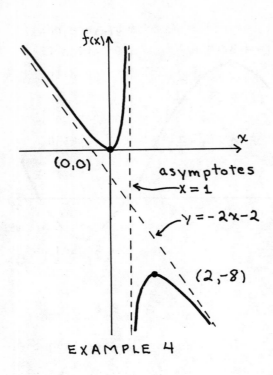

EXAMPLE 4

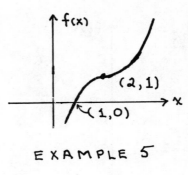

EXAMPLE 5

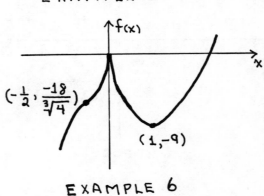

EXAMPLE 6

SECTION 4.8

The differentiation of $y = x^2$ with respect to t yields by the chain rule

$$\frac{dy}{dt} = 2x \frac{dx}{dt}.$$ (1)

The rates of change: $\frac{dy}{dt}$ and $\frac{dx}{dt}$, are related by equation (1). Appropriately, problems involving such equations are called "related rates" problems.

Equation (1) may have raised a question in your mind. Namely, how can we differentiate an equation with respect to a variable that doesn't appear in that equation? To answer that question, we will look at the area formula for a rectangle. Differentiating A = LW with respect to t (where t is presumably time in some units like seconds), we obtain

$$\frac{dA}{dt} = L\frac{dW}{dt} + W\frac{dL}{dt} .$$ (2)

The variable t is not apparent in the equation A = LW, but yet equation (2) makes sense. There are two possibilities. It could be that A, W, and L are not functions of t. That is, the area, width, and length of a rectangle are not dependent on time. In this case $\frac{dA}{dt}$, $\frac{dW}{dt}$, and $\frac{dL}{dt}$ are all zero and equation (2) simply says that $0 = 0$, which is undeniably true. Recall from Chapter 3 that if A and t are independent, then $\frac{dA}{dt} = 0$.

Now, what of the other possibility where A, W, and L are dependent on (are functions of) t? Equation (2) is that expression which relates the rates of change of these variables measured relative to time. Let us examine this possibility with a specific example.

Example 1. Suppose that at an instant when the length of a rectangle is 6 inches and its width is 8 inches, the length is increasing at 2 inches per second and the width is decreasing at 1 foot per minute.   How fast is the area of the rectangle changing?

Solution. We need to evaluate $\frac{dA}{dt}$. The conditions of the problem tell us that L = 6 in., W = 8 in., $\frac{dL}{dt}$ = 2 in./sec., and $\frac{dW}{dt}$ = -1/5 in./sec. Substituting these values in equation (2), we arrive at $\frac{dA}{dt}$ = 14.8 in.^{2}/sec. Consequently, the area is increasing at the rate of 14.8 square inches per second.

Notice that we made some numerical adjustments before we recorded a value for $\frac{dW}{dt}$. One adjustment was needed because not all the quantities were specified in the same units. Related rates are derivatives taken relative to the same variable. Here we chose to do everything in inch-seconds, with the variable t in terms of seconds. You may omit the units from your calculations as long as they are not mixed, like seconds and minutes or feet and inches. But, just as with your work in the physical sciences, the proper units should be attached to the final result.

The second adjustment in $\frac{dW}{dt}$ (a minus sign) was needed since W was decreasing rather than increasing. The statements: "x is increasing at 2 feet per second" and "x is decreasing at -2 feet per second," are equivalent. Both are written as $\frac{dx}{dt}$ = 2 ft./sec. Use of the terms increasing and decreasing enables us to specify change as a non-negative measurement. It is clearer to say that y is decreasing at 3 feet per second than it is to say that y is increasing at -3 feet per second.

Aside from the notion of differentiating an equation with respect to a variable perhaps only implicitly present, related rates problems are just like ones you have done before. In spirit they are like those of Section 3.4. In setting they are like extremum word problems. The question posed by the problem tells you what variable you want to differentiate (say A) and what you want to differentiate with respect to (say t). After determining these two variables, you should find an equation involving "A" and differentiate this equation with respect to "t". The information in the problem should enable you to evaluate the other quantities so that you may solve for $\frac{"dA"}{dt}$.

Example 2. A rectangle is inscribed in a circle of radius 5 inches. If the length of the rectangle is decreasing at 2 inches per second, how fast is the area changing at the instant when the length is 6 inches?
Solution. We need to find $\frac{dA}{dt}$ given L = 6" and $\frac{dL}{dt}$ = -2 in./sec.

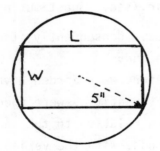

$$A = LW,$$

so $\frac{dA}{dt} = W \frac{dL}{dt} + L \frac{dW}{dt} = -2W + 6 \frac{dW}{dt}$.

Since
$$L^2 + W^2 = 10^2 \text{ (Why?)}, \qquad (3)$$
W = 8. Differentiation of equation (3)

with respect to t gives
$$2L \frac{dL}{dt} + 2W \frac{dW}{dt} = 0 \quad \text{or} \quad \frac{dW}{dt} = -\frac{L}{W} \cdot \frac{dL}{dt} .$$
Thus, $\frac{dW}{dt} = \frac{3}{2}$ in./sec. and $\frac{dA}{dt}$ = -7 in.^{2}/sec. Consequently, the area of the rectangle is decreasing at 7 square inches per second.

Example 3. The perimeter of a rectangle is fixed at 24 inches. If the length is increasing at 1 inch per second, when is the area of the rectangle decreasing?
Solution. We are given that P = 24 = 2(L + W), $\frac{dP}{dt}$ = 0 and $\frac{dL}{dt}$ = 1 in./sec. We

need to determine when $\frac{dA}{dt} < 0$. Since $\frac{dP}{dt} = 0 = 2(\frac{dL}{dt} + \frac{dW}{dt})$, it follows that

$\frac{dW}{dt} = -\frac{dL}{dt} = -1$ in./sec. Next, $\frac{dA}{dt} = L\frac{dW}{dt} + W\frac{dL}{dt} = -L + W$, so $\frac{dA}{dt} < 0$ iff $-L + W < 0$ or $W < L$. Since $P = 24$, $W = 12 - L$, and $\frac{dA}{dt} < 0$ if $12 - L < L$ or $6 < L$. But, we must have both W and L positive, so $L < 12$. Consequently, the area of the rectangle is decreasing when $6 \leq L < 12$.

Remembering the theorems on increasing and decreasing functions, we have included $L = 6$ since A is a continuous function. We exclude $L = 12$ since it is not in the domain of A (why?). Notice that A is increasing when $0 < L \leq 6$. Thus, A has its maximum when $L = W = 6$, as noted in Section 4.5.

The following example is not easy. Make sure you understand the problem and the first sentence of the solution before reading on.

Example 4. A rubber ball is thrown vertically downward with a certain initial velocity from a height of 4 feet. If the ball rebounds with one-half its impact speed and returns exactly to its original height before falling again, how fast was it thrown initially?

Solution. After some thought, it is apparent that in order to determine the initial velocity we must determine the "initial velocity" required to rise from the ground to a precise height of 4 feet. We will use the formulas for vertical motion as given by the text.

$$x(t) = \frac{1}{2}gt^2 + v_0 t + x_0,$$
$$v(t) = gt + v_0,$$

and

$$a(t) = g,$$

where $g = -32$ ft./sec.^{2} is the acceleration due to gravity, v_0 is the initial velocity (plus or minus depending on direction), and x_0 is the initial height relative to some reference point which we take here as the ground. Our solution is in two parts.

First we will consider the motion of the ball after it strikes the ground. For this rebound, we need to find v_0 if $v(t) = 0$ when $x(t) = 4$.

$$v(t) = 0 = -32t + v_0$$

and

$$x(t) = 4 = -16t^2 + v_0 t + 0.$$

Substituting the first equation ($v_0 = 32t$) in the second, we find that $4 = 16t^2$ or $t = 1/2$ second. And, if $t = 1/2$, $v_0 = 32t = 16$ ft./sec. Therefore, the impact velocity was -32 feet per second (why?).

Second we need to determine the initial velocity v_0 which produces this impact velocity of -32 ft./sec. That is, we want to find v_0 if $x_0 = 4$ and if $x(t) = 0$ when $v(t) = -32$.

$$v(t) = -32 = -32t + v_0$$

and

$$x(t) = 0 = -16t^2 + v_0 t + 4.$$

(Only for the purpose of illustration will we solve this set of equations differently from the approach taken above.)

From the first equation, we have that

$$t = \frac{v_0 + 32}{32}$$

so that

$$0 = -16(\frac{v_0 + 32}{32})^2 + v_0(\frac{v_0 + 32}{32}) + 4$$

and after some algebra

$$v_0 = \pm 16\sqrt{3} .$$

Consequently, the ball is initially thrown downward ($v_0 = -16\sqrt{3}$) with a speed of $16\sqrt{3}$ feet per second. How do you interpret the result $v_0 = +16\sqrt{3}$ ft./sec.?

The ball's trip was a quick one, approximately .64 seconds. Its maximum speed was 32 ft./sec. and, of course, it finished its trip at 0 ft./sec. Was its speed ever 20 ft./sec.? Be careful!

<u>Example 5.</u> A right circular cone of height 6 inches and radius 4 inches is full of water. The cone springs a leak and loses water at the rate of 2 cu.in./min.

   (a) How long will it take for the depth of the water to drop to 3 inches?

   (b) At that instant, how fast is the water level falling?

<u>Solution.</u> (a) The formula for the volume of a cone is $V = \frac{1}{3}\pi r^2 h$. Since $r = 4$ when $h = 6$, we may reason from similar triangles that $r = 2$ when $h = 3$. Thus,

$V = \frac{1}{3}\pi(4)^2 6 - \frac{1}{3}\pi(2)^2 3 = 28\pi$ cubic inches of water have leaked out when $h = 3$

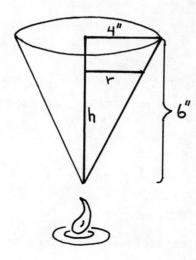

inches. Since $\frac{dV}{dt}$ = -2 cu.in./min.,
it will take 14π minutes for the
depth of the water to drop to
3 inches. (Notice that no Calculus
was needed to answer this question.)

(b) Since $\frac{r}{4} = \frac{h}{6}$ at all times (by similar triangles), it follows that $\frac{dr}{dt} = \frac{2}{3} \frac{dh}{dt}$. Further,

$$\frac{dV}{dt} = \frac{1}{3}\pi(r^2 \frac{dh}{dt} + 2rh \frac{dr}{dt}).$$

Thus,
$$-2 = \frac{1}{3}\pi(4 \cdot \frac{dh}{dt} + 2(2)(3)(\frac{2}{3} \frac{dh}{dt}))$$

and
$$\frac{dh}{dt} = -1/(2\pi).$$

Consequently, the water level is falling at the rate of $1/(2\pi)$ inches per minute.

Example 6. The perimeter of a rectangle is fixed at 24 inches. How fast is the
area changing when it is 32 square inches?

Solution. We are given that P = 24 = 2(L + W) and $\frac{dP}{dt}$ = 0. We need to evaluate
$\frac{dA}{dt}$ when A = LW = 32. First, we may rewrite

$$\frac{dA}{dt} = L \frac{dW}{dt} + W \frac{dL}{dt}$$

as
$$\frac{dA}{dt} = L \frac{dW}{dt} + \frac{32}{L} \cdot \frac{dL}{dt} . \qquad (4)$$

Since $\frac{dP}{dt}$ = 0 = 2($\frac{dW}{dt} + \frac{dL}{dt}$), we have $\frac{dW}{dt} = - \frac{dL}{dt}$ and we may rewrite equation (4) as follows:

$$\frac{dA}{dt} = \frac{dL}{dt}\left(\frac{32}{L} - L\right). \tag{5}$$

The only piece of given information we have yet to use is $P = 24 = 2(L + W)$. Combining this fact with $A = 32 = LW$, we obtain $L = 8$ or $L = 4$. Correspondingly we have from equation (5) that

$$\frac{dA}{dt} = -4\frac{dL}{dt} \text{ when } L = 8$$

and

$$\frac{dA}{dt} = 4\frac{dL}{dt} \text{ when } L = 4. \tag{6}$$

Literally, the area is decreasing or increasing, respectively, 4 times as fast the length is increasing. (We can go no further towards finding a specific numerical value for $\frac{dA}{dt}$; we have completely used all the available information.)
   Suppose this problem had asked instead for $\frac{dA}{dL}$. Can you supply the details to show that $\frac{dA}{dL} = -4$ when $L = 8$ and $\frac{dA}{dL} = 4$ when $L = 4$? You should note that these results tell us neither more nor less than equation (6).

SECTION 4.9

Once you have memorized the formula

$$f(x + h) \simeq f(x) + h \cdot f'(x) \qquad \text{(for small } h) \tag{1}$$

and understand its derivation, there are only two other points to remember in order to make approximations using the differential.

   First, as discussed when the derivative was defined, "h" may be negative as well as positive. Second, when making an approximation for $f(c)$, you should choose a value for x which makes $f(x)$ and $f'(x)$ easy to evaluate and does not yield "too large" a value for h. Notice that then $c = x + h$. The question of "too large" is relative to f and x. It involves a more detailed discussion of error. The following example will implicitly, but clearly, illustrate how the error is affected by the choices for x and h. Error analysis is part of a higher level course.

Example 1. Using differentials, approximate $\sqrt{22}$.

Solution. Quite clearly the function involved here is $f(x) = \sqrt{x}$ with $f'(x) = 1/(2\sqrt{x})$. Thus, equation (1) becomes

$$\sqrt{22} \simeq \sqrt{x} + \frac{h}{2\sqrt{x}} \text{ with } 22 = x + h.$$

If we choose $x = 25$, we need $h = -3$. Thus,

$$\sqrt{22} \simeq \sqrt{25} + \frac{-3}{2\sqrt{25}} = 4.7 \;.$$

The relative error: $(f(x + h) - f(x))/f(x)$, is approximately $h \cdot f'(x)/f(x)$ by equation (1). In this example, the relative error = -.06. The percentage error is 100 times the relative error, here -6%. The results for other choices of x and h are compiled in the following table.

x	h	$f(x) + h \cdot f'(x) \simeq \sqrt{22}$	% error
1	21	11.50	1050
4	18	6.50	225
9	13	5.17	72
16	6	4.75	19
25	-3	4.70	-6
36	-14	4.83	-20
49	-27	5.07	-28
64	-42	5.37	-33

Example 2. A square of side 4 inches is incorrectly measured to have side 4.3 inches. Compare the true area of the square with the actual and approximate values for the area found with the erroneous measurement. Draw a diagram.

Solution. $A(s) = s^2$ is the formula for the area of a square of side s. Hence $A(4) = 16$ and $A(4.3) = 18.49$. Using differentials, $A(x + h) \simeq A(x) + h \cdot A'(x)$. If $h = .3$ and $x = 4$, we have $A(4.3) \simeq 16 + (.3)(2)(4) = 18.40$. Notice that the missing corner in the diagram

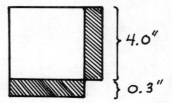

has area of 0.09 = 18.49 - 18.40 square inches. The area of the unshaded region is A(4).

In Section 3.3, we mentioned that $\frac{dy}{dx}$ = f'(x) when y = f(x) and that the symbols dy and dx may be individually given a meaning. Some texts do so by writing dx = h and dy = h·f'(x) ≃ f(x + h) - f(x). This practice stems from the historical setting of Δy = f(x + h) - f(x) and Δx = (x + h) - (x) = h. The symbol Δ is the Greek capital letter d, standing here for "difference" or "change in."

It is legitimate to write dy = f'(x)dx. We may think of this statement as representing the derivative of y = f(x) with respect to an unspecified variable. We thus have options like $\frac{dy}{dt}$ = f'(x)$\frac{dx}{dt}$ and $\frac{dy}{dx}$ = f'(x)$\frac{dx}{dx}$ = f'(x) and $\frac{dy}{dz}$ = f'(x)$\frac{dx}{dt}$ · $\frac{dt}{dz}$.

Neither we nor the text attempt to justify the usage of "dx" when it is alone. There will be instances when we will employ the dx notation on a stand alone basis. In such cases, the usage is purely for certain notational conveniences and will be so explained.

Integration

SECTIONS 5.1-3

Aside from gaining an intuitive feeling for the integral and comfort with the notation, your main objective in these Sections is to be able to solve problems like the one in the first example below.

It is not always the case that the extrema (M_k and m_k) of the function occur at the endpoints of each subinterval of the partition. Further, even if they do, it is not always the case that the maxima, say, occur at the right-hand endpoints of the subintervals. Exercises addressing this point appear after Example 3.

Example 1. Calculate $U_f(P)$ and $L_f(P)$ for $f(x) = x^2 - 2x + 2$ on $[0,3]$ if

$$P = \{0, \frac{3}{4}, \frac{3}{2}, 2, 3\}.$$

Solution. $U_f(P) = M_1(x_1 - x_0) + M_2(x_2 - x_1) + M_3(x_3 - x_2) + M_4(x_4 - x_3)$

$$= M_1(\frac{3}{4} - 0) + M_2(\frac{3}{2} - \frac{3}{4}) + M_3(2 - \frac{3}{2}) + M_4(3 - 2)$$

$$= \frac{3}{4} M_1 + \frac{3}{4} M_2 + \frac{1}{2} M_3 + M_4$$

Also,

$$L_f(P) = \frac{3}{4} m_1 + \frac{3}{4} m_2 + \frac{1}{2} m_3 + m_4.$$

Using the curve sketching techniques of Chapter 4 we can produce the following diagram. We have $M_1 = 2$, $M_2 = \frac{5}{4}$, $M_3 = 2$, $M_4 = 5$, $m_1 = \frac{17}{16}$, $m_2 = 1$, $m_3 = \frac{5}{4}$, $m_4 = 2$, which you should carefully verify. Consequently, $U_f(P) = 8 \frac{7}{16}$ and $L_f(P) = 4 \frac{11}{64}$.

(Thus, $4 \frac{11}{64} \le I \le 8 \frac{7}{16}$. The actual value of I is 6.)

95

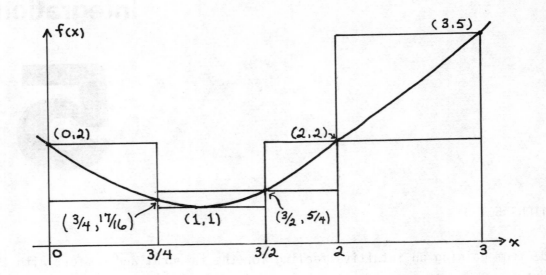

Of course, we knew that the function in this example was integrable since
it was continuous on a closed interval. The theorem proving that these
conditions are sufficient to insure a function is integrable is verified in
Appendix B.4 of the text. Try reading it over if you want a real challenge, the
details are by no means easy! In fact, much of the theory behind integration is
normally treated in a Junior Level course.

The purpose of the next two examples is to enhance your intuitive apprecia-
tion of the integral via the model of rectangles. The first example shows that
it is not necessary for a function to be continuous on a closed interval in order
for it to be integrable. The second example is that of a function which is not
integrable.

Example 2. $f(x) = x + [x] - 2$ for $1 \leq x \leq 5/2$.
     Notice first that f is discontinuous at $x = 2$. Nevertheless, f is integrable
on the interval [1,5/2]. To avoid messy and detailed algebra, we present the
argument at an intuitive level. Mathematicians frequently call this type of
argument "hand-waving."
     A partition P of the interval [1,5/2] leads to two possibilities. Either
$x = 2$ is not a point of the partition P (Diagram 1B) or it is a point of the

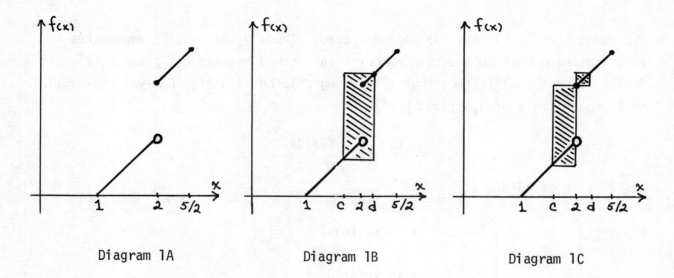

Diagram 1A   Diagram 1B   Diagram 1C

partition P (Diagram 1C). Points c and d are successive points of the partition P
which bracket x = 2. That is, either

$$P = \{1, \ldots, c, d, \ldots, 5/2\} \tag{1}$$

or

$$P = \{1, \ldots, c, 2, d, \ldots, 5/2\} \tag{2}$$

with c < 2 < d.

Since f is continuous on the intervals [1,c] and [d,5/2] , we know that

$$\int_1^c f(x)dx \qquad\text{and}\qquad \int_d^{5/2} f(x)dx$$

exist for all c < 2 < d. Apparently, any difficulties we may encounter will occur
on the interval [c,d]. We are considering all partitions P of [1,5/2] so the
length of the interval [c,d] approaches zero. As this happens, the areas of the
shaded rectangles in Diagrams 1B and 1C approach zero. Symbolically, we are
saying that

$$U_f(P) - L_f(P)$$

for the partitions in (1) and (2) are roughly

$$[f(d) - f(c)](d - c) \tag{3}$$

and

$$[f(d) - f(2)](d - 2) + [f(2) - f(c)](2 - c) \tag{4}$$

respectively. As c approaches 2 from the left and d approaches 2 from the right, the quantities in (3) and (4) approach zero. Thus, $U_f(P) - L_f(P)$ approaches zero and there must be a unique number I sandwiched between $U_f(P)$ and $L_f(P)$. It follows from the definition that f is integrable on $[1,5/2]$. Can you show that this common limit of $U_f(P)$ and $L_f(P)$:

$$I = \int_1^{5/2} f(x)dx,$$

has a value of 13/8?

Example 3.
$$f(x) = \begin{cases} 1 & x \text{ is rational} \\ 0 & x \text{ is irrational} \end{cases}$$

This is the Dirichlet function first encountered on p. 54 of the text. Let us consider $\int_0^1 f(x)dx$. This problem is easy once we bear in mind that between

any two real numbers there are rationals and there are irrationals. Consequently, no matter how the partition P is chosen, $M_k = 1$ and $m_k = 0$ for all k. Thus, $U_f(P) = M_1(x_1 - x_0) + \cdots + M_n(x_n - x_{n-1}) = (x_1 - x_0) + \cdots + (x_n - x_{n-1}) = x_n - x_0 = 1 - 0 = 1$; and, $L_f(P) = 0$ for all partitions P of $[0,1]$. This function is not integrable on $[0,1]$ since there is not a unique number I such that $0 = L_f(P) \le I \le U_f(P) = 1$ for all partitions P of $[0,1]$.

EXERCISES. Find $L_f(P)$ and $U_f(P)$ for each of the following.

1. $f(x) = x^3 - 4x$, $x \in [-2,3]$; $P = \{-2, \frac{1}{2}, 2, 3\}$.

2. $f(x) = |x^2 - 4|$, $x \in [-3,3]$; $P = \{-3, -\frac{5}{2}, -\frac{1}{2}, 1, 3\}$.

3. $f(x) = x^3 - 4x$, $x \in [-2,3]$; $P = \{-2, -\frac{1}{2}, \frac{1}{2}, 2, 3\}$.

4. $f(x) = \begin{cases} x^2, & -2 \le x < 1; \\ 3 - x, & 1 \le x \le 3 \end{cases}$ $P = \{-2, -1, \frac{1}{2}, 1, 3\}$.

ANSWERS.

1. $L_f(P) = -\dfrac{147}{16}$, $U_f(P) = \dfrac{45}{2}$ 3. $L_f(P) = -\dfrac{51}{8}$, $U_f(P) = \dfrac{171}{8}$

2. $L_f(P) = \dfrac{45}{8}$, $U_f(P) = \dfrac{205}{8}$ 4. $L_f(P) = \dfrac{9}{8}$, $U_f(P) = \dfrac{21}{2}$

SECTION 5.4

The following theorem fully lives up to its name in practice and theory.

The Fundamental Theorem of Calculus. Let f be continuous on [a,b]. If G is an antiderivative of f on [a,b], then

$$\int_a^b f(t)dt = G(b) - G(a).$$

Some texts present two versions of The Fundamental Theorem of Calculus. Theorem 5.3.5 is usually the other version.

Theorem 5.3.5.   If f is continuous on [a,b], the function F defined on [a,b] by setting

$$F(x) = \int_a^x f(t)dt$$

is continuous on [a,b], differentiable on (a,b) and satisfies $F'(x) = f(x)$ for all x in (a,b).

When we refer to the Fundamental Theorem of Calculus (F.T.C.), it will be the one so named above.

It was shown in Section 4.2 that if two functions have identical derivatives on an open interval, then those functions differ by no more than an additive constant. Taken together with the two above theorems, it is sometimes loosely and incorrectly asserted that differentiation and integration are inverse processes. Such is not exactly the case. For example, suppose $f(x) = 3x^2$. An antiderivative for f is of the form $G(x) = x^3 + c$ since $G'(x) = 3x^2 = f(x)$.

Thus, the derivative of an antiderivative of f is f. <u>But</u>, an antiderivative of the derivative of f is not necessarily f:

$$f'(x) = 6x \text{ and } \int_a^x f'(t)dt = \int_a^x 6tdt = 3x^2 - 3a^2 \text{ which is } f(x) \text{ only if } a = 0.$$

To apply the F.T.C. to a function f, it is necessary to find an anti-derivative for f. Chapter 8 is devoted to developing "techniques of integration." We will discuss one such technique at this time. It is the most sophisticated, but least remembered.

We call the method: "Drop back and punt." To determine an antiderivative G for the function f by this method,

(1) Make a guess of what G is. (Note that G stands for guess.)

(2) Calculate G'.

(3) Is G' = f? If not, return to step 1. If yes, you are done.

This procedure also contains a valuable suggestion. Since you are now a whiz at differentiation, you can always completely check your calculation of an anti-derivative.

<u>Example 1</u>. Calculate $\int_1^2 (x^3 + \frac{1}{x^2}) dx$.

An antiderivative for $f(x) = x^3 + x^{-2}$ is $G(x) = \frac{x^4}{4} - x^{-1} + 5$ since

$$\frac{d}{dx}(\frac{x^4}{4} - \frac{1}{x} + 5) = x^3 + \frac{1}{x^2} . \text{ Thus,}$$

$$\int_1^2 (x^3 + \frac{1}{x^2}) dx = (\frac{2^4}{4} - \frac{1}{2} + 5) - (\frac{1}{4} - \frac{1}{1} + 5) = 4\frac{1}{4}.$$

<u>Example 2</u>. Calculate $\int_1^3 x(x^2 - 4)^3 dx$.

Since $\frac{d}{dx}(x^2 - 4)^4 = 8x(x^2 - 4)^3$, an antiderivative of $x(x^2 - 4)^3$ is

$\frac{1}{8}(x^2 - 4)^4$. Thus, $\int_1^3 x(x^2 - 4)^3 dx = (\frac{1}{8} 5^4) - (\frac{1}{8}(-3)^4) = 68$.

We could have "multiplied out" $x(x^2 - 4)^3$, but that's too much work when you can see a shorter and slicker way.

Example 3. Determine two antiderivatives of $f(t) = (3t - 2)^{1/5}$.

Since $\frac{d}{dt}(3t - 2)^{6/5} = \frac{18}{5}(3t - 2)^{1/5}$, one antiderivative of f is

$$\frac{5}{18}(3t - 2)^{6/5}.$$

Another is $\frac{5}{18}(3t - 2)^{6/5} + 7$.

We will use the dx which appears in $\int_a^b f(x)dx$ only to tell us which is the "variable of integration." Suppose the variables x and t are <u>independent</u>. Then, $\frac{d}{dx}(xt^2) = t^2$ and $\frac{d}{dt}(xt^2) = 2tx$. In much the same way, we now assert for independent x and t that

$$\int_a^b xt^2 dx = \frac{x^2}{2} t^2 \Big|_a^b = t^2 \frac{(b^2 - a^2)}{2}$$

and

$$\int_a^b xt^2 dt = x \frac{t^3}{3} \Big|_a^b = x \frac{(b^3 - a^3)}{3} .$$

SECTION 5.5

The area of a region in the plane is independent of that region's location in the plane. Regions R1, R2, and R3 in Diagram 1 have the same area because they are congruent.

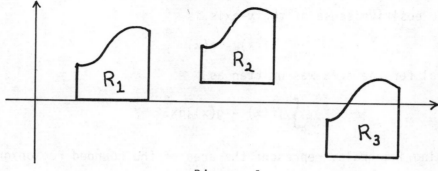

Diagram 1

Further, it is not necessary that the region be a rectangle save for "one wavy side."

Example 1. Using integrals, represent the area of the bounded region graphed below in Diagram 2A.

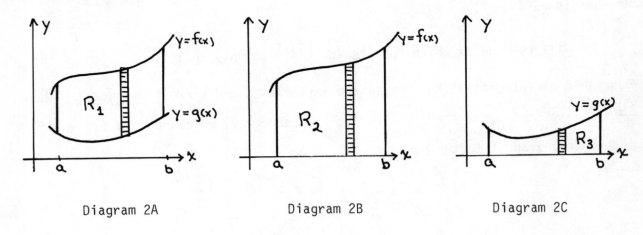

Diagram 2A   Diagram 2B   Diagram 2C

Solution. Notice that Area (R_1) = Area (R_2) - Area (R_3). Thus

$$\text{Area } (R_1) = \int_a^b f(x)dx - \int_a^b g(x)dx$$

which simplifies to

$$\text{Area } (R_1) = \int_a^b [f(x) - g(x)]dx.$$

It is useful to use "representative rectangles" such as those drawn in the diagrams above. Notice in Diagram 2A that the height of the "representative rectangle" measured in the positive sense of the y-axis is

$$f(x) - g(x)$$

and our integral for the area was written as

$$\int_a^b [f(x) - g(x)]dx.$$

Example 2. Using integrals, represent the area of the bounded region graphed in Diagram 3A.

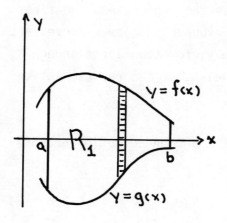

Diagram 3A

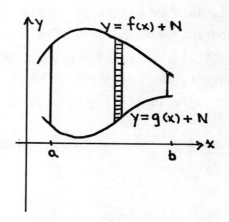

Diagram 3B

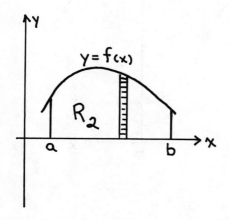

Diagram 3C

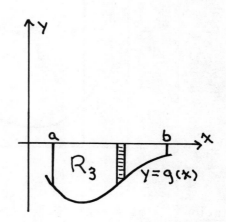

Diagram 3D

<u>Solution</u>. This time Area (R_1) = Area (R_2) + Area (R_3). But notice carefully

that $\int_a^b g(x)dx$ is not Area (R_3). It is the negative of Area (R_3). Thus,

Area (R_3) = $-\int_a^b g(x)dx = \int_a^b (-g(x))dx$. This happens because the height of the

representative rectangle in Diagram 3D is ($-g(x)$). As in the last example,

Area (R_1) = $\int_a^b [f(x) - g(x)]dx$.

Another way to justify this result is to raise the region into the first quadrant (Diagram 3B). The area is unchanged. The upper and lower curves are given by $y = f(x) + N$ and $y = g(x) + N$, respectively, for some large enough N. However, the reasoning of Example 1 tells us the height of the representative rectangle is $(f(x) + N) - (g(x) + N) = f(x) - g(x)$.

<u>Example 3</u>. Using integrals, represent the area of the bounded region graphed below.

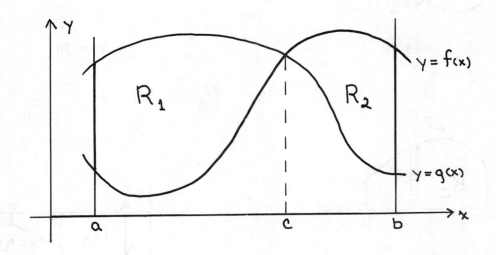

<u>Solution</u>. Area = Area (R_1) + Area (R_2)

$$= \int_a^c [g(x) - f(x)]\,dx + \int_c^b [f(x) - g(x)]\,dx$$

$$\neq \int_a^b [g(x) - f(x)]\,dx$$

This example points up the one principle that may be applied to all problems of this type; <u>determine the curves which serve as upper and lower boundaries of the region and the intervals over which those curves are indeed the upper and lower boundaries</u>.

Example 4. Using integrals, represent the area of the bounded region graphed below.

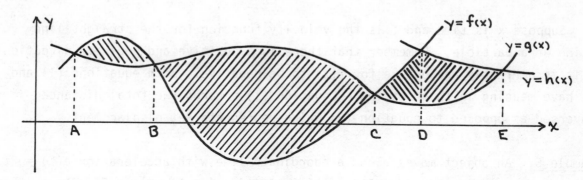

Solution. Area $= \int_A^B [f(x) - g(x)]dx + \int_B^C [g(x) - f(x)]dx$

$\qquad\qquad + \int_C^D [f(x) - g(x)]dx + \int_D^E [h(x) - g(x)]dx.$

It is not always the case that $\int_a^b f(x)dx$ represents the area of some region.

Consider the following graph.

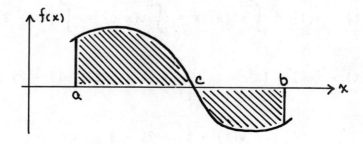

The area of the shaded region is given by

$$\int_a^c f(x)dx + \int_c^b (-f(x))dx = \int_a^b |f(x)|dx \qquad (1)$$

which is not the same here as

$$\int_a^b (f(x)dx. \qquad (2)$$

An obvious, but incomplete, check on your calculation of some area is that your answer should never be negative. Such a check _might_ catch the use of equation (2) rather than (1) in an area problem like this one.

Suppose x is time and f is the velocity function for the straight line motion of a particle. Remember that there is a positive and negative direction so it is perfectly reasonable for f(x) to be negative. Both equations (1) and (2) have meaning in this setting. Equation (1) records the total distance traversed as opposed to equation (2) which represents net displacement.

Example 5. An object moves along a coordinate line with acceleration a(t) = 2t - 2 units per second per second. Its initial position (at t = 0) is 5 units to the left of the origin. One second later the object is moving at -4 units per second.

    (a) Determine the initial velocity of the object.

    (b) Find the position of the object when t = 4.

    (c) How far has the object traveled during these 4 seconds?

Solution. If d(t) and v(t) denote position and velocity, respectively, at time t, then d"(t) = v'(t) = a(t). We are given that d(0) = -5 and v(1) = -4. Since v'(t) = a(t), we know from the F.T.C. that

$$v(t) - v(1) = \int_1^t v'(x)dx = \int_1^t a(x)dx = \int_1^t (2x - 2)dx$$

or

$$v(t) - (-4) = (x^2 - 2x) \Big|_1^t = (t^2 - 2t) - (1 - 2)$$

Therefore,

$$v(t) = t^2 - 2t - 3$$

and v(0) = 0 - 0 - 3 = -3 units/sec.   (Answer to part (a))

Again from the F.T.C., we may write

$$d(t) - d(0) = \int_0^t d'(x)dx = \int_0^t v(x)dx$$

so that

$$d(t) = d(0) + \int_0^t (x^2 - 2x - 3)dx$$

$$= d(0) + \left(\frac{x^3}{3} - x^2 - 3x\right)\Big|_0^t$$

$$= \frac{t^3}{3} - t^2 - 3t - 5.$$

Thus, $d(4) = \frac{64}{3} - 16 - 12 - 5 = \frac{-35}{3}$. The object is $\frac{35}{3}$ units to the left of the origin when $t = 4$.

(Answer to part (b))

In order to compute the total distance traveled rather than the net displacement just calculated, we examine

$$s = \int_0^4 |v(t)|dt = \int_0^4 |t^2 - 2t - 3|dt.$$

We may write this integral as

$$s = \int_0^3 (t^2 - 2t - 3)dt + \int_3^4 (-t^2 + 2t + 3)dt.$$

You should verify that $s = \frac{34}{3}$ units.

(Answer to part (c))

Example 6. A point moves about the plane such that its x-coordinate is changing at the rate of 2 units/sec. and its y-coordinate is changing at the rate of $t^{1/3}$ units per second (t = time). If the particle is at the point (3,-5) when t = 1 second, where is it 7 seconds later?

Solution. Let (f(t), g(t)) be the position of the particle at time t. The F.T.C. tells us that

$$f(8) - f(1) = \int_1^8 (2)dt$$

and

$$g(8) - g(1) = \int_1^8 t^{1/3}dt.$$

We are given that f(1) = 3 and g(1) = -5. You should finish the calculations to show that when t = 8 the particle is at $(17, 6\frac{1}{4})$. Is the value of the x-coordinate surprising?

SECTION 5.8

The heart of the Substitution Principle is the chain rule. Frequently, the aim is to make the integrand easier to integrate by reexpressing it as the sum of powers of some quantity all multiplied by the derivative of that quantity. Symbolically, if we begin with

$$\int (\text{some expression in } x) \, dx$$

and then substitute

$$u = \text{some quantity in } x \tag{1}$$
$$du = (\text{derivative of that quantity}) \, dx$$

our goal is to obtain

$$\int (u^k + u^\ell + \cdots + u^m) du \qquad (k, \ell, m \text{ constants});$$

which is very easy to evaluate:

$$\frac{u^{k + 1}}{k + 1} + \frac{u^{\ell + 1}}{\ell + 1} + \cdots + \frac{u^{m + 1}}{m + 1} + C \tag{2}$$

We may then use equation (1) to rewrite equation (2) in terms of x. The key step of course is the <u>choice</u> of "u = an expression in x." There very well may be some trial and error needed here, but there is no substitute for experience when applying the Substitution Principle. The more problems you do, the more proficient you will become at integration. Some simple examples are in order before we take a formal look at the Substitution Principle.

<u>Example 1.</u> Evaluate $\int 2x(x^2 - 1)^4 dx$.

<u>Solution.</u>   Let $u = x^2 - 1$.

Then $du = 2x dx$.

So, $\int 2x(x^2 - 1)^4 dx = \int u^4 du = \dfrac{u^5}{5} + C = \dfrac{(x^2 - 1)^5}{5} + C$

<u>Example 2.</u> Evaluate $\int (6x + 3)\sqrt{2x^2 + 2x} \, dx$.

<u>Solution.</u>   Let $u = 2x^2 + 2x$.

Then $du = (4x + 2)dx$ or $dx = \dfrac{du}{4x + 2}$.

So, $\int (6x + 3)\sqrt{2x^2 + 2x}\ dx = \int (6x + 3)\sqrt{2x^2 + 2x}\ \dfrac{du}{4x + 2}$

$$= \int \frac{3}{2}\sqrt{2x^2 + 2x}\ du = \frac{3}{2} \int u^{1/2} du = u^{3/2} + C$$

$$= (2x^2 + 2x)^{3/2} + C.$$

You should verify that the substitution $t = \sqrt{2x^2 + 2x}$ also works resulting in $\int 3t^2 dt$.

Example 3. Evaluate $\int (x^5 + x^3)dx$.

Solution.   Let $u = x^2$.

Then $du = 2xdx$ or $dx = \dfrac{du}{2x}$.

So, $\int (x^5 + x^3)dx = \int (x^5 + x^3)\dfrac{du}{2x} = \dfrac{1}{2} \int (x^4 + x^2)du$

$$= \frac{1}{2} \int (u^2 + u)du = \frac{1}{2}[\frac{u^3}{3} + \frac{u^2}{2}] + C = \frac{u^3}{6} + \frac{u^2}{4} + C$$

$$= \frac{x^6}{6} + \frac{x^4}{4} + C, \text{ which should be no surprise!}$$

Example 4. Evaluate $\int (3 + \dfrac{1}{(2x - 1)^2})dx$.

Solution.   Let $u = 2x - 1$.

Then $du = 2dx$.

So, $\int (3 + \dfrac{1}{(2x - 1)^2})dx = \int (3 + \dfrac{1}{u^2})\dfrac{du}{2} = \dfrac{3}{2}u - \dfrac{1}{2u} + C$

$$= \frac{3}{2}(2x - 1) - \frac{1}{2(2x - 1)} + C = 3x - \frac{3}{2} - \frac{1}{2(2x - 1)} + C.$$

The need for the Substitution Principle depends on how well you can see the structure of the integrand. You might well have visualized the integrand in Example 1 as the fourth power of some quantity, $x^2 - 1$, times the derivative of that quantity, $2x$, before you even read the solution. Formally, what you may have seen is $\int [g(x)]^4 g'(x)dx$, where $g(x) = x^2 - 1$. The Substitution Principle simply serves to condense and simplify this to $\int u^4 du$, where $u = g(x)$. Clearly, the

Substitution Principle was not needed in Example 3, for the integrand was already in that basic and simple form for which we shoot.

Perhaps a better first step in Example 4 would have been to write

$$\int (3 + \frac{1}{(2x - 1)^2})dx = 3x + \int \frac{1}{(2x - 1)^2} dx.$$

If you complete this calculation, you will find that

$$\int (3 + \frac{1}{(2x - 1)^2})dx = 3x - \frac{1}{2(2x - 1)} + C.$$

But, where is that "$-\frac{3}{2}$" that appears in the answer given in Example 4? The answer lies in your understanding the meaning of "+ C" in the general form for the antiderivative of a function. One should read F(x) + C as F(x) + "any real number." Thus, "F(x) $-\frac{3}{2}$ + any real number" and "F(x) + any real number" represent and include the same set of functions; namely, all of those whose derivative is F'(x).

How do we apply the Substitution Principle? Well,

(1) Select a change of variable u = g(x) as guided by your experience and extreme good luck.

(2) Calculate du = g'(x)dx.

(3) Substitute $\frac{du}{g'(x)}$ for dx.

(4) Rewrite the integrand in terms of u, using u = g(x).

(5) If you can see it, calculate the antiderivative and rewrite the result in terms of x, the original variable. Otherwise, you have two options: Abandon the selection u = g(x) and try another or iterate the process with a selection t = h(u) in the event there seems to be a multiple application of the chain rule that needs to be unraveled.

Some questions may come to mind. Does this procedure always work? Answer: yes, if you are fortunate in your choice of u = g(x). Can steps (3) and (4) be done at the same time? Answer: one normally does just that, but is careful not to simply replace dx by du unless dx = du. What does "du = g'(x)dx" mean? Answer: read on.

A formal examination of the Substitution Principle may be instructive. First, we need to remember that the "dx" in $\int f(x)dx$ is used to indicate the variable of integration. It also indicates the last level or link in the chain rule differentiation that led from an antiderivative of f to f itself. That is,

$$\int f(u)du = F(u) + C$$

may be written as

$$\int F'(u)du = F(u) + C.$$

Suppose now that $u = g(x)$. Consequently,

$$F(u) = F(g(x)) \text{ and } F'(u) = F'(g(x)) \cdot g'(x)$$

by the chain rule. Thus,

$$\int F'(g(x))du = \int F'(u)du = F(u) + C = F(g(x)) + C$$

$$= \int \frac{d}{dx}(F(g(x)))dx = \int F'(g(x)) \cdot g'(x)dx.$$

Omitting the intervening steps, we have that

$$\int F'(g(x))du = \int F'(g(x))g'(x)dx, \tag{3}$$

which suggests, notationally at least, that

$$du = g'(x)dx. \tag{4}$$

Proper application of equation (4) is critical. If we make the substitution or "change of variable" $u = g(x)$ and obtain

$$\int F'(u)du = \int F'(g(x))dx,$$

we have failed to take into account how the changes in u and x are related. We should have written

$$\int F'(u)du = \int F'(g(x)) \cdot \underline{g'(x)dx}.$$

A specific example should serve to permanently remind you of this point.

<u>Example 5</u>. Evaluate $\int x^4 dx$ using the substitution $u = x^2$.

<u>Incorrect Solution</u>. Let's see. If $u = x^2$, then $u^2 = x^4$. Ah! Thus,

$$\int x^4 dx = \int u^2 dx .$$

But wait, I better change that dx to a du since u is the new variable. So,

$$\int u^2 du = \frac{u^3}{3} + C = \frac{(x^2)^3}{3} + C = \frac{x^6}{3} + C.$$

Just to prove that I'm reading carefully, I'll apply the Drop Back and Punt

philosophy. Now, $\frac{d}{dx}(\frac{x^6}{3} + C) = 2x^5 \neq x^4$!! Boy is this a dumb problem.

<u>Correct solution</u>. If $u = x^2$, then $du = 2xdx$. So,

$$\int x^4 dx = \int x^4 \frac{du}{2x} = \frac{1}{2} \int x^3 du = \frac{1}{2} \int (u^{1/2})^3 du = \frac{1}{2} \int u^{3/2} du$$

$$= \frac{1}{5} u^{5/2} + C = \frac{1}{5}(x^2)^{5/2} + C = \frac{x^5}{5} + C,$$

which <u>is</u> right. Ho, Hum!

     In addition to forgetting the chain rule through the omission of equation (4)
and steps 3 and 4 of the procedure, a common error is to factor a variable out
of the integral sign.

<u>Example 6</u>. Evaluate $\int x^4 dx$ using the substitution $u = x^2$.
<u>Incorrect solution</u>. Ah, here's that same problem again. Think chain rule! If

$u = x^2$, $du = 2x\ dx$ and $dx = \frac{du}{2x}$. I've got it this time.

$$\int x^4 dx = \int x^4 \frac{du}{2x} = \frac{1}{2} \int x^3 du = \frac{1}{2} \int x \cdot x^2 \cdot du = \frac{1}{2} \int xu\ du$$

What do I do with that extra x? It isn't a "u", so let's factor it out.

$$\frac{x}{2} \int u du = \frac{x}{2} \frac{u^2}{2} + C = \frac{x}{2} \frac{x^4}{2} + C = \frac{x^5}{4} + C.$$

<u>Comment</u>. "It isn't a u" is the source of the error here. Since $u = x^2$, x is
related to the variable of integration u in $\int \dots du$. In this case x and u are
<u>dependent</u>. Thus, "factoring out" x is just as gross an error as "factoring out"

an expression in <u>u</u> from $\int \dots du$. You wouldn't write

$$\int x\sqrt{1 - x^2}\ dx = x \int \sqrt{1 - x^2}\ dx,$$

so be careful not to let the change in variable lead you into this trap. To be on
the safe side, only "factor out" <u>constants</u> as supported by the rule:

$$\int cf(x)dx = c \int f(x)dx.$$

There is an occasion, though, when symbols other than numerals may be "factored out" from under the integral sign. Namely, suppose the variables x and y are independent; then

$$\int x^2 y^3 dx = y^3 \int x^2 dx \text{ and } \int x^2 y^3 dy = x^2 \int y^3 dy.$$

The u and x of Example 6 were not independent since we had u = x^2.

If you avoid these two errors, then the Substitution Principle should prove very useful. All you need is some good fortune in selecting the substitution. That comes with experience.

There is but one point left to cover. How do we apply the Substitution Principle to definite integrals? There are two procedures, each with its own virtues. First, we could evaluate the definite integral: $\int_a^b f(x)dx$, as an indefinite integral to obtain

$$\int f(x)dx = F(x) + C,$$

where of course F'(x) = f(x). It follows then from the Fundamental Theorem of Calculus that

$$\int_a^b f(x)dx = (F(x) + C)\Big|_a^b = F(b) - F(a).$$

Example 7. Evaluate $\int_1^3 x(1 - x^2)^{1/3}dx.$

Solution.   Let u = 1 - x^2.
             Then du = -2xdx.

So, $\int x(1 - x^2)^{1/3}dx = -\frac{1}{2}\int u^{1/3}du = -\frac{3}{8}u^{4/3} + C = -\frac{3}{8}(1 - x^2)^{4/3} + C$ and

$$\int_1^3 x(1 - x^2)^{1/3}dx = \left(-\frac{3}{8}(1 - x^2)^{4/3}\right)\Big|_1^3 = \left(-\frac{3}{8}(-2)^4\right) - (0) = -6.$$

It is standard procedure to omit the + C when calculating a definite integral. Can you justify this omission?

Example 8. Evaluate $\int_1^8 x^{-1/3}(1 + 3x^{2/3})^3 dx$.

Solution.   Let $u = x^{1/3}$.

   Then $du = \frac{1}{3}x^{-2/3} dx$ and $dx = 3u^2 du$.

So, $\int x^{-1/3}(1 + 3x^{2/3})^3 dx = \int 3u(1 + 3u^2)^3 du$.

   Let $t = 1 + 3u^2$.

   Then $dt = 6u\, du$.

So, $\int 3u(1 + 3u^2)^3 du = \int \frac{t^3}{2} dt = \frac{t^4}{8} + C$

$\qquad = \frac{(1 + 3u^2)^4}{8} + C = \frac{(1 + 3x^{2/3})^4}{8} + C,$

and $\int_1^8 x^{-1/3}(1 + 3x^{2/3})^3 dx = \frac{(1 + 3x^{2/3})^4}{8}\bigg|_1^8 = \frac{(13)^4}{8} - \frac{(4)^4}{8} = 3531\,\frac{1}{8}.$

Another purpose of this example is to suggest that there may be occasion for you to iterate the application of the Substitution Principle. The choice "$u = 1 + 3x^{2/3}$" could have led to the final result in one application.

   Our second technique for evaluating definite integrals by using the Substitution Principle is embodied in the equation:

$$\int_a^b f(g(x))g'(x)dx = \int_{g(a)}^{g(b)} f(u)du,$$

which is proven on p. 216 of the text. Examples 7 and 8 will be reworked in the context of this equation.

Example 9. Evaluate $\int_1^3 x(1 - x^2)^{1/3} dx$.

Solution.   Let $u = 1 - x^2$.   Further, if $x = 1$, $u = 0$
             Then $du = -2x dx$.   and if $x = 3$, $u = -8$.

So, $\displaystyle\int_1^3 x(1 - x^2)^{1/3} dx = \int_0^{-8} x(1 - x^2)^{1/3} \frac{du}{-2x} = -\frac{1}{2} \int_0^{-8} u^{1/3} du$

$$= -\frac{3}{8} u^{4/3} \Big|_0^{-8} = -\frac{3}{8}(-8)^{4/3} + 0 = -6.$$

<u>Example 10.</u> Evaluate $\displaystyle\int_1^8 x^{-1/3}(1 + 3x^{2/3})^3 dx.$

<u>Solution.</u> Let $u = x^{1/3}$. Then $du = \frac{1}{3} x^{-2/3} dx$ and $dx = 3u^2 du$. Further, $u = 2$ when $x = 8$, and $u = 1$ when $x = 1$. Thus,

$$\int_1^8 x^{-1/3}(1 + 3x^{2/3})^3 dx = \int_1^2 3u(1 + 3u^2)^3 du.$$

Next, let $t = 1 + 3u^2$ so that $dt = 6u\,du$. Notice that $t = 13$ when $u = 2$ and that $t = 4$ when $u = 1$. So,

$$\int_1^2 3u(1 + 3u^2)^3 du = \frac{1}{2} \int_4^{13} t^3 dt = \frac{1}{8} t^4 \Big|_4^{13} = \frac{1}{8}(13)^4 - \frac{1}{8}(4)^4 = 3531 \frac{1}{8}.$$

As noted earlier, each of these two techniques has its virtues. By first evaluating the integral as if it were indefinite, we are in a good position to check our work by differentiation. The second technique still admits such a worthwhile check but since we don't have to rewrite the antiderivative in terms of the original variable the amount of algebra involved is frequently reduced quite a bit.

Our closing example is presented as a matter of interest.

<u>Example 11.</u> Evaluate $\displaystyle\int_{-1}^1 t\sqrt{t^2 + 1}\, dt.$

<u>Solution.</u> Let $x = t^2 + 1$ so that $dx = 2t\, dt$. When $t = -1$, $x = 2$ and when $t = 1$ $x = 2$. Therefore,

$$\int_{-1}^1 t\sqrt{t^2 + 1}\, dt = \frac{1}{2} \int_2^2 \sqrt{x}\, dx = 0.$$

Hmm!

EXERCISES. Evaluate the following integrals using the Substitution Principle.

1. $\int \dfrac{x}{\sqrt{x + 2}} \, dx$ [try u = x + 2]

2. $\int \dfrac{x^3}{\sqrt{x^2 + 1}} \, dx$

3. $\int \sqrt{x^4 + x^2} \, dx \quad (x > 0)$

4. $\int \dfrac{x^2}{\sqrt{x - 1}} \, dx$

5. $\int_1^2 (x + 2)^3 (x - 1)^7 dx$ [try u = x - 1]

6. $\int \dfrac{x^2}{(x - 1)^4} \, dx$

ANSWERS.

1. $(x - 2)\sqrt{x + 2} + C$

2. $\dfrac{1}{3}(x^2 - 2)\sqrt{x^2 + 1} + C$

3. $\dfrac{1}{3}(x^2 + 1)^{3/2} + C$

4. $\dfrac{\sqrt{x - 1}}{15}(6x^2 + 8x + 16) + C$

5. $\dfrac{3241}{880}$

6. $\dfrac{-(3x^2 - 3x + 7)}{(x - 1)^3} + C$

SECTION 5.9

If $H(x) = \displaystyle\int_2^x f(t)dt$, what is $H'(x)$? By Theorem 5.3.5, $H'(x) = f(x)$. If

$H(x) = \displaystyle\int_8^x f(t)dt$, what is $H'(x)$? Again, $H'(x) = f(x)$. If

$$H(x) = \int_3^{x^2} f(t)dt,$$

formula 5.9.7 on p. 219 of the text informs us that
$$H'(x) = f(x^2) \cdot 2x$$
by virtue of the chain rule. Modeling the argument in the text, we find that

$$\text{if } F(x) = \int_3^x f(t)dt, \text{ then } H(x) = F(x^2).$$

Consequently, as $F'(x) = f(x)$,

$$H'(x) = \frac{d}{dx}F(x^2) = F'(x^2) \cdot \frac{d}{dx}(x^2) = f(x^2) \cdot 2x.$$

The lower limit for each of the above problems was a constant. Now, if

$$H(x) = \int_{x}^{3} f(t)dt,$$

what is H'(x)? Noting that

$$H(x) = -\int_{3}^{x} f(t)dt,$$

we easily see that H'(x) = -f(x). Similarly, if

$$G(x) = \int_{x^3}^{7} f(t)dt,$$

then

$$G'(x) = -3x^2 \cdot f(x^3).$$

Finally, suppose that

$$F(x) = \int_{\sqrt{x}}^{x^3} f(t)dt.$$

In order to calculate F'(x), we select a number $\underline{a}$ from the domain of f and write

$$F(x) = \int_{\sqrt{x}}^{a} f(t)dt + \int_{a}^{x^3} f(t)dt$$

so that

$$F'(x) = - \frac{1}{2\sqrt{x}} \cdot f(\sqrt{x}) + 3x^2 \cdot f(x^3).$$

EXERCISES. Compute.

1. F'(x) for $F(x) = \int_{x^3}^{2} \frac{dt}{1 + t^2}$.

2. F'(x) for $F(x) = \int_{x^3}^{x^2} \sqrt{1 + t} \; dt$.

3. $F'(4)$ for $F(x) = \displaystyle\int_{\sqrt{x}}^{x} \frac{dt}{t}$.

4. $H'(x)$ for $H(x) = x \cdot \displaystyle\int_{x^2}^{1} f(t)dt$.

5. $H'(x)$ for $H(x) = \displaystyle\int_{1}^{x^2} x \cdot f(t)dt$.

6. $F'(2)$ for $F(x) = \displaystyle\int_{2}^{x^2} t \cdot F(t)dt$.

7. $H'(2)$ for $H(x) = x \cdot \displaystyle\int_{1}^{x} [1 + H'(t)]dt$.

ANSWERS.

1. $-3x^2/(1 + x^6)$ 2. $2x\sqrt{1 + x^2} - 3x^2\sqrt{1 + x^3}$

3. $1/8$ 4. $\displaystyle\int_{x^2}^{1} f(t)dt - 2x^2 f(x^2)$

5. $2x^2 f(x^2) - \displaystyle\int_{x^2}^{1} f(t)dt$ 6. $16 \displaystyle\int_{2}^{16} t \cdot F(t)dt$ 7. -1

SECTION 5.10

It might help at this time to review Examples 1-4 in Section 5.5 of the Supplement. The principle used in setting up integrals for area problems bears repeating.

> Determine the curves which serve as upper and lower boundaries of the region and the intervals over which those curves are indeed the upper and lower boundaries.

Recall also the use of representative rectangles.

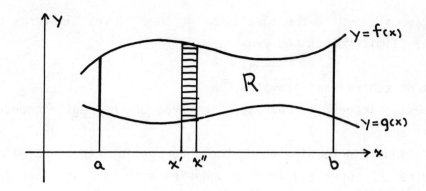

If x' and x" are successive points from a partition of [a,b], then

$$\int_a^b [f(x) - g(x)]dx \tag{1}$$

is the limit of a sum of terms looking like

$$[f(x_0) - g(x_0)] \cdot \Delta x, \tag{2}$$

where $x' \leq x_0 \leq x"$ and $\Delta x = x" - x'$ (Equations 5.1.1-2 in text). The factors in (2) are the dimensions of our representative rectangle. In this limit process the integral sign ($\int$) was historically derived as a skinny "ess" abbreviation for the word "sum." In like fashion, dx symbolically replaces delta x (Δx), after the limit has been formed. It remains useful to informally regard [f(x) - g(x)] and dx as the "dimensions" of the "representative rectangle." We do not attach any formal meaning to the dx. Nevertheless, it is useful for indicating the variable of integration. Thus, by positioning our representative rectangle horizontally, we know we want one or more integrals of the form:

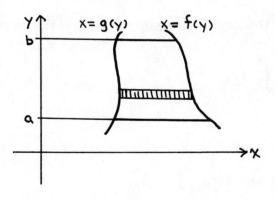

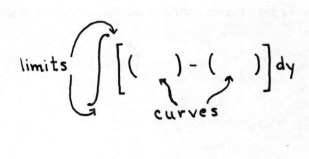

The "dy" tells us to write the equations for the "curves" in terms of y and to determine the "limits" from the y-axis.

We present two related examples.

Example 1. Using integrals, represent the area of the region bounded by the curves $x = y^2$ and $x - y = 2$.

Solution. We should determine where these curves intersect and with the aid of a graph determine the upper and lower boundaries with corresponding intervals. Solving $x = y^2$ and $x - y = 2$ simultaneously, we find $(1,-1)$ and $(4,2)$ as the points of intersection.

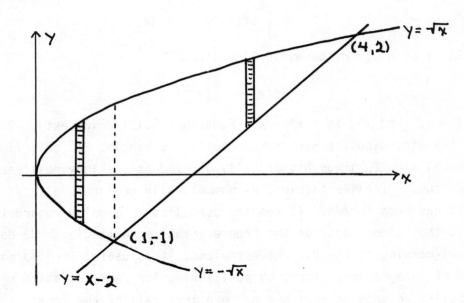

Representative rectangles have been added to the graph to help you visualize that

$$\text{Area} = \int_0^1 [(\sqrt{x}) - (-\sqrt{x})]dx + \int_1^4 [(\sqrt{x}) - (x - 2)]dx.$$

Example 2. Redo Example 1 using integrals of the form $\int \cdots dy$.

Solution.

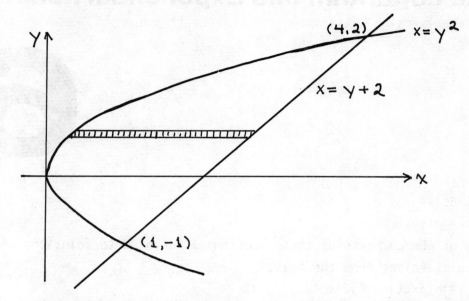

The dimensions of the rectangle are measured in the positive senses of the axes. Thus, y changes from -1 to 2. The "lower curve" is $x = y^2$ and the "upper curve" is $x = y + 2$.

$$\int_{-1}^{2} [(y + 2) - (y^2)]dy.$$

The Logarithm and Exponential Functions

6

SECTIONS 6.1-4

The bulk of our comments on these Sections is devoted to solutions of the following exercises from the text.

p. 241 (Section 6.2): 10, 17, 18

p. 248 (Section 6.3): 5, 8, 25, 31, 32, 39, 43b, 45f

p. 256 (Section 6.4): 11, 28, 44, 49, 65

First we highlight some of the critical facts from these Sections.

You must memorize the following derivatives and the companion integrals.

$$\frac{d}{dx}(e^{f(x)}) = e^{f(x)}f'(x) \quad \text{and} \quad \int e^{f(x)}f'(x)dx = e^{f(x)} + C, \tag{1}$$

and for $f(x) \neq 0$

$$\frac{d}{dx}[\log|f(x)|] = \frac{f'(x)}{f(x)} \text{ and } \int \frac{f'(x)}{f(x)} dx = \log|f(x)| + C. \tag{2}$$

Presumably, you recognize that equations (1) and (2) are chain rule complications of the special case of $f(x) = x$. Namely,

$$\frac{d}{dx}(e^x) = e^x \text{ and } \int e^x dx = e^x + C, \tag{3}$$

and

$$\frac{d}{dx}[\log|x|] = \frac{1}{x} \text{ and } \int \frac{dx}{x} = \log|x| + C. \tag{4}$$

With the Substitution Principle under your belt you should find the integrals stated above a bit easier. Only one comment can be added. Whenever the integrand is a quotient you should be alert for the possibility that the numerator, save for multiplication by some constant, is the derivative of the denominator. Equation (2) then applies. This observation will be more meaningful later on when problems of this nature don't stand alone, but are mixed and hidden with problems requiring other formulas and techniques.

Many of the facts about logarithms and exponentials may be recalled from their graphs.

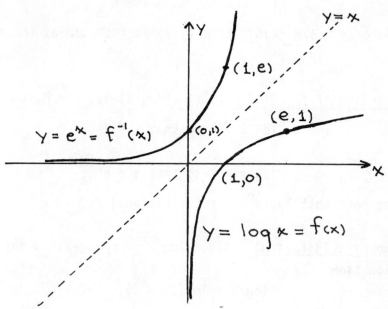

For instance, $y = e^x$ and $y = \log x$ are inverses as they are symmetric with respect to the line $y = x$. $\text{Dom}(f) = \text{Range}(f^{-1}) = (0, \infty)$. $\text{Range}(f) = \text{Dom}(f^{-1}) = (-\infty, \infty)$. Both f and f^{-1} are increasing, $\log 1 = 0$, $\log x > 0$ for $x > 1$, and $\log x < 0$ for $0 < x < 1$, etc., etc. Further, the proofs of these facts are all easy to establish: f is increasing on $(0, \infty)$ since $f'(x) = \frac{1}{x} > 0$ for $x > 0$. As for facts like $\log(xy) = \log x + \log y$, $e^{x + y} = e^x e^y$, $\log_2 8 = 3$, and $\log e^{-4} = -4$, we presume you know something about "logs and antilogs" or at least read the book.

Finally, you are probably a bit uneasy about that number e. You might be wondering whatever happened to that ole log to the base 10 number system. The emergence of e resulted from seeking a solution to $\int x^n dx$ for $n = -1$, which is both natural and unavoidable. As amply developed by the text, the real number e, whatever we call it, is thus inescapable. The fascinating quality of this number is its continual appearance in mathematics and natural sciences in very curious ways.

Solution for Exercise 10 (p. 241). Since $\int_a^b \frac{dx}{x} = \log b - \log a$ represents the area under the curve $y = 1/x$ from $x = a$ to $x = b$ for $1 \leq a < b$, we have from

$$\log n = \log mn - \log m$$

that the area under $y = 1/x$ from 1 to $n \; (> 1)$ is the same as the area from $m \; (\geq 1)$ to mn.

Solution for Exercise 17 (p. 241). If $(2 - \log x) \log x = 0$, then either

$$\log x = 0 \text{ so that } x = 1$$

or

$$\log x = 2 \text{ so that } x = e^2.$$

Here we used the fact that $\log(e^r) = r$ for rational r.

Solution for Exercise 18 (p. 241). Since $\log x^r = r \log x$, we may rewrite the original equation as

$$\log\sqrt{x} = \log(2x - 1).$$

Now, the logarithm function is one-to-one so

$$\sqrt{x} = 2x - 1.$$

Thus,

$$x = (2x - 1)^2 = 4x^2 - 4x + 1$$

or

$$(4x - 1)(x - 1) = 0.$$

The possible solution $x = 1/4$ does not fit the original equation. Remember that $\log T$ is defined only for $T > 0$. (Where did this extraneous solution come from?) The only solution is $x = 1$.

Solution for Exercise 5 (p. 248). $f(x) = \log \sqrt{1 + x^2}$. As the log function is only defined for positive argument, we need $\sqrt{1 + x^2} > 0$. But, $\sqrt{1 + x^2} \geq 0$ implies $1 + x^2 > 0$ which is true for all real x. Thus, $\text{Dom}(f) = (-\infty, \infty)$.

$$f'(x) = \frac{1}{\sqrt{1 + x^2}} \cdot \frac{d}{dx}(\sqrt{1 + x^2}) = \frac{x}{1 + x^2}.$$

Solution for Exercise 8 (p. 248). Since the domain of the logarithm function consists precisely of all positive real numbers, the domain of f(x) = log(log x) is found by solving log x > 0. Easily, x > 1. Thus, Dom(f) = (1,∞).

$$f'(x) = \frac{1}{\log x} \cdot \frac{d}{dx}(\log x) = \frac{1}{x \log x}.$$

Solution for Exercise 25 (p.248). Noticing that $\frac{d}{dx}(1 + x\sqrt{x}) = \frac{3}{2}\sqrt{x}$ greatly simplifies this problem, for then we may rewrite the problem as:

$$\frac{2}{3}\int \frac{\frac{d}{dx}(1 + x\sqrt{x})}{1 + x\sqrt{x}} \, dx.$$

Otherwise, one might start with the substitution u = $\sqrt{x}$ (or x = u^2) in hopes of simplifying the integrand. Thus, dx = 2u du and

$$\int \frac{\sqrt{x}}{1 + x\sqrt{x}} \, dx = \int \frac{u \cdot 2u}{1 + u^3} \, du = 2\int \frac{u^2}{1 + u^3} \, du$$

$$= \frac{2}{3} \log|1 + u^3| + C = \frac{2}{3} \log|1 + x\sqrt{x}| + C.$$

Solution for Exercise 31 (p. 248). Let u = x^2 - 1. Then du = 2x dx. Next, u = 15 when x = 4 and u = 24 when x = 5. Thus,

$$\int_4^5 \frac{x}{x^2 - 1} \, dx = \frac{1}{2} \int_{15}^{24} \frac{du}{u} = \frac{1}{2} \log u \Big|_{15}^{24} = \frac{1}{2}(\log 24 - \log 15) = \log\sqrt{\frac{8}{5}}.$$

Solution for Exercise 32 (p. 248). Let u = log x, then du = dx/x.

So, $\int \frac{\log x}{x} \, dx = \int u \, du = \frac{1}{2} u^2 + C = \frac{1}{2}(\log x)^2 + C$

and $\int_1^e \frac{\log x}{x} \, dx = \frac{1}{2}(\log x)^2 \Big|_1^e = (\frac{1}{2}(\log e)^2) - (\frac{1}{2}(\log 1)^2) = \frac{1}{2}.$

Solution for Exercise 39 (p. 248).

$$\log g(x) = \frac{1}{2}[\log(x - 1) + \log (x - 2) - \log (x - 3) - \log (x - 4)].$$

$$\frac{g'(x)}{g(x)} = \frac{1}{2}[\frac{1}{x - 1} + \frac{1}{x - 2} - \frac{1}{x - 3} - \frac{1}{x - 4}].$$

$$g'(x) = \frac{-2x^2 + 10x - 11}{(x - 1)(x - 2)(x - 3)(x - 4)} \sqrt{\frac{(x - 1)(x - 2)}{(x - 3)(x - 4)}} \ .$$

Solution for Exercise 43b (p. 248).

Let $f(x) = \log (1 - x)$.

Then $f'(x) = -(1 - x)^{-1}$,

$$f''(x) = -(1 - x)^{-2},$$

$$f'''(x) = -2(1 - x)^{-3},$$

$$f^{(iv)}(x) = -2 \cdot 3(1 - x)^{-4},$$

$$f^{(v)}(x) = -2 \cdot 3 \cdot 4(1 - x)^{-5}.$$

It seems reasonable now to guess that

$$f^{(n)}(x) = \frac{d^n}{dx^n}[\log(1 - x)] = -(n - 1)! \ (1 - x)^{-n}.$$

We can test our guess by applying the principle of mathematical induction to the proposition

$$P(n) : f^{(n)}(x) = -(n - 1)! \ (1 - x)^{-n} \qquad \text{for } n \geq 1.$$

P(1) is true, as shown above. Next, assume P(n) is true for some integer $n = k \geq 1$:

$$f^{(k)}(x) = -(k - 1)! \ (1 - x)^{-k}.$$

Differentiating this equation with respect to x, we have

$$f^{(k + 1)}(x) = -(k - 1)! \ (-k)(1 - x)^{-k-1}(-1)$$

$$= - k! \ (1 - x)^{-k-1}$$

and the truth of P(n) is established for all $n \geq 1$.

Solution for Exercise 45f (p. 248).

$$f(x) = \log (8x - x^2) \quad , \quad 0 < x < 8.$$

$$f'(x) = \frac{2(4 - x)}{x(8 - x)} \quad , \quad 0 < x < 8.$$

$$f''(x) = \frac{-2(x^2 - 8x + 32)}{x^2(8 - x)^2} \quad , \quad 0 < x < 8.$$

[handwritten marginalia: $\log x(8-x)$ $f'(x) = (\log x)\left[(8-x)(-1)\right] + (8-x)\left(\frac{1}{x}\right)$ $(\log x)(x-8) + \frac{8-x}{x}$]

Dom(f) = (0,8). As x(8 - x) > 0 for 0 < x < 8, f'(x) < 0 for 4 - x < 0 and f'(x) > 0 for 4 - x > 0. Thus, f is increasing on (0,4] and decreasing on [4,8). The only critical point for f is x = 4. As f"(x) < 0 for all x in (0,8), the local maximum at (4, log 16) is the absolute maximum. As $\lim_{x \downarrow 0} f(x) = -\infty$, there is no absolute minimum. With f"(x) < 0 on the domain of f there are no points of inflection and f is everywhere concave down.

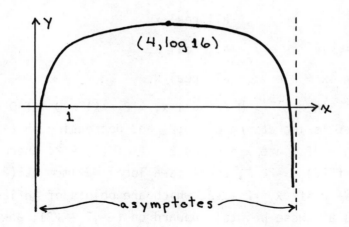

Solution for Exercise 11 (p. 256).

$$\frac{dy}{dx} = e^{\sqrt{x}} \cdot \frac{d}{dx} (\log\sqrt{x}) + \log\sqrt{x} \cdot \frac{d}{dx} (e^{\sqrt{x}}) = \frac{e^{\sqrt{x}}}{2x} + \frac{e^{\sqrt{x}}\log\sqrt{x}}{2\sqrt{x}} .$$

Solution for Exercise 28 (p. 256). Let u = 1/x, so du = $- \frac{dx}{x^2}$.

$$\int \frac{e^{1/x}dx}{x^2} = - \int e^u du = -e^u + C = -e^{1/x} + C .$$

Solution for Exercise 44 (p. 256). Let $u = x^2$, so $du = 2x\, dx$. When $x = 0$, $u = 0$; and, when $x = 1$, $u = 1$.

$$\int_0^1 x(e^{x^2} + 2)\,dx = \frac{1}{2} \int_0^1 (e^u + 2)\,du = \left.\frac{e^u + 2u}{2}\right|_0^1$$

$$= \frac{e + 2}{2} - \frac{1 + 0}{2} = \frac{e + 1}{2}\ .$$

Solution for Exercise 49 (p. 256).

$$e^{-2.1} = \frac{1}{e^{2.1}} = \frac{1}{e^2 \cdot e^{0.1}} \approx \frac{1}{(7.39)(1.11)} \approx 0.12\ .$$

Solution for Exercise 65 (p. 256).

$$f(x) = x^2 e^{-x} \qquad\qquad \text{, all real } x.$$

$$f'(x) = x(2 - x)e^{-x} \qquad\quad \text{, all real } x.$$

$$f''(x) = (x^2 - 4x + 2)e^{-x}\ , \text{ all real } x.$$

$\text{Dom}(f) = (-\infty, \infty)$. Since $e^{-x} > 0$ for all x, $x(2 - x) > 0$ iff $0 < x < 2$ and since f is continuous, f is increasing on $[0,2]$ and decreasing on $(-\infty,0]$ and on $[2,\infty)$. The critical points of f are $x = 0$ and 2. As $f''(0) = 2$, there is a local minimum at $(0,0)$. Since $f''(2) = -2e^{-2}$, there is a local maximum at $(2, 4e^{-2})$. Note also that $x^2 - 4x + 2 = 0$ iff $x = 2 \pm \sqrt{2}$, which are points of inflection since the concavity changes at these points: upward on $(-\infty, 2 - \sqrt{2}]$, downward on $[2 - \sqrt{2},\ 2 + \sqrt{2}]$, upward on $[2 + \sqrt{2}, \infty)$

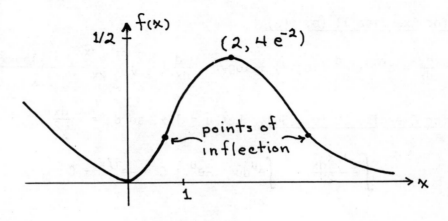

SECTION 6.5

In a moment solutions for problems 1j, 1ℓ, 2a, 3e, 4, 6e, 6h, 7d, 9c, 9d, 10a, 10f, 10g, 11c, and 15 on p. 266 of the text will be given. First we make some general comments.

If you bear in mind that $f(x) = \log x$ and $g(x) = e^x$ are inverse functions:

$$e^{\log x} = x \text{ and } \log(e^x) = x, \tag{1}$$

you should be able to derive all the essential formulas needed to do the problems in this Section. In fact, the logarithm and exponential functions to any given base a > 0 are inverses. That is,

$$\log_a b = u \text{ if and only if } b = a^u.$$

There is <u>no need</u> to memorize all the formulas for derivatives and integrals of logarithms and exponentials to a base other than e. You only need to realize that

$$\log_A B = \frac{\log B}{\log A} \quad \text{(natural logarithms base e)} \tag{2}$$

and

$$A^B = e^{B \log A}. \tag{3}$$

We suggest that rather than memorizing that

$$\frac{d}{dx}(a^x) = a^x \log a \qquad \text{for a = constant,} \tag{4}$$

you take but a moment to derive the result as follows:

$$\frac{d}{dx}(a^x) = \frac{d}{dx}(e^{x \log a}) = e^{x \log a} \cdot \log a = a^x \log a. \tag{5}$$

There are many advantages to this approach. First, it is not uncommon to "forget" the log a in equation (4). You are less likely to do so with the procedure in equation (5), which is a direct result of formula (3). Second, you are reinforcing your awareness that $\log x$ and e^x are inverse functions. Third, you can only use equation (4) if a is a constant. You need the approach of (5) to handle $\frac{d}{dx}(x^{3x})$:

$$\frac{d}{dx}[x^{3x}] = \frac{d}{dx}[(e^{\log x})^{3x}] = \frac{d}{dx}[e^{3x \cdot \log x}]$$

$$= e^{3x \log x} \{\frac{d}{dx}(3x \log x)\} = x^{3x} \{3 + 3 \log x\}.$$

$$(3x)(\tfrac{1}{x}) + (\log x)(3)$$

$$\frac{3x}{x} + \log x\, 3$$

$$3 + \log x 3$$

You can't avoid needing to know (1), (2), and (3), but it is unnecessary to memorize (4).

Solution for Exercise 1j (p. 266). Since $\log_A(B^t) = t \log_A(B)$, we may write

$$\log_{100}(10^{-4/5}) = -\frac{4}{5}[\log_{100}(10)] = -\frac{4}{5}[\frac{1}{2}] = -\frac{2}{5}.$$

Notice that $\log_{100}(10) = \frac{1}{2}$, as $10 = 100^{1/2}$.

Solution for Exercise 1ℓ (p. 266). Let $\log_9(\sqrt{3}) = t$. By equation (2), $t = \frac{\log\sqrt{3}}{\log 9}$.
But, $\log\sqrt{3} = \frac{1}{2} \log 3$ and $\log 9 = 2 \log 3$, so that $t = 1/4$. Alternately, we could have gone from $\log_9(\sqrt{3}) = t$ to $\sqrt{3} = 9^t$ or $3^{1/2} = 9^t = 3^{2t}$. Since $y = 3^x$ is a one-to-one function, $3^{1/2} = 3^{2t}$ tells us that $\frac{1}{2} = 2t$ or $t = \frac{1}{4}$ as before.

Solution for Exercise 2a (p. 266). By the Theorem on p. 236,

$$\log_p xy = \frac{\log xy}{\log p} = \frac{\log x + \log y}{\log p}.$$

Therefore,

$$\log_p xy = \frac{\log x}{\log p} + \frac{\log y}{\log p} = \log_p x + \log_p y.$$

Solution for Exercise 3e (p. 266).

$$\int_2^x \frac{dt}{t} = \log t \Big|_2^x = \log x - \log 2 \text{ and } \log_2 x = \frac{\log x}{\log 2}.$$

Thus,

$$\log x - \log 2 = \frac{\log x}{\log 2}$$

which we may solve for $\log x$. This gives us

$$\log x = \frac{(\log 2)^2}{\log 2 - 1} = C \text{ and } x = e^C.$$

Solution for Exercise 4 (p. 266).

(i) Since $y = \log x$ is an increasing function for all $x > 0$, if $0 < A < B$, then $\log A < \log B$. Thus, if $e^{t_1} < a < e^{t_2}$, $\log(e^{t_1}) < \log a < \log(e^{t_2})$ and $t_1 < \log a < t_2$.

(ii) As $y = e^x$ is an increasing function for all x, $\log x_1 < b < \log x_2$ implies $x_1 < e^b < x_2$.

Solution for Exercise 6e (p. 266). Following the above suggestion we may write 10^{-x^2} as $e^{-x^2 \log 10}$. Thus, if $u = -x^2 \log 10$, $du = -2x \log 10\, dx$ and

$$\int x\, 10^{-x^2} dx = \int \frac{e^u du}{-2 \log 10} = \frac{-e^u}{2 \log 10} + C = \frac{-10^{-x^2}}{2 \log 10} + C.$$

Solution for Exercise 6h (p. 266). From equation (2), $\log_5 x = \frac{\log x}{\log 5}$. Thus, if $u = \log x$, $du = \frac{1}{x}dx$ and

$$\int \frac{\log_5 x}{x}\, dx = \int \frac{\log x}{x \log 5}\, dx = \frac{1}{\log 5} \int u\, du = \frac{(\log x)^2}{2 \log 5} + C$$

$$= \frac{\log 5}{2} (\log_5 x)^2 + C.$$

Solution for Exercise 7d (p. 266).

$$f(x) = \log_3(\log_2 x) = \frac{\log(\log_2 x)}{\log 3} = \frac{\log(\frac{\log x}{\log 2})}{\log 3}.$$

$$f'(x) = \frac{1}{\log 3} \cdot \frac{1}{\frac{\log x}{\log 2}} \cdot \frac{d}{dx}\left(\frac{\log x}{\log 2}\right) = \frac{1}{x \log 3 \log x}.$$

$$f'(e) = \frac{1}{e \log 3}.$$

Solution for Exercise 9c (p. 266).

$$(x^2 + 2)^{\log x} = (e^{\log(x^2 + 2)})^{\log x} = e^{\log(x^2 + 2)\log x}.$$

So,
$$\frac{d}{dx}(x^2 + 2)^{\log x} = e^{\log(x^2 + 2)\log x}\frac{d}{dx}(\log(x^2 + 2)\log x)$$

$$= (x^2 + 2)^{\log x}\left(\frac{\log(x^2 + 2)}{x} + \frac{2x\,\log x}{x^2 + 2}\right).$$

Solution for Exercise 9d (p. 266). There is a slightly different way from that used in the last exercise to solve "$\frac{d}{dx}(f(x))^{g(x)}$" problems. We begin by setting

$$y = \left(\frac{1}{x}\right)^x$$

so that

$$\log y = \log\left[\left(\frac{1}{x}\right)^x\right] = x \cdot \log\left(\frac{1}{x}\right) = -x\,\log x. \tag{6}$$

Differentiation of this last equation yields

$$\frac{1}{y}\frac{dy}{dx} = -(1 + \log x)$$

or

$$\frac{dy}{dx} = -\left(\frac{1}{x}\right)^x(1 + \log x).$$

Had we used the technique of the last exercise instead, it would help to first write

$$y = \left(\frac{1}{x}\right)^x = x^{-x} = (e^{\log x})^{-x} = e^{-x\,\log x}.$$

Taking logs of both sides, we have equation (6).

Solutions for Exercises 10a, 10f, 10g (p. 266).

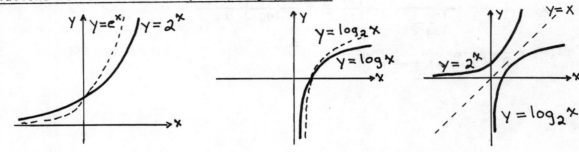

Solution for Exercise 11c (p. 266).

$$f(x) = 10^{\sqrt{1 - x^2}} \qquad \text{for } -1 \leq x \leq 1.$$

$$f'(x) = \frac{-x(\log 10)10^{\sqrt{1 - x^2}}}{\sqrt{1 - x^2}} \qquad \text{for } -1 < x < 1.$$

So, f is increasing on [-1,0] and decreasing on [0,1]. Thus, f has a local maximum at (0,10) and endpoint minima at the points (-1,1) and (1,1).

Solution for Exercise 15 (p. 266). From $\frac{d}{dx}(\log x) = \frac{1}{x}$ it follows that

$$\lim_{h \to 0} \frac{\log(1 + h) - \log 1}{h} = 1,$$

for this is the derivative of log x evaluated at 1. Thus,

$$\lim_{h \to 0} \frac{1}{h} \log(1 + h) = 1$$

or

$$\lim_{h \to 0}[\log (1 + h)^{1/h}] = 1.$$

Since the logarithm function is continuous, $\log [\lim_{h \to 0} (1 + h)^{1/h}] = 1$. But, log[e] = 1 and the log is a 1-1 function, so the result follows.

SECTION 6.6

In this section you should follow the algorithm for word problems given in Section 4.5 of the Supplement. Carefully read the text and each example to gain an understanding of the application of the exponential function to physical and economic situations. A solution to exercise twelve (p. 273) is given below for your gratification.

From the problem on p. 269 of the text, we know that if $A'(t) = k \cdot A(t)$, then $A(t) = A(0)e^{kt}$.

(a) (i) We are given that $A(4) = \frac{3}{4} A(0)$ and want to determine t such that $A(t) = \frac{1}{2} A(0)$. First, we determine k.

$$\frac{3}{4} A(0) = A(4) = A(0)e^{4k}$$

$$\frac{3}{4} = e^{4k} \text{ so that } k = \frac{1}{4} \log \frac{3}{4}.$$

Thus, if

$$A(t) = \frac{1}{2} A(0) = A(0)e^{(\frac{1}{4} \log \frac{3}{4})t}$$

$$\log \frac{1}{2} = \frac{t}{4} \log \frac{3}{4}$$

$$-4 \log 2 = t \log \frac{3}{4}$$

and

$$t = \frac{4 \log 2}{\log 4 - \log 3} \simeq \frac{2.76}{1.39 - 1.10} \simeq 9.5 \text{ years.}$$

(ii) The initial condition is now $A(3) = \frac{3}{4} A(0)$. This gives $k = \frac{1}{3} \log \frac{3}{4}$ and

$$t = \frac{3 \log 2}{\log 4 - \log 3} \simeq 7.1 \text{ years.}$$

(b) Let us choose this year (i.e., now) as our reference point. We are given that $A(0) = 3$ and $A(-1) = 4$. Substituting in $A(t) = A(0)e^{kt}$, we find that $4 = 3e^{-k}$ and $k = \log \frac{3}{4}$. Thus, $A(t) = 3e^{t \cdot \log(3/4)} = 3(\frac{3}{4})^t$. Therefore,

(i) there were $3(\frac{3}{4})^{-2} = 16/3$ lbs. 2 years ago.

(ii) there were $3(\frac{3}{4})^{-10}$ lbs. 10 years ago.

(iii) there will be $3(\frac{3}{4})^{10}$ lbs. left 10 years from now.

SECTION 6.7

Application of the equation

$$\int f(x) \cdot g'(x)dx = f(x) \cdot g(x) - \int f'(x) \cdot g(x)dx \qquad (1)$$

is referred to as the technique of integration by parts. This equation is frequently written in the form

$$\int u \ dv = uv - \int v \ du, \qquad (2)$$

where we have made the identification

$$u = f(x) \qquad \text{and} \qquad dv = g'(x)dx$$

so that $\qquad du = f'(x)dx \qquad$ and $\qquad v = g(x).$

You should organize your work as exampled below.

<u>Example 1</u>. Evaluate $\int xe^{-2x}dx.$

<u>Solution</u>. We set

$$u = x \qquad \text{and} \qquad dv = e^{-2x}dx$$

so that $\qquad du = dx \qquad$ and $\qquad v = -\frac{1}{2}e^{-2x}.$

Substituting in equation (2), we have

$$\int xe^{-2x}dx = -\frac{x}{2}e^{-2x} - \int -\frac{1}{2}e^{-2x}dx$$

$$= -\frac{x}{2}e^{-2x} - \frac{1}{4}e^{-2x} + C$$

<u>Example 2</u>. Evaluate $\int x \cdot \log x \ dx.$

<u>Solution</u>. We set

$$u = \log x \qquad \text{and} \qquad dv = x \ dx$$

so that $\qquad du = \frac{1}{x} dx \qquad$ and $\qquad v = \frac{x^2}{2}.$

Substituting these selections in equation (2), viz.,

$$\int u \ dv = uv - \int v \ du,$$

we have that

$$\int x \, \log x \, dx = \frac{x^2}{2} \log x - \int \frac{1}{x} \frac{x^2}{2} \, dx$$

$$= \frac{x^2}{2} \log x - \frac{x^2}{4} + C.$$

On the basis of just these two examples some observations are immediate. Application of the technique of integration by parts is simple once you know it is the technique to apply, once you know the choices for u and dv, and if the integral occurring on the right side of the equation is easier to handle than the original integral. Making problems easy is a matter of recognizing through continued experience which techniques best resolve which types of problems. Naturally, a problem is easy if and only if _you_ can solve it. Since most problems are inherently easy, the burden of effort lies with you.

Two specific comments will clarify the second point noted above. First, the choices for u and dv may be anything, as long as their product is the original integrand. Thus, the selections $u = x^{-2}$ and $dv = x^3 \cdot \log x \, dx$ are legitimate for Example 2. Naturally, though, you should choose u and dv so that you can evaluate du and v with relative ease. These last selections, though legitimate, are consequently inappropriate. Second, the choices for u and dv that lead to a resolution of the problem are not necessarily unique. In terms of relative ease and speed there may well be a best selection for u and dv, but you should not conclude that there is only one selection that is appropriate. Our next two examples are intended to reinforce this point.

Example 3. Evaluate $\int x \cdot \log x \, dx$.

Solution. We set $u = x \cdot \log x$ and $dv = dx$
so that $du = (1 + \log x)dx$ and $v = x.$
Substituting these selections in

$$\int u \, dv = uv - \int v \, du$$

we have that

$$\int x \cdot \log x \, dx = x^2 \cdot \log x - \int x(1 + \log x)dx.$$

It would seem at first that we have managed to derive a more complicated integral on the right, but we may rewrite this last equation as:

$$\int x \cdot \log x \, dx = x^2 \cdot \log x - \int x \, dx - \int x \cdot \log x \, dx$$

or
$$2 \int x \cdot \log x \, dx = x^2 \log x - \frac{x^2}{2} + C.$$

Consequently,
$$\int x \cdot \log x \, dx = \frac{x^2}{2} \log x - \frac{x^2}{4} + C.$$

This last example also illustrates an important, perchance fortuitous, feature that occurs sometimes when you integrate by parts. Namely, some constant multiple of the original integral occurs on the right hand side of the equation. In that event you do exactly what we did here. Further, notice that in the last line of the solution, we have written C rather than C/2 as the algebra would seem to dictate. This notational looseness is fair here as long as we read the C (and thus the C/2) to represent "any real number." Whether we call it C or C/2 is of no importance. Could we, though, say that C^2 is any real number? (Why not?)

Example 4. Evaluate $\int x \cdot \log x \, dx$.

Solution. We set $u = x$ and $dv = \log x \, dx$
so that $du = dx$ and $v = x \cdot \log x - x$.
Notice that the evaluation of v in this case is not as easy as before. In fact it is necessary to use integration by parts to evaluate v (see p. 276 of the text). Substituting our selections in

$$\int u \, dv = uv - \int v \, du$$

we have that

$$\int x \cdot \log x \, dx = x^2(\log x - 1) - \int (x \log x - x) \, dx.$$

You should (be able to) complete this solution on your own.

There is one other apparently obvious choice we could make for u and dv in order to evaluate $\int x \cdot \log x \, dx$ by parts. Namely, we could let u = 1 and dv = x·log x dx. Why is this selection absurd, here and in general? (Substitute in equation (2) to see why.)

Notice how we needed to use integration by parts twice in the solution for

Example 4.   We remind you that the initial reason for Examples 2 through 4 was to
point out that the selection of u and dv is not unique but often leads to solutions
of varying degrees of difficulty to execute.

You should notice that in each of the above solutions we have actually written
down the equation into which we are substituting. This is always a good practice
in any setting where you are substituting values or expressions into some formula.
Next to algebraic sloppiness the most common error is the misuse of a standard
formula. When integrating by parts, it is not uncommon to overlook the minus
sign preceding the integral on the right. Such an error in Example 3 would give us

$$\int x \cdot \log x \, dx = x^2 \cdot \log x + \int x(1 + \log x) dx$$

which reduces to the absurdity:

$$0 = x^2 \log x + \int x \, dx.$$

We are lucky here in that the result signals an error. But, as is more frequently
the case, we are the last to recognize our own mistakes!

A technical question may have occurred to you. If not, read on and it will.
Why, as in Example 2, have we written a particular antiderivative:

$\frac{x^2}{2}$, rather than the more general antiderivative: $\frac{x^2}{2} + C$? Should there be a "+C"
involved when we determine v? A look at equation (1) will answer this point.

$$\int f(x) \cdot g'(x) \, dx = f(x)(g(x) + C) - \int f'(x)(g(x) + C) dx$$

$$= f(x) \cdot g(x) + C \cdot f(x) - C \int f'(x) dx - \int f'(x) \cdot g(x) dx$$

$$= f(x) \cdot g(x) - \int f'(x) \cdot g(x) dx.$$

Clearly, with this justification, the answer is no. The choice of v = g(x) may be
any particular function as long as its derivative is the expression selected for
dv = g'(x)dx.

How are definite integrals evaluated using integration by parts? Answer:
with the same amount of luck and grace required by indefinite integrals. One could
apply the equation

$$\int_a^b f(x) \cdot g'(x) \, dx = f(x) \cdot g(x) \Big|_a^b - \int_a^b f'(x) \cdot g(x) \, dx$$

$$= f(b) \cdot g(b) - f(a) \cdot g(a) - \int_a^b f'(x) \cdot g(x) \, dx,$$

but we would recommend against this practice in general. If one reserves the use of "a and b" to the very end and evaluates the integral as if it were indefinite, then one always has open the option of checking the result by differentiation.

Thus, to evaluate $\int_1^e x \cdot \log x \, dx$, our suggestion is that you first obtain

$$\int x \cdot \log x \, dx = \frac{x^2}{2} \log x - \frac{x^2}{4} + C$$

and <u>then</u> insert the limits of integration to find that

$$\int_1^e x \cdot \log x \, dx = \left(\frac{x^2}{2} \log x - \frac{x^2}{4}\right) \Big|_1^e = \left(\frac{e^2}{2} - \frac{e^2}{4}\right) - \left(0 - \frac{1}{4}\right) = (e^2 + 1)/4.$$

<u>Example 5.</u> Evaluate $\int x^3 e^x dx$.

<u>Solution.</u> We set $u = x^3$ and $dv = e^x dx$

so that $du = 3x^2 \, dx$ and $v = e^x$.

Thus,

$$\int x^3 e^x dx = x^3 e^x - \int 3x^2 e^x dx.$$

Next, we set $u = 3x^2$ and $dv = e^x dx$

so that $du = 6x \, dx$ and $v = e^x$.

Thus,

$$\int 3x^2 e^x dx = 3x^2 e^x - \int 6x e^x dx$$

and

$$\int x^3 e^x dx = x^3 e^x - 3x^2 e^x + \int 6x e^x dx.$$

Now, we set $u = 6x$ and $dv = e^x dx$

so that $du = 6 \, dx$ and $v = e^x$.

Thus,

$$\int x^3 e^x dx = x^3 e^x - 3x^2 e^x + 6x e^x - \int 6 e^x dx$$

$$= x^3 e^x - 3x^2 e^x + 6x e^x - 6e^x + C.$$

It was necessary to integrate by parts three times to complete this example. Notice that in each instance u was selected as the polynomial part of the integrand. Suppose that you hadn't noted the reduction in the degree of the polynomial part of the integrand, where presumably you would eventually reach the point of a constant times e^x for the integrand. What if you had chosen for the third application

$$u = e^x \qquad \text{and} \qquad dv = 6x\, dx$$
$$du = e^x dx \qquad \text{and} \qquad v = 3x^2$$

so that

$$\int x^3 e^x dx = x^3 e^x - 3x^2 e^x + \left(3x^2 e^x - \int 3x^2 e^x dx\right)$$

or

$$\int x^3 e^x dx = x^3 e^x - \int 3x^2 e^x dx.$$

This alternate third application of integration by parts would simply have reciprocated the second application and you would have worked your way back to the previous juncture. This "reversal of field" is an error of wasted effort that can occur in solutions involving multiple applications of integration by parts. It is easy enough to guard against this error, but admittedly the heated excitement of integrating by parts dulls your sensitivities. You should note that if we select $u = e^x$ and $dv = 3x^2 dx$ at the second juncture, you will derive the comforting statement that $0 = 0$. This statement is comforting only because you underlined it and memorized it at a very tender age.

Example 6. Evaluate $\displaystyle\int \frac{x^3 dx}{\sqrt{x^2 - 4}}$

Solution. We set

$$u = x^2 \qquad \text{and} \qquad dv = \frac{x}{\sqrt{x^2 - 4}}\, dx$$

so that

$$du = 2x\, dx \qquad \text{and} \qquad v = \sqrt{x^2 - 4}.$$

Thus,

$$\int \frac{x^3}{\sqrt{x^2 - 4}} \, dx = x^2\sqrt{x^2 - 4} - \int 2x\sqrt{x^2 - 4} \; dx$$

$$= x^2\sqrt{x^2 - 4} - \frac{2}{3}(x^2 - 4)^{3/2} + C.$$

It is possible that you might gain an inflated sense of confidence in the technique of integration by parts because that is the focal point of integration problems in this part of the course. Before reading on see if you can solve Example 6 using the Substitution Principle. (Try $t = x^2 - 4$.)

You should be aware that the most difficult part of integrating is trying to decide which technique to apply rather than how to apply the respective procedures. It is with a small sense of relief that you may note that you won't fully experience the impact of such decision making until Chapter 8.

The Trigonometric and Hyperbolic Functions

7

If your trigonometric skills reside towards the bleaker end of the spectrum, you may have some very dark moments ahead of you. The following exercise is intended to lighten any impending edge of darkness. It is <u>critical</u> that you complete all parts of this exercise.

    <u>Review exercise on trigonometry</u>: Consider the following diagram where angles OAE, OBC, EDC, and OCE are right angles and where the length of segment OE is

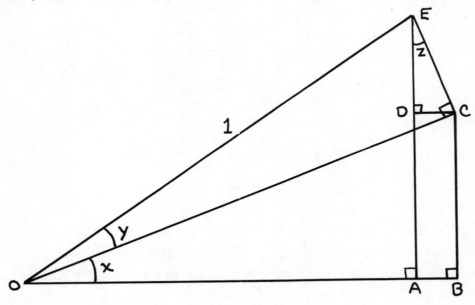

1 unit. Verify that angle z equals angle x. Now determine the lengths of all the line segments in the diagram. (For example EC = sin y and AB = DC = sin x sin y.)

142

Having done so, you should be able to conclude that

$$\sin(x + y) = AE = \sin x \cos y + \sin y \cos x \qquad (1)$$

and

$$\cos(x + y) = OA = \cos x \cos y - \sin x \sin y. \qquad (2)$$

Using these two results, the identity

(*) $$\sin^2 x + \cos^2 x = 1 \qquad (3)$$

and the definitions of the other trig functions in terms of sine and cosine, you should derive the following identities:

(*) $1 + \tan^2 x = \sec^2 x$ $\qquad\qquad 1 + \cot^2 x = \csc^2 x$

$\qquad \sin(x - y) = \sin x \cos y - \cos x \sin y$ $\qquad\qquad$ (Replace y by -y in (1).)

$\qquad \cos(x - y) = \cos x \cos y + \sin x \sin y$

$\qquad \tan(x + y) = \dfrac{\tan x + \tan y}{1 - \tan x \tan y}$ $\qquad\qquad$ (Combine (1) and (2).)

$\qquad \tan(x - y) = $?

(*) $\sin 2x = 2 \sin x \cos x$ $\qquad\qquad$ (Replace y by x in (1).)

(*) $\cos 2x = \cos^2 x - \sin^2 x$

(*) $\cos^2 x = \dfrac{1}{2}(1 + \cos 2x), \qquad \sin^2 x = \dfrac{1}{2}(1 - \cos 2x)$ $\quad$ (Combine (3) and the last identity to obtain these.)

$\qquad \tan 2x = \dfrac{2}{\cot x - \tan x}$

Next, evaluate each of the six trigonometric functions at 0, $\pi/6$, $\pi/4$, $\pi/3$, $\pi/2$, π, and $5\pi/3$ and sketch the graphs of each of the functions on the interval $[-\pi, 2\pi]$. Finally, sketch the graphs of $y = \sin x$, $y = \sin 2x$, $y = 2 \sin x$, and $y = 3 \sin 4x$ for $-\pi/4 \le x \le 2\pi$ on the same set of coordinate axes. Some of this work has been done for you in Figures 7.1.1-3 and Table 7.1.2 in the text.

Only if you have faithfully executed all parts of the above exercise and checked them with a competent friend and only if the end of the semester has not yet arrived should you continue. You will need the identities marked on the left by an asterisk (*) to solve many of the problems in the next few Chapters. Memorize these identities.

It is imperative that you memorize the differentiation formulas for the six trigonometric functions.

$$\frac{d}{dx}(\sin x) = \cos x \qquad\qquad \frac{d}{dx}(\cos x) = -\sin x$$

$$\frac{d}{dx}(\tan x) = \sec^2 x \qquad\qquad \frac{d}{dx}(\cot x) = -\csc^2 x$$

$$\frac{d}{dx}(\sec x) = \sec x \tan x \qquad\qquad \frac{d}{dx}(\csc x) = -\csc x \cot x.$$

The derivative of each "co" function carries a minus sign. Notice that certain pairings occur: sine with cosine, tangent with secant, and cotangent with cosecant.

Two limits are developed in this Section, primarily for the purpose of deriving the differentiation formulas for the trigonometric functions.

$$\lim_{x \to 0} \frac{\sin x}{x} = 1 \qquad \text{and} \qquad \lim_{x \to 0} \frac{1 - \cos x}{x} = 0. \tag{4}$$

Frequently these limits are useful when calculating the limit of a more complicated expression. You should be alert for the possibility of obtaining the expressions in (4) by algebraic manipulation. The following examples illustrate this point. Additional limit problems appear after the second example.

Example 1. Evaluate $\displaystyle\lim_{x \to 0} \frac{2x}{\sin 3x}$.

At first this limit appears undefined. However we can write this as $\displaystyle\lim_{x \to 0} \frac{2}{3} \cdot \frac{3x}{\sin 3x}$. Let $u = 3x$. As $x \to 0$, we have $3x \to 0$ so $u \to 0$. Hence,

$$\lim_{x \to 0} \frac{2x}{\sin 3x} = \lim_{u \to 0} \frac{2}{3} \frac{u}{\sin u} = \frac{2}{3} \lim_{u \to 0} \frac{1}{\frac{\sin u}{u}} = \frac{2}{3} \ .$$

Example 2. Evaluate $\displaystyle\lim_{x \to 0} \left(x \csc 2x - \frac{2}{x \cot 3x} + \frac{\sin^2 x}{x(1 + \cos x)} \right)$.

We use some basic identities to rephrase the problem so that the results in (4) may be used as follows. Which identities are used in the first step?

$$\lim_{x \to 0} \left(\frac{x}{\sin 2x} - \frac{2 \sin 3x}{x \cos 3x} + \frac{1 - \cos^2 x}{x(1 + \cos x)} \right)$$

$$= \lim_{x \to 0} \left(\frac{1}{2} \cdot \frac{2x}{\sin 2x} - \frac{6}{\cos 3x} \cdot \frac{\sin 3x}{3x} + \frac{1 - \cos x}{x} \right)$$

$$= \frac{1}{2} \cdot 1 - \frac{6}{1} \cdot 1 + 0 = -5\frac{1}{2} .$$

EXERCISES. Evaluate the following limits.

1. $\displaystyle \lim_{x \to 0} \frac{\sin^2 x}{x(1 + \cos x)}$.

2. $\displaystyle \lim_{x \to 0} x \cot x$.

3. $\displaystyle \lim_{x \to 0} \frac{\tan(3x)}{x^2 + 2x}$.

4. $\displaystyle \lim_{x \to 0} \frac{\sec x - 1}{x \sec x}$.

5. $\displaystyle \lim_{x \to \pi/2} \frac{\cos x}{x - \pi/2}$.

6. $\displaystyle \lim_{x \to 0} x^2 (\cot^2(3x) + 1)$.

7. $\displaystyle \lim_{x \to 0} \frac{1 - \cos(6x)}{x^2}$.

8. $\displaystyle \lim_{x \to \pi/4} \frac{\sin(x + \frac{\pi}{4}) - 1}{x - \frac{\pi}{4}}$.

ANSWERS.

1. 0 [See Example 2]   2. 1   3. 3/2   4. 0   5. -1

6. 1/9   7. 18 [Use $2 \sin^2 A = 1 - \cos 2A$]   8. -1

SECTION 7.2

There are six integrals associated with the derivatives given in Table 7.2.1 of the text.

$$\int \cos x \, dx = \sin x + C \qquad\qquad \int \sin x \, dx = -\cos x + C$$

$$\int \sec^2 x \, dx = \tan x + C \qquad\qquad \int \csc^2 x \, dx = -\cot x + C$$

$$\int \sec x \tan x \, dx = \sec x + C \qquad\qquad \int \csc x \cot x \, dx = -\csc x + C$$

These six formulas should also be remembered. Note that <u>you have already done so</u> when you memorized the table of derivatives. The following integrals should also be at your fingertips:

$$\int \tan x \, dx, \qquad \int \cot x \, dx, \qquad \int \sec x \, dx, \qquad \int \csc x \, dx.$$

See Table 7.2.2 if you can't evaluate them.

If you have an integral involving $\sin x$ and/or $\cos x$ and you are unable to see how to evaluate the integral, then it may help to rewrite the integrand in terms of $\sec x$ and $\tan x$ or in terms of $\csc x$ and $\cot x$. What you are doing is relying on the natural groupings these functions inherit from their differentiation formulas. For example,

$$\int \frac{\sin^2 x}{\cos^4 x} \, dx = \int \tan^2 x \sec^2 x \, dx = \frac{1}{3} \tan^3 x + C.$$

Again, $\int \sec^2 x \cot x \, dx$ may be solved as

$$\int \frac{\sec^2 x}{\tan x} \, dx = \log |\tan x| + C$$

or, in terms of $\sin x$ and $\cos x$, as

$$\int \frac{dx}{\sin x \cos x} = \int \frac{2 \, dx}{\sin 2x} = 2 \int \csc(2x) dx$$

$$= \log|\csc(2x) - \cot(2x)| + C.$$

A significant part of evaluating trigonometric integrals is using the identities to rephrase, and perhaps simplify, the problems.

Of course, you should not forget the chain rule. As a reminder, notice that

$$\frac{d}{dx}(\sin 2x) = 2 \cos 2x$$

$$\frac{d}{dx}(\cos x^2) = -2x \sin(x^2)$$

$$\frac{d}{dx}(\tan(\sec 3x)) = \sec^2(\sec 3x) \cdot 3 \sec 3x \tan 3x$$

and

$$\frac{d}{dx}(\csc(\frac{1}{x})) = \frac{1}{x^2} \csc(\frac{1}{x}) \cot(\frac{1}{x}).$$

Hopefully, you are not tempted now to claim that $\csc(\frac{1}{x}) = \sin x$. Where is the error in this temptation and why is it ridiculous? More generally is $f(\frac{1}{x}) = \frac{1}{f(x)}$? Of course not!

Integration by parts is frequently useful in evaluating integrals which mix the three types of expressions: polynomial, trigonometric, and exponential (including logarithmic). For example, the integrals

$$\int e^x \cos x \, dx, \qquad \int x^2 \sin x \, dx, \qquad \int x^3 \cos(x^2) dx$$

will succumb to integration by parts. Not all such "mixed integrands", however, require integration by parts. For example,

$$\int x \cos(x^2) dx = \frac{1}{2} \sin(x^2) + C \qquad \text{(Substitute } u = x^2.)$$

and

$$\int x^2 e^{x^3} dx = \frac{1}{3} e^{x^3} + C. \qquad \text{(Substitute } u = x^3.)$$

SECTION 7.3

For what value of x is $\sin x = 1/2$? The answer to this question is not unique, since $f(x) = \sin x$ is not a one-to-one function. Since it is not one-to-one, $f(x) = \sin x$ does not have an inverse. The same is true of the other five trigonometric functions. However, we can restrict the domain of each trigonometric function so that each function is one-to-one on its restricted domain. In this context we may and do refer to the inverse trigonometric functions. Arcsin x and arctan x are nicely developed by the text by restricting $x = \sin y$ and $x = \tan y$

to y in $[-\pi/2, \pi/2]$ and y in $(-\pi/2, \pi/2)$ respectively. You should question why these restricted domains are chosen instead of others. After all, $x = \sin y$ is one-to-one on $[\pi/2, 3\pi/2]$, on $[3\pi/2, 5\pi/2]$, etc. We will answer this question indirectly by treating $y = \arccos x$. Incidentally, $\sin x = 1/2$ if and only if $x = \pi/6 + 2n\pi$, where n is any integer. However, $x = \arcsin(1/2)$ iff $x = \pi/6$.

A few examples will help to clarify the relationship between the trigonometric functions and their inverses.

Example 1. Find $\cos(\arcsin 1/2)$.

We are asked to find the cosine of that angle (or number) whose sine is 1/2. We will call this angle θ. A convenient way to think of trigonometric functions is in the very context in which we all first learned about them: as ratios of the <u>directed lengths</u> of the sides of right triangles. In particular, the sine of either acute angle in a right triangle may be thought of as the ratio of "opposite to hypotenuse."

Clearly, θ is that angle in the diagram whose sine is 1/2. Another way to say this is $\theta = \arcsin(1/2)$. Since the range of the arcsine function has been restricted to $[-\pi/2, \pi/2]$, $\theta = \pi/6$.

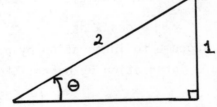

Therefore, $\cos(\arcsin(1/2)) = \cos(\pi/6) = \sqrt{3}/2$. Notice that this result agrees with our diagram, since the length of the other leg is $\sqrt{3}$.

Example 2. Find $\tan(\arcsin x)$.

As above we construct a diagram with $\theta = \arcsin x$.

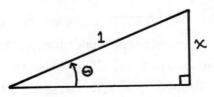

Since the length of the third side of the right triangle is $\sqrt{1 - x^2}$ (why?), we have that

$$\tan(\arcsin x) = \tan(\theta) = \frac{x}{\sqrt{1 - x^2}} \ .$$

Example 3. Find $\tan(\arcsin(-3/5))$.

Again we consult a diagram with $\theta = \arcsin(-3/5)$.

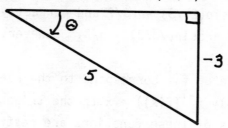

Thus,

$$\tan(\arcsin(-3/5)) = \tan(\theta) = -3/4.$$

This refreshing diagram with one side labeled with a negative directed length may cause you to wonder why the third side here was computed as 4 rather than -4, or more generally, why the third side in the last example was computed as $\sqrt{1 - x^2}$ rather than $-\sqrt{1 - x^2}$. The reason why we may consistently choose the positive radical is that if both legs of the triangle are negative then θ lies in the third quadrant; viz., $\pi \leq \theta \leq 3\pi/2$, which is disjoint from the designated ranges for the inverse trigonometric functions. In plotting the ratios $\frac{\text{opposite}}{\text{hypotenuse}}$, $\frac{\text{opposite}}{\text{adjacent}}$, etc., the "hypotenuse" is always positive. "Opposite" is measured parallel to the y-axis and "adjacent" is measured parallel to the x-axis. Depending on the signs of "opposite" and "adjacent," there are three possible orientations for the resulting triangle.

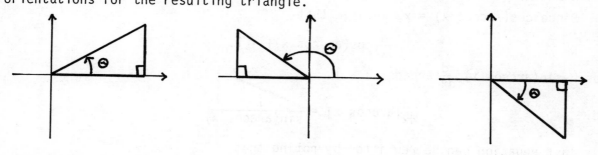

no sides negative   only "adjacent" negative   only "opposite" negative

Example 4. Find arcsin(sin(2π/3)).

The fact that $f^{-1}(f(x)) = x$ suggests here that arcsin(sin(2π/3)) = 2π/3. This is troublesome, however, since the range of the arcsine function, [-π/2,π/2], does not include 2π/3. Therefore, arcsin(sin(2π/3)) cannot be 2π/3. What do we do?

We could note that sin(2π/3) = √3/2 and correctly conclude with great joy that arcsin(sin(2π/3)) = arcsin(√3/2) = π/3. However, our enthusiasm is avoiding the more general question.

The domain and range of f^{-1} correspond to the range and domain, respectively, of f. Before we may apply $f^{-1}(f(x)) = x$ to the trigonometric functions, we must remember that the domains of these functions are restricted to enable the restricted functions to be one-to-one and have inverses. The choice for the restricted domain of the sine function was [-π/2,π/2]. Therefore, we need to find a number y such that

y is in [-π/2,π/2] <u>and</u> sin y = sin 2π/3.

Then it will be true that arcsin(sin y) = y. As you might suspect, y = π/3.

We didn't need to know that sin 2π/3 = √3/2 = sin π/3 to solve this problem. We simply needed to know that sin 2π/3 = sin π/3 and π/3 is in [-π/2,π/2]. Can you now show that arcsin(sin(37π/18)) = π/18? Is arccos(cos 5π/4) = 3π/4?

We define y = arccos x iff y is in [0,π] and cos y = x. In the following derivation of $\frac{d}{dx}$[arccos x] you should see if you can determine why [0,π] is chosen as the restricted domain for cos y = x.

Since cos(arccos x) = x, we have that

$$\frac{d}{dx}[\cos(\arccos x)] = 1.$$

Thus, $-\sin(\arccos x) \frac{d}{dx}(\arccos x) = 1$

and

$$\frac{d}{dx}[\arccos x] = \frac{-1}{\sin(\arccos x)}.$$

This last equation can be rewritten by noting that

$$\sin(\arccos x) = \pm \sqrt{1 - \cos^2(\arccos x)} = \pm \sqrt{1 - x^2}.$$

Since we may select the + sign for all values of x (why?), we find that

$$\frac{d}{dx}(\arccos x) = \frac{-1}{\sqrt{1 - x^2}} \ .$$

The answer to the "why" is that arccos x is <u>by definition</u> between 0 and π, and on this interval the sine function is never negative. With this thought in mind, look at the text's development of the derivatives for arcsine and arctangent. In the first instance, we need to assert that

$$\cos(\arcsin x) = +\sqrt{1 - \sin^2(\arcsin x)} \ .$$

The assertion is legitimate since the range of arcsine has been restricted to $[-\pi/2, \pi/2]$ and the cosine function is not negative on this interval. The case of the arctangent is left for you.

You probably have noticed with some surprise that

$$\frac{d}{dx}(\arcsin x) = -\frac{d}{dx}(\arccos x).$$

Is this result significant and what does it mean? First of all, this result and others explain in part why the text only develops the inverse trigonometric functions for sine and tangent. Arcsine and arctangent are sufficient for most applications that may arise.

We can explain our surprising equation by rewriting it as

$$\frac{d}{dx}(\arcsin x + \arccos x) = 0,$$

which implies that

$$\arcsin x + \arccos x = C$$

where C is a constant. The value of C is $\pi/2$ as may be found by substituting any value, say zero, for x. A quick glance at the diagram below will reveal that we didn't need to differentiate to show that

$$\arccos(x) + \arcsin(x) = \pi/2 \ .$$

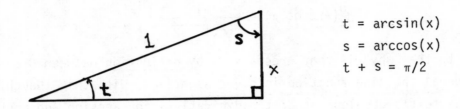

$$t = \arcsin(x)$$
$$s = \arccos(x)$$
$$t + s = \pi/2$$

We will now derive the derivative of $y = \text{arcsec}(x)$. Our procedure will differ slightly from that taken by us and the text for the other inverse functions. We do so because we want to suggest that it is _not_ necessary to memorize the derivatives (and associated integration formulas) for the inverse trigonometric functions.

We adopt the view that by simply remembering the one or two procedures for deriving the desired formulas and by following these through for each function, one ends up remembering (as opposed to memorizing) the formulas and learns something in the meantime. We recommend this procedure to you only if it appeals to you. Admittedly some of you would much prefer a deck of 3 x 5 "flash cards" and a six-pack.

Suppose, then, that you are given $y = \text{arcsec } x$ and are asked to write down $\frac{dy}{dx}$. Let us suppose further that you have a normal memory and are accordingly flustered by this request. Calling into play your many talents, you note that if

$$y = \text{arcsec } x$$

then

$$\sec y = x$$

and

$$\frac{d}{dx}(\sec y) = 1.$$

Thus,

$$\sec y \tan y \frac{dy}{dx} = 1,$$

$$\frac{dy}{dx} = \frac{1}{\sec y \tan y},$$

which may be written as

$$\frac{dy}{dx} = \frac{1}{x\sqrt{x^2 - 1}}$$

by consulting the diagram at the top of the next page.

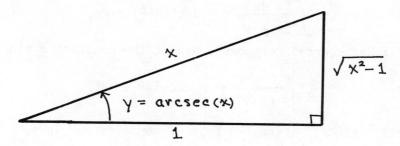

Can you examine this derivation to suggest why we <u>might</u> restrict the domain of $x = \sec y$ to y in $[-\pi, -\pi/2) \cup [0, \pi/2)$ in order for the inverse arcsecant function to be well defined? Usually the domain is restricted differently to $[0, \pi/2) \cup (\pi/2, \pi]$. (Why is $\pi/2$ excluded?) This restriction is preferred, but it forces us to note that sometimes $\tan y = \sqrt{\sec^2 y - 1}$ and sometimes $\tan y = -\sqrt{\sec^2 y - 1}$. The net effect is that we must then write

$$\frac{dy}{dx} = \frac{1}{|x|\sqrt{x^2 - 1}}, \qquad |x| > 1.$$

Notice the absolute value symbol.

Some texts denote the inverse trigonometric functions as $\sin^{-1} x$. This notation is inconsistent with $\sin^n x = (\sin x)^n$. There is a vast difference between $\sin^{-1} x = \arcsin x$ and $\sin^{-1} x$ read as $\frac{1}{\sin x}$.

As noted earlier, you will have no real use for arccos, arccot, and arccsc. However, you should be able to evaluate, say, $\frac{d}{dx}(\arccos(1 - 2x))$. Your best approach is that taken above. You should verify that the result of this differentiation is $1/\sqrt{x - x^2}$. Similarly, you will rarely have use for arcsec. However, arcsin(x) and arctan(x) should become close friends, even though a method for sometimes circumventing this acquaintance is given in Chapter 8. This method is hinted at in the cautionary note below.

You will probably remember the integration formula

$$\int \frac{dx}{1 + x^2} = \arctan(x) + C,$$

but you will just as likely misuse the more general formula

$$\int \frac{dx}{a^2 + x^2} = \frac{1}{a} \arctan(\frac{x}{a}) + C \; , \qquad\qquad \text{(See text, p. 307)}$$

Namely, you will be uncertain of how to handle the constant _a_ since

$$\int \frac{dx}{\sqrt{a^2 - x^2}} = \arcsin(\frac{x}{a}) + C$$

lacks the coefficient 1/a. Even the best of memories may lapse in such a situation. Again, our suggestion is to replace some memory by technique. An example will suffice.

To evaluate $\int \frac{dx}{9 + 16x^2}$, begin with the substitution $t = \frac{4}{3} x$. Therefore,

$$\int \frac{dx}{9 + 16x^2} = \int \frac{3/4 \, dt}{9 + 9t^2} = \frac{1}{12} \int \frac{dt}{1 + t^2} = \frac{1}{12} \arctan(t) + C$$

$$= \frac{1}{12} \arctan(\frac{4x}{3}) + C.$$

The material in Chapter 8 will take you one step further. The substitution $4x = 3 \tan(t)$ gives us

$$\int \frac{dx}{9 + 16x^2} = \int \frac{3/4 \, \sec^2 t \, dt}{9 + 9 \tan^2 t} = \frac{1}{12} \int dt = \frac{1}{12} t + C$$

$$= \frac{1}{12} \arctan(\frac{4x}{3}) + C.$$

You might find it helpful to mimic each of these substitutions in verifying that

$$\int \frac{dx}{\sqrt{9 - 16x^2}} = \frac{1}{4} \arcsin(\frac{4x}{3}) + C.$$

SECTION 7.4

Solutions for Exercises 7, 21, 22, 23, 24, and 28 on p. 311 of the text are given below. Additional exercises are given after the solution for Exercise 28.

Solution for Exercise 7 (p. 311). The function $y = a \sin x + b \cos x$ has period 2π so it is sufficient to examine the function for x in $[0, 2\pi]$. We then calculate

$y' = a \cos x - b \sin x$. Let us assume for the moment that $ab \neq 0$. The condition $y' = 0$ implies that $a \cos x = b \sin x$ or $\tan x = \frac{a}{b}$. This equation has two

solutions in the interval $[0, 2\pi]$, i.e., where $\cos x = \pm b / \sqrt{a^2 + b^2}$. These two values together with the endpoints 0 and 2π give the critical points of y. By the Maximum-Minimum Theorem we know we must have a maximum on $[0, 2\pi]$. Computing

the values of y at each of the four critical points yields $\sqrt{a^2 + b^2}$ as the desired maximum. Notice that this expression also covers the excluded cases when $ab = 0$.

Solution for Exercise 21 (p. 311). A clearly labelled diagram helps.

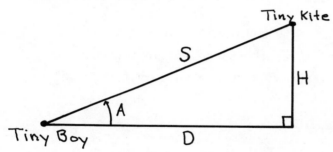

S = 100 feet (not constant)

H = 60 feet (constant)

$\frac{dD}{dt} = -10$ feet/min.

What is $\frac{dA}{dt}$ in radians/min.?

We may write the solution in two ways.

$$\sin A = H/S = 60/S$$

$$\cos A \, \frac{dA}{dt} = \frac{-60}{S^2} \, \frac{dS}{dt}$$

so

$$\frac{dA}{dt} = \frac{-60 \sec A}{S^2} \, \frac{dS}{dt} \, .$$

As

$$D^2 + H^2 = S^2,$$

$$2D \, \frac{dD}{dt} + 0 = 2S \, \frac{dS}{dt}$$

and

$$\frac{dA}{dt} = \frac{-60 \sec A}{S^2} \cdot \frac{D}{S} \, \frac{dD}{dt} \, .$$

At the required instant,

$$\frac{dA}{dt} = .06 \text{ radians/min.}$$

OR

$$A = \arctan\left(\frac{H}{D}\right)$$

$$= \arctan\left(\frac{60}{D}\right)$$

Thus,

$$\frac{dA}{dt} = \frac{1}{1 + (60/D)^2} \cdot \frac{-60}{D^2} \cdot \frac{dD}{dt}$$

$$= \frac{-60}{D^2 + (60)^2} \cdot \frac{dD}{dt} \, .$$

Since D = 80 feet at the required instant, we have

$$\frac{dA}{dt} = .06 \text{ radians/min.}$$

Solution for Exercise 22 (p. 311). Again, a sketch is enlightening.

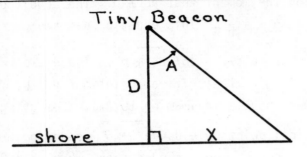

D = 1/2 mile (constant)

X = 1 mile (not constant)

$\frac{dA}{dt}$ = 2π radians/min.

What is $\frac{dX}{dt}$?

Starting with X = D tan A, we have

$$\frac{dX}{dt} = D \sec^2 A \frac{dA}{dt} = D(1 + (\frac{X}{D})^2) \frac{dA}{dt} .$$

At the required instant $\frac{dX}{dt} = \frac{1}{2}(1 + 4)2\pi = 5\pi$ miles/min. This rate is approximately 9420 miles per hour - slow, as the speed of light goes. But notice that as X gets arbitrarily large the light will be moving down the shoreline faster than it can arrive at the shore!

Solution for Exercise 23 (p. 311).

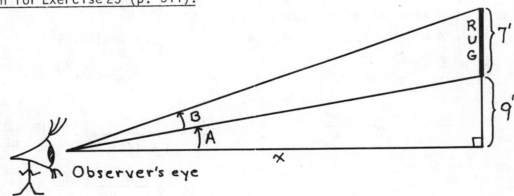

We have that B = arctan($\frac{16}{X}$) - arctan($\frac{9}{X}$) for 0 < x < ∞. With a minimum of effort we find

$$\frac{dB}{dx} = \frac{7(144 - x^2)}{(x^2 + 81)(x^2 + 256)} .$$

Since $\frac{dB}{dx}$ = 0 for x = 12 and since B is increasing for 0 < x < 12 and decreasing for 12 < x, B attains an absolute maximum when x = 12. Thus, the most favorable

position is 12 feet from the wall. How do you interpret the solution x = -12?
Had the tapestry been one-sided and suspended in space, one interpretation would
call x = -12 the least favorable position, depending on your point of view of
tapestries.

Partial solution for Exercise 24 (p. 311). The analysis on this one is left for
you, but here is the set-up (A = θ).

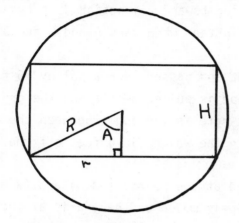

R = 6 inches

Vol. = $\pi r^2 H$

r = R sin A = 6 sin A

H = 2R cos A = 12 cos A

with 0 < A < π/2.

Solution for Exercise 28 (p. 311).

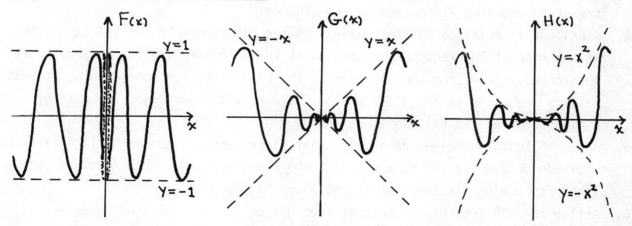

    F is discontinuous at x = 0 since F(x) gets arbitrarily close to each value
between -1 and 1 as x approaches 0. That is, $\lim_{x \to 0} F(x)$ does not exist. On the
other hand, G and H are continuous at x = 0 since

$$\lim_{x \to 0} G(x) = \lim_{x \to 0} H(x) = 0 = G(0) = H(0).$$

These limits are easily verified by use of the Pinching Theorem. (How?) Only H is differentiable at 0. (Try using the limit definition of the derivative.)

EXERCISES.

1. Car A is traveling east at 30 miles per hour and car B is traveling north at 22.5 miles per hour. Both cars are traveling toward a junction J of two roads. At what rate is angle JAB changing at the instant when cars A and B are 300 ft. and 400 ft., respectively, from the junction?

2. A rope 32 feet long is attached to a weight and passed over a pulley 16 feet above the ground. The other end of the rope is pulled away along the ground at the rate of 3 feet per second. At what rate is the angle between the rope and the ground changing at the instant when the weight is 4 feet off the ground?

3. The two ends of a rubber strip are fastened at the points (-3,0) and (3,0), respectively. If the midpoint of the strip is moved up the y-axis at the rate of 4 units per minute at what rate is the angle between the two parts of the strip changing at the end of two minutes?

4. A balloon is released 500 feet away from our observer. If the balloon rises at the rate of 100 feet per minute and at the same time a wind is carrying it horizontally away from the observer at the rate of 75 feet per minute, at what rate is the angle of elevation of the observer's line of sight increasing, 6 minutes after the balloon is released?

5. A searchlight is trained on a plane that flies directly above the light at an altitude of 2 miles and at a speed of 400 miles per hour. How fast must the light be turning 2 seconds after the plane passes directly overhead?

6. A flag pole 15 feet high stands at the edge of the roof of a building so that the foot of the pole is 60 feet above an observer's eye level. How far from the building should the observer stand so that the pole subtends the maximum possible angle at his eye?

ANSWERS.

1. .0308 radians/sec.   2. -0.12 rad./sec.   3. -24/73 rad./min.

4. 0.05 rad./min.   5. 9/164 rad./sec.   6. $30\sqrt{5}$ feet

SECTIONS 7.6-7

Some of the technical aspects of hyperbolic functions are treated by the text.
The purpose of the following comments is to give you in non-technical terms some
feeling for the importance of these topics.

 A complex number is any number of the form a + bi, where a and b are real
numbers and i is the symbol chosen to represent one of the roots of the equation
x^2 = -1. That is, i = $\sqrt{-1}$, which is not a real number. The other root of the
equation x^2 = -1 is of course -i. It is quite likely that you have encountered
these numbers before. To help refresh your memory, we simply mention the Argand
diagram used for representing complex numbers:

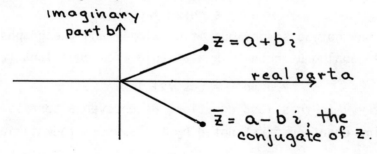

and, DeMoivre's formula used for computing the nth roots of any real number x:

$$\sqrt[n]{x} = |x|^{1/n}[\cos(\frac{2k\pi}{n}) + i \sin(\frac{2k\pi}{n})], \qquad 1 \le k \le n.$$

Now that your memory has been restored we will use the complex numbers to state
a very close relationship between the circular and hyperbolic functions. Rigorous
study of these and many other concepts of complex function theory are left for an
upper level course.

 The hyperbolic sine and cosine are defined in terms of the exponential
function. These definitions are stated below with the "slight" change that the
variable z is a complex number. By definition, z = a + bi for real numbers a and

b, so z is real iff b = 0.

$$sinh(z) = \frac{e^z - e^{-z}}{2} \text{ and } cosh(z) = \frac{e^z + e^{-z}}{2}. \tag{1}$$

With no desire or attempt to justify the statement, we remark that it is also true that

$$sin(z) = \frac{e^{iz} - e^{-iz}}{2i} \text{ and } cos(z) = \frac{e^{iz} + e^{-iz}}{2}. \tag{2}$$

The equations in (1) and (2) are strikingly similar. In fact, they permit us to conclude that

$$sin(z) = \frac{1}{i} sinh(iz) \text{ and } cos(z) = cosh(iz).$$

With a little thought it might become apparent to you now why the derivatives of the hyperbolic functions so closely resemble those of their mnemonic counterparts. Aside from the ever present "h" at the end, only some signs are changed. The changes in sign are brought about by the presence of "i."

The equations in (1) may be combined to yield

$$e^z = sinh(z) + cosh(z),$$

a result that may not be so obvious if we simply add the graphs of $y = sinh\ x$ and $y = cosh\ x$. An analogous result is available from equations (2):

$$e^{iz} = cos\ z + i\ sin\ z.$$

Substituting $z = \pi$ in this last equation, we uncover a terrifyingly simple, but ever fascinating equation first noted by the Swiss mathematician Leonard Euler (1707-1783):

$$e^{\pi i} = -1.$$

Most of the real work in physics and other applied areas of mathematics is done in terms of functions of complex variables. The above remarks suggest that in such a setting the hyperbolic and circular functions are on equal footing.

We turn now to two closely related physical problems where the hyperbolic functions arise in a natural way. Our first problem is that of the hanging cable. A presumably homogeneous, inelastic cable is suspended between two points (A and B) that need not be at the same height. Let us place our cable on coordinate axes with the y-axis being the "vertical" line through the lowest point on the curve.

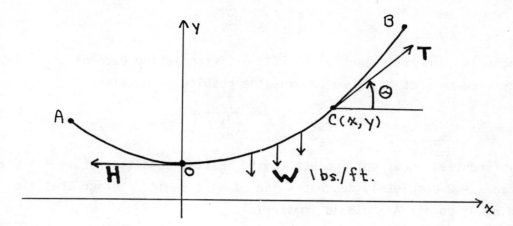

We assume that the cable is stable so that the forces acting upon it are in equilibrium. There are three forces acting on the arbitrary section OC of the cable. Focusing our attention on just this section of the cable with the rest removed, the following forces would maintain the stable state of the cable.

(1) H, the horizontal tension pulling on the cable at the point O.

(2) T, the tangential tension pulling on the cable at the point C.

(3) WL, the weight of the section OC which is L feet long and weighs
       W lbs. per foot.

As the vertical forces on section OC must balance,

$$WL = T \sin \theta.$$

Similarly, the equilibrium of the cable requires the horizontal forces to be in balance:

$$H = T \cos \theta.$$

Consequently,

$$\tan \theta = \frac{W}{H} L = \frac{1}{a} L, \text{ where } a = \frac{H}{W} \text{ a constant.}$$

We assume the y-coordinate of point O to be a. This legitimate assumption will simplify subsequent work. Now, if y = f(x) describes the cable, the fact that $\frac{dy}{dx} = \tan \theta$ permits us to write

$$\frac{dy}{dx} = \frac{1}{a} L.$$

This is a simple formula until we note that the arc length L of section OC is given by

$$L = \int_0^x \sqrt{1 + (\frac{dy}{dt})^2} \; dt.$$

(Arc length is covered in Section 11.7.) Differentiating each of these last two equations with respect to x, we combine the results to obtain

$$\frac{d^2y}{dx^2} = \frac{1}{a}\sqrt{1 + (\frac{dy}{dx})^2} \; . \tag{3}$$

This "differential equation" subject to the initial condition that $\frac{dy}{dx} = 0$ and y = a when x = 0 completely describes the hanging cable. Notice that the positioning of points A and B is immaterial.

To solve equation (3), let $z = \frac{dy}{dx}$.

Then,
$$\frac{dz}{dx} = \frac{1}{a} \sqrt{1 + z^2}$$

and
$$\int \frac{dz}{\sqrt{1 + z^2}} = \int \frac{dx}{a}$$

or
$$\text{arcsinh}(z) = \frac{1}{a} x + C.$$

As x = 0 implies z = 0 and arcsinh(0) = 0, we have that C = 0. Therefore,

$$\frac{dy}{dx} = z = \sinh(\frac{x}{a}).$$

Easily then

$$y = a \cosh(\frac{x}{a}) + k.$$

Since y = a when x = 0 and cosh(0) = 1, we have

$$y = a \cosh(\frac{x}{a})$$

as the description of the hanging cable. You should remember that a > 0 is the "elevation" of the lowest point on the cable. The shape taken by the hanging cable is called a catenary. It is, of course, familiar to builders of suspension bridges as well as other contraptions that suggest no visible means of support.

We begin our second application by rotating our catenary about the x-axis to generate a "surface of revolution." This particular surface is called a catenoid. Points A and B will generate circles which lie on parallel planes and, of course, have their centers on the x-axis. Suppose that _a_ soap film is permitted to form on these two circles. Quite curiously the surface of the soap film joining the two bounding rings will be the aforementioned catenoid.

We can but will not carry this discussion further. Hopefully, it will not leave you thinking: "Cosh almighty, I've been left hanging again!"

The Technique of Integration

8

In dealing with the more complicated integration problems of this Section and later Sections, you will find that completing the square is an algebraic necessity. The following steps mimic the general approach. Starting with

$$Ax^2 + Bx, \quad (A \neq 0),$$

we write

$$A(x^2 + \frac{B}{A} x).$$

We want to determine a value of C so that

$$x^2 + \frac{B}{A} x + C$$

may be written as a perfect square. Since $(x + D)^2 = x^2 + 2Dx + D^2$ we note that $C = (\frac{1}{2} \frac{B}{A})^2$ is the proper choice. Carefully adding and substracting constants so as not to change the value of the original expression, we have

$$Ax^2 + Bx = A(x + \frac{B}{2A})^2 - A(\frac{B}{2A})^2.$$

For example, $x^2 + 6x$ becomes $(x + 3)^2 - 9$ and $x^2 - 2x + 3$ becomes $(x - 1)^2 + 2$. Further,

$$x^2 + 4x - 3y^2 + 4y + 5$$

may be written as

$$(x + 2)^2 - 3(y - \frac{2}{3})^2 + \frac{7}{3} .$$

164

EXERCISES. "Complete the squares" in the following.

1. $x^2 + 6x$

2. $x^2 + 6x - 7$

3. $x^2 - 6x - 7$

4. $x^2 + 4x + 9$

5. $x^2 + 4x - 9$

6. $9x^2 + 6x$

7. $4x^2 - 8x + 7$

8. $4x^2 - 6x + 5$

9. $x^2 - 9y^2 + 4x - 12y$

ANSWERS.

1. $(x + 3)^2 - 9$

2. $(x + 3)^2 - 16$

3. $(x - 3)^2 - 16$

4. $(x + 2)^2 + 5$

5. $(x + 2)^2 - 13$

6. $9(x + \frac{1}{3})^2 - 1$

7. $4(x - 1)^2 + 3$

8. $4(x - \frac{3}{4})^2 + \frac{11}{4}$

9. $(x + 2)^2 - 9(y + \frac{2}{3})^2$

SECTION 8.2

The method of partial fractions applies to all integrals where the integrand is a rational function; that is, the quotient of two polynomials. There are two stages in this method. The first stage is strictly algebraic and is made up of three steps.

Step one consists of reexpressing the rational function as the sum of a polynomial and a rational function whose numerator is of a smaller degree than the denominator. The remaining steps in our method presume this standard form (degree of numerator less than degree of denominator) which is obtained by long division. For example,

$$
\begin{array}{r}
x + 3 \\
x^3 + x \overline{\smash{\big)}\ x^4 + 3x^3 + 3x^2 \qquad\quad + 1} \\
\underline{x^4 \qquad\quad + x^2} \\
3x^3 + 2x^2 \qquad\quad + 1 \\
\underline{3x^3 \qquad\quad + 3x} \\
2x^2 - 3x + 1
\end{array}
$$

so that

$$\int \frac{x^4 + 3x^3 + 3x^2 + 1}{x^3 + x} \, dx = \int (x + 3 + \frac{2x^2 - 3x + 1}{x^3 + x}) \, dx$$

$$= \frac{x^2}{2} + 3x + \int \frac{2x^2 - 3x + 1}{x^3 + x} \, dx \, .$$

The second step consists of factoring the denominator into linear and quadratic expressions, where the quadratics are not reducible; that is, not factorable into two linear terms using only real numbers. Note that $4x^2 + 9 = (2x + 3i)(2x - 3i)$ is not a legitimate factorization as $i = \sqrt{-1}$ is not a real number. The Fundamental Theorem of Algebra assures us that every polynomial with real coefficients can be factored into linear and irreducible quadratic factors.

The third step consists of decomposing the rational function that remains after step one into a sum of partial fractions by considering <u>all</u> fractions of the form

$$\frac{A}{(ax + b)^k} \quad \text{and} \quad \frac{Bx + C}{(cx^2 + dx + e)^\ell} \, , \tag{1}$$

where these denominators divide the denominator of the rational function. You must remember to consider all possibilities. For example,

$$\frac{N(x)}{(x + 1)^n} \qquad \deg N < n$$

gives rise to n possible terms:

$$\frac{A_i}{(x + 1)^i} \qquad 1 \le i \le n,$$

where each A_i is a constant.

For example, suppose we had

$$RF = \frac{P(x)}{x(x + 1)^3(x^2 + 1)^2} \tag{2}$$

after steps 1 and 2, where $P(x)$ is a polynomial of degree less than 8 (why?). We would write

$$RF = \frac{A}{x} + \frac{B}{x+1} + \frac{C}{(x+1)^2} + \frac{D}{(x+1)^3} + \frac{Ex+F}{(x^2+1)} + \frac{Gx+H}{(x^2+1)^2} \qquad (3)$$

Notice that we have <u>allowed</u> (some of the values A through H may turn out later to be zero) for <u>all</u> partial fractions of form (1) whose denominators <u>divide</u> the denominator of RF. We next substitute values for x and obtain a system of equations in the variables A through H. (Since there are 8 variables here, we will need to substitute 8 different values for x to obtain a system of 8 equations.) Solving this system to find the values of A through H, we will be ready for stage two of the method of partial fractions: the integration of expression (3).

Once you have decomposed a rational function into partial fractions, you come up with a number of integrals to be evaluated, each of which is represented by one of the following four forms:

I. $\int \frac{A}{Bx+D} \, dx, \ B \neq 0$ II. $\int \frac{A}{(Bx+D)^k} \, dx, \ B \neq 0, \ k > 1$

III. $\int \frac{Ax+B}{Dx^2+Ex+F} \, dx, \ D \neq 0$ IV. $\int \frac{Ax+B}{(Dx^2+Ex+F)^k} \, dx, \ D \neq 0, \ k > 1$

Remember that $Dx^2 + Ex + F$ is irreducible.

This is one instance where a discussion of the general case is more revealing than a horde of examples. Naturally, it is not our intent that you memorize these formulas. It is the technique you should observe. You should also notice that only three of the basic formulas from the last Section are used.

<u>Solution of Type I</u>:

$$\int \frac{A}{Bx+D} \, dx = \frac{A}{B} \int \frac{B}{Bx+D} \, dx = \frac{A}{B} \log|Bx+D| + C.$$

<u>Solution of Type II</u>:

$$\int \frac{A}{(Bx+D)^k} \, dx = \frac{A}{B} \int \frac{B}{(Bx+D)^k} \, dx = \frac{A}{B(-k+1)}(Bx+D)^{-k+1} + C.$$

Solution of Type III: This one will take a little longer. Our solution is motivated by the observation that if the numerator is a multiple of $2Dx + E$, say $Ax + B = k(2Dx + E)$, then we could quickly ascertain that:

$$\int \frac{Ax + B}{Dx^2 + Ex + F} \cdot dx = k \log|Dx^2 + Ex + F| + C.$$

Therefore, we write

$$Ax + B = \frac{A}{2D}(2Dx + E) + (B - \frac{AE}{2D})$$

so that

$$\int \frac{Ax + B}{Dx^2 + Ex + F} dx = \frac{A}{2D} \log|Dx^2 + Ex + F| + (B - \frac{AE}{2D}) \int \frac{dx}{Dx^2 + Ex + F} . \tag{4}$$

We can handle the remaining integral by "completing the square" and appealing to the arctangent, viz.,

$$\int \frac{du}{u^2 + a^2} = \frac{1}{a} \arctan(\frac{u}{a}) + C. \tag{5}$$

First,

$$Dx^2 + Ex + F = D[(x + \frac{E}{2D})^2 + (\frac{F}{D} - \frac{E^2}{4D^2})].$$

Since $Dx^2 + Ex + F$ is irreducible, the quantity $\frac{F}{D} - \frac{E^2}{4D^2}$ is positive (verify this using the quadratic formula). We may set

$$a = \sqrt{\frac{F}{D} - \frac{E^2}{4D^2}} . \tag{6}$$

Using the substitution $u = x + \frac{E}{2D}$, we may simplify equation (4) to

$$\int \frac{Ax + B}{Dx^2 + Ex + F} dx = \frac{A}{2D} \log|Dx^2 + Ex + F| + (\frac{B}{D} - \frac{AE}{2D^2}) \int \frac{du}{u^2 + a^2}$$

and use formula (5) to complete the work. Assuredly, a special case will look simpler.

$$\int \frac{x + 7}{x^2 + 2x + 5} \, dx = \frac{1}{2} \int \frac{2x + 2 + 12}{x^2 + 2x + 5} \, dx$$

$$= \frac{1}{2} \int \frac{2x + 2}{x^2 + 2x + 5} \, dx + 6 \int \frac{dx}{(x + 1)^2 + 4}$$

$$= \frac{1}{2} \log|x^2 + 2x + 5| + 3 \arctan(\frac{x + 1}{2}) + C.$$

Notice in the general case that if $A = 0$, we obtain only the arctangent; the logarithm does not come into play (see Equation (4)).

Solution of Type IV: This one will take even longer than the last because a technique from the _next_ Section is needed. We will take the solution up to that point. The remaining step is easy to state and that is all we will do.

Suppose first that $A = 0$. We complete the square as before. If we let

$$u = x + \frac{E}{2D} \quad \text{and} \quad a^2 = \frac{F}{D} - \frac{E^2}{4D^2} \, ,$$

then

$$\int \frac{B}{(Dx^2 + Ex + F)^k} \, dx = \frac{B}{D^k} \int \frac{du}{(u^2 + a^2)^k} \, .$$

We propose the substitution $u = a \tan(t)$ so that $du = a \sec^2 t \, dt$. The integral is changed to:

$$\frac{B}{D^k} \int \frac{a \sec^2 t \, dt}{a^{2k} (\sec^2 t)^k} = \frac{B}{Da^{2k - 1}} \int \cos^{2k - 2} t \, dt \, .$$

It is at this point that we need the next Section.

In the event that $A \neq 0$, we again may mimic the solution under Type III.

Thus, $\displaystyle \int \frac{Ax + B}{(Dx^2 + Ex + F)^k} \, dx = \frac{A}{2D} \int \frac{2Dx + E}{(Dx^2 + Ex + F)^k} \, dx + \int \frac{B - AE/2D}{(Dx^2 + Ex + F)^k} \, dx$.

Of the two integrals arising on the right-hand side of this equation, the right-most has just been discussed, whereas the other easily evaluates as

$$\frac{A}{2D(-k + 1)} (Dx^2 + Ex + F)^{-k + 1} + C \, .$$

The purpose of the following example is to provide comment on techniques for solving the systems of equations that involve the undetermined coefficients.

<u>Example</u>. Evaluate the integral $\int \dfrac{x^4 - 4x^2 + 2x - 10}{(x + 1)^3(x^2 + 4)}\,dx$.

<u>Solution</u>. By some lucky coincidence the integrand is already in standard form for the application of partial fractions. That is, the degree of the numerator is less than that of the denominator and the denominator has been factored. Therefore, we write

$$\frac{x^4 - 4x^2 + 2x - 10}{(x + 1)^3(x^2 + 4)} = \frac{A}{x + 1} + \frac{B}{(x + 1)^2} + \frac{C}{(x + 1)^3} + \frac{Dx + E}{x^2 + 4} \tag{7}$$

or

$$x^4 - 4x^2 + 2x - 10 = A(x + 1)^2(x^2 + 4) + B(x + 1)(x^2 + 4) + C(x^2 + 4)$$
$$+ (Dx + E)(x + 1)^3 . \tag{8}$$

We wish to determine values of A, B, C, D, and E that will result in equation (8) being true for all values of x. Equation (7) will then hold as well for $x \neq -1$. Since there are 5 undetermined coefficients (A, B, C, D, E) we need to substitute 5 different values for x. Excellent choices seem to be x = -1 and x = 0 (why?). The best next choices are x = 1, 2, and -2 because they will keep the numbers in our calculations within the limits of finger addition. Thus,

$$
\begin{array}{llll}
(x = -1) & -15 = & 5C \\
(x = 0) & -10 = & 4A + 4B + 4C & + E \\
(x = 1) & -11 = & 20A + 10B + 5C + 8D + 8E \\
(x = 2) & -6 = & 72A + 24B + 8C + 54D + 27E \\
(x = -2) & -14 = & 8A - 8B + 8C + 2D - E .
\end{array}
$$

The first equation tells us C = -3, so we substitute this value in the other four equations and simplify wherever possible.

$$
\begin{aligned}
2 &= 4A + 4B + E \\
2 &= 10A + 5B + 4D + 4E \\
6 &= 24A + 8B + 18D + 9E \\
10 &= 8A - 8B + 2D - E \;.
\end{aligned}
$$

Our next move is to solve any one of these equations for any one of the variables present and then substitute that expression in the remaining equations. The result will be a system of 3 equations involving 3 variables. Solution of the first equation for E:

$$
E = 2 - 4A - 4B, \tag{9}
$$

will give rise to the least amount of algebra. Substituting this expression for E in the other 3 equations and regrouping terms, we find

$$
\begin{aligned}
6 &= 6A + 11B - 4D \\
6 &= 6A + 14B - 9D \\
6 &= 6A - 2B + D \;.
\end{aligned}
$$

This time we select D and the third equation.

$$
D = 6 - 6A + 2B \;. \tag{10}
$$

Substituting this expression in the other two equations, we find that

$$
\begin{aligned}
10 &= 10A + B \\
15 &= 15A - B \;.
\end{aligned}
$$

Further,

$$
B = 10 - 10A \tag{11}
$$

and
$$
15 = 15A - (10 - 10A) \text{ or } 25 = 25A.
$$

Therefore, A = 1 and, by substituting back in equations (11), (10), and (9); B = 0, D = 0, and E = -2. Therefore

$$
\int \frac{x^4 - 4x^2 + 2x - 10}{(x + 1)^3(x^2 + 4)}\, dx = \int \frac{dx}{x + 1} + \int \frac{-3dx}{(x + 1)^3} + \int \frac{-2dx}{x^2 + 4}
$$

$$
= \log|x + 1| + \frac{3}{2}(x + 1)^{-2} - \arctan(\tfrac{x}{2}) + C.
$$

Unfortunately, the drama of the algebra involved distracts one from the simplicity of the method of partial fractions.

SECTIONS 8.3-4

Consider the indefinite integral

$$\int \frac{\csc x \, \sin^2 x}{\sec^3 x} \, dx.$$

You should verify that it may be rewritten as

$$\int \cos^3 x \, \sin x \, dx \quad \text{or} \quad \int \frac{\tan x}{\sec^4 x} \, dx \quad \text{or} \quad \int \frac{\cot^3 x}{\csc^4 x} \, dx.$$

We would most likely evaluate one of these three variations in order to handle the original problem. The reason for this was suggested when we studied the derivatives of the trigonometric functions; they appear in certain natural pairings. For example, $\frac{d}{dx}(\tan x) = \sec^2 x$ and $\frac{d}{dx}(\sec x) = \tan x \, \sec x$. Consequently, it is sufficient to focus our attention on integrals of the form:

$$\int \sin^m x \, \cos^n x \, dx,$$
$$\int \tan^m x \, \sec^n x \, dx,$$
$$\int \cot^m x \, \csc^n x \, dx,$$

where m and n are integers.

The text gives explicit techniques for integrating several of these cases and the remaining cases are treated similarly.

You should remember that for the case of

$$\int \sin^m x \, \cos^n x \, dx$$

we not only can try the substitutions u = cos x (if n is odd) or u = sin x (if m is odd), but we may resort to the identities

$$\sin^2 x = \frac{1 - \cos 2x}{2} \quad \text{and} \quad \cos^2 x = \frac{1 + \cos 2x}{2}$$

if m and n are both even.

It is most important that you realize that there is not a unique approach to any integral (involving trigonometric functions), even though some approaches may be quicker or easier than others. We conclude with 8 examples to make this very point. Examples 1 through 6 will evaluate

$$\int \frac{\csc x \, \sin^2 x}{\sec^3 x} \, dx$$

and the last two examples will focus on

$$\int \sin^2 x \, \cos^2 x \, dx.$$

Example 1.

$$\int \frac{\csc x \, \sin^2 x}{\sec^3 x} \, dx = \int \cos^3 x \, \sin x \, dx$$

$$= - \int u^3 du = - \frac{1}{4} u^4 + C = - \frac{1}{4} \cos^4 x + C.$$

Let $u = \cos x$
$du = - \sin x \, dx$

Example 2.

$$\int \frac{\csc x \, \sin^2 x}{\sec^3 x} \, dx = \int \cos^3 x \, \sin x \, dx$$

$$= \int (\cos^2 x \, \sin x) \, \cos x \, dx$$

$$= \int (1 - u^2) \cdot u \cdot du = \int (u - u^3) du$$

$$= \frac{u^2}{2} - \frac{u^4}{4} + C = \frac{\sin^2 x}{2} - \frac{\sin^4 x}{4} + C.$$

Let $u = \sin x$
$du = \cos x \, dx$

Example 3.

$$\int \frac{\csc x \, \sin^2 x}{\sec^3 x} \, dx = \int \frac{\tan x}{\sec^4 x} \, dx$$

$$= \int \frac{\sec x \, \tan x \, dx}{\sec^5 x} = \int u^{-5} du$$

$$= - \frac{1}{4} u^{-4} + C = \frac{-1}{4 \sec^4 x} + C.$$

Let $u = \sec x$
$du = \sec x \, \tan x \, dx$

Example 4.

$$\int \frac{\csc x \, \sin^2 x}{\sec^3 x} \, dx = \int \frac{\tan x}{\sec^4 x} \, dx$$

$$= \int \frac{\tan x}{\sec^6 x} \, \sec^2 x \, dx = \int \frac{u}{(1 + u^2)^3} \, du \qquad \text{Let } u = \tan x$$

$$\qquad\qquad\qquad\qquad\qquad\qquad\qquad\qquad\qquad\qquad du = \sec^2 x \, dx$$

$$= -\frac{1}{4}(1 + u^2)^{-2} + C = \frac{-1}{4 \sec^4 x} + C.$$

Example 5.

$$\int \frac{\csc x \, \sin^2 x}{\sec^3 x} \, dx = \int \frac{\cot^3 x}{\csc^4 x} \, dx$$

$$= \int \frac{\cot^2 x}{\csc^5 x} \, \cot x \, \csc x \, dx \qquad \text{Let } u = \csc x$$

$$\qquad\qquad\qquad\qquad\qquad\qquad\qquad\qquad\qquad du = \csc x \, \cot x \, dx$$

$$= -\int \frac{u^2 - 1}{u^5} \, du = \int (u^{-5} - u^{-3}) \, du$$

$$= -\frac{1}{4} u^{-4} + \frac{1}{2} u^{-2} + C = \frac{1}{2 \csc^2 x} - \frac{1}{4 \csc^4 x} + C.$$

Example 6.

$$\int \frac{\csc x \, \sin^2 x}{\sec^3 x} \, dx = \int \frac{\cot^3 x}{\csc^4 x} \, dx \qquad \text{Let } u = \cot x$$

$$\qquad\qquad\qquad\qquad\qquad\qquad\qquad\qquad\qquad du = -\csc^2 x \, dx$$

$$= -\int \frac{u^3 \, du}{(1 + u^2)^3} = -\int \frac{v - 1}{v^3} \cdot \frac{dv}{2} \qquad \text{Let } v = u^2 + 1$$

$$\qquad\qquad\qquad\qquad\qquad\qquad\qquad\qquad\qquad dv = 2u \, du$$

$$= \frac{-1}{2} \int (v^{-2} - v^{-3}) dv$$

$$= \frac{1}{2} v^{-1} - \frac{1}{4} v^{-2} + C$$

$$= \frac{1}{2(u^2 + 1)} - \frac{1}{4(u^2 + 1)^2} + C = \frac{\sin^2 x}{2} - \frac{\sin^4 x}{4} + C.$$

It should be noted that it is not that often that all six substitutions work out so easily. However, it is always easy to <u>execute</u> each substitution.

Have you memorized the identities marked with an asterisk on p. 143 of the Supplement?

<u>Example 7.</u>

$$\int \sin^2 x \cos^2 x \, dx = \int (\frac{1 - \cos 2x}{2})(\frac{1 + \cos 2x}{2}) \, dx$$

$$= \frac{1}{4} \int (1 - \cos^2 2x)dx = \frac{1}{4} \int dx - \frac{1}{4} \int \cos^2 2x \, dx$$

$$= \frac{x}{4} - \frac{1}{8} \int (1 + \cos 4x)dx = \frac{x}{4} - \frac{x}{8} - \frac{\sin 4x}{32} + C.$$

<u>Example 8.</u>

$$\int \sin^2 x \cos^2 x \, dx = \int (\frac{\sin 2x}{2})^2 \, dx = \frac{1}{4} \int \sin^2 2x \, dx$$

$$= \frac{1}{8} \int (1 - \cos 4x)dx = \frac{x}{8} - \frac{\sin 4x}{32} + C.$$

<u>EXERCISES.</u>

1. $\int \tan x \cot x \, dx$ 2. $\int \sin x \tan x \, dx$ 3. $\int \sec x \cot x \, dx$

4. $\int \frac{\cos^2 x}{1 - \cos 2x} \, dx$ 5. $\int \sin x \sec^2 x \, dx$ 6. $\int \frac{\csc x}{\sec x} \, dx$

7. $\int (1 + \cos 2x)\tan x \, dx$ 8. $\int (\sin x - \cos x)^2 dx$ 9. $\int \frac{\csc^2 x + 1}{\sec x} \, dx$

<u>ANSWERS.</u>

1. $x + C$ 2. $\log|\sec x + \tan x| - \sin x + C$ 3. $\log|\csc x - \cot x| + C$

4. $-\frac{1}{2}[x + \cot x] + C$ 5. $\sec x + C$ 6. $\log|\sin x| + C$

7. $\sin^2 x + C$ 8. $x + \frac{\cos 2x}{2} + C$ 9. $\sin x - \csc x + C$

SECTION 8.5

It is due to the identities $\sin^2 x + \cos^2 x = 1$ and $1 + \tan^2 x = \sec^2 x$ that trigonometric substitutions are particularly well-suited for integrals involving $\sqrt{x^2 + a^2}$, $\sqrt{x^2 - a^2}$, or $\sqrt{a^2 - x^2}$. (Why isn't $\sqrt{-a^2 - x^2}$ included in this list?)

If an expression of the form $\sqrt{x^2 + a^2}$ occurs in the integral, the substitution $x = a \tan t$ is usually useful. For then we have

$$\sqrt{a^2 + x^2} = \sqrt{a^2 + a^2 \tan^2 t} = \sqrt{a^2(1 + \tan^2 t)} = \sqrt{a^2 \sec^2 t} = a \sec t$$

where we have assumed without loss of generality that a is positive (so that $\sqrt{a^2} = a$). Our algebra is supported by Diagram 1.

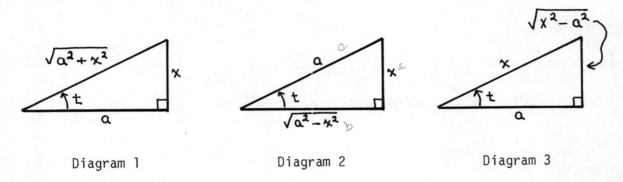

Diagram 1   Diagram 2   Diagram 3

If the integrand contains an expression of the form $\sqrt{a^2 - x^2}$, then the substitution $x = a \sin t$ is called for. Thus, $\sqrt{a^2 - x^2} = a \cos t$ as shown in Diagram 2.

For an expression of the form $\sqrt{x^2 - a^2}$ the substitution $x = a \sec t$ is usually useful, as $\sqrt{x^2 - a^2} = a \tan t$. (See Diagram 3.)

Many students draw these diagrams to help them with the substitutions. We shall draw them for most of the examples.

Example 1. Evaluate $\displaystyle \int \frac{x^3}{\sqrt{x^2 + 4}}\, dx$.

<u>Solution</u>. Let x = 2 tan t, then dx = 2 sec^{2}t dt.

$$\int \frac{x^3}{\sqrt{x^2 + 4}}\, dx = \int \frac{8\ \tan^3 t}{2\ \sec t}\, 2\ \sec^2 t\, dt$$

$$= 8 \int \tan^2 t(\sec t\ \tan t)dt.$$

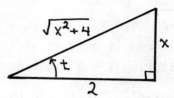

Now we make the substitution u = sec t, so du = sec t tan t dt.

$$\int \frac{x^3 dx}{\sqrt{x^2 + 4}} = 8 \int (u^2 - 1)du = \frac{8u^3}{3} - 8u + C = \frac{8\ \sec^3 t}{3} - 8\ \sec t + C.$$

From the diagram we can see that sec t = $\frac{1}{2}\sqrt{x^2 + 4}$ so that

$$\int \frac{x^3 dx}{\sqrt{x^2 + 4}} = \frac{(x^2 + 4)^{3/2}}{3} - 4(x^2 + 4)^{1/2} + C.$$

<u>Example 2</u>. Evaluate $\int \frac{\sqrt{x^2 - 25}}{x}\, dx$.

<u>Solution</u>. Let x = 5 sec t, then dx = 5 sec t tan t dt.

$$\int \frac{\sqrt{x^2 - 25}}{x}\, dx = \int \frac{5\ \tan t}{5\ \sec t} \cdot 5\ \sec t\ \tan t\, dt$$

$$= 5 \int \tan^2 t\, dt = 5 \int (\sec^2 t - 1)dt$$

$$= 5\ \tan t - 5t + C$$

$$= \sqrt{x^2 - 25} - 5\ \text{arcsec}(\tfrac{x}{5}) + C$$

or

$$= \sqrt{x^2 - 25} - 5\ \arctan(\frac{\sqrt{x^2 - 25}}{5}) + C,$$

where we have consulted the diagram to rewrite the final result in terms of x.

<u>Example 3</u>. Evaluate $\int_{-2}^{2\sqrt{3}} \sqrt{16 - x^2}\, dx$.

Solution. Let x = 4 sin t. Then, dx = 4 cos t dt, t = π/3 when x = 2√3, and
t = -π/6 when x = -2. It should be noted that
our diagram is only an aid. When making the
substitution x = 4 sin t or equivalently

t = arcsin($\frac{x}{4}$), it should be remembered that

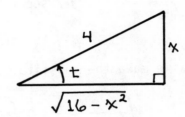

the range of the inverse sine function (here, t)

is restricted to $[-\frac{\pi}{2}, \frac{\pi}{2}]$. This is why the new

limits of integration were set at π/3 and -π/6, even though there are other
solutions to both sin t = √3/2 and sin t = -1/2. Continuing with the solution, we
obtain

$$\int_{-2}^{2\sqrt{3}} \sqrt{16 - x^2}\ dx = \int_{-\pi/6}^{\pi/3} 4\cos t \cdot 4\cos t\ dt = 16 \int_{-\pi/6}^{\pi/3} \frac{1 + \cos 2t}{2}\ dt$$

$$= 8(t + \frac{\sin 2t}{2})\Big|_{-\pi/6}^{\pi/3} = 4\pi + 4\sqrt{3}.$$

This problem can also be done by the method of integration by parts. Can you
provide the details?

Example 4. Evaluate $\int_{3\sqrt{2}}^{6} \frac{x}{\sqrt{x^2 - 9}}\ dx.$

Solution. Let x = 3 sec t, then dx = 3 sec t tan t dt.

$$\int_{3\sqrt{2}}^{6} \frac{x}{\sqrt{x^2 - 9}}\ dx = \int_{\pi/4}^{\pi/3} \frac{3\sec t}{3\tan t} \cdot 3\sec t\tan t\ dt$$

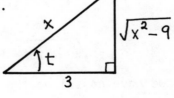

$$= 3 \int_{\pi/4}^{\pi/3} \sec^2 t\ dt = 3\tan t\Big|_{\pi/4}^{\pi/3} = 3(\sqrt{3} - 1).$$

Our substitution of x = 3 sec t was not necessary to simplify this problem, as

$$\int \frac{x}{\sqrt{x^2 - 9}}\ dx = \int x(x^2 - 9)^{-1/2}\ dx = (x^2 - 9)^{1/2} + C$$

is about as easy as they come.

The use of trigonometric substitution is not limited to integrals involving the expressions:

$$\sqrt{x^2 + a^2}, \qquad \sqrt{x^2 - a^2}, \qquad \sqrt{a^2 - x^2} .$$

We can also handle the expressions:

$$\sqrt{b^2x^2 + a^2}, \qquad \sqrt{b^2x^2 - a^2}, \qquad \text{and } \sqrt{a^2 - b^2x^2}, \tag{1}$$

by using the substitutions

$$bx = a \tan t, \qquad bx = a \sec t, \qquad \text{and } bx = a \sin t,$$

respectively. The corresponding triangular aids are just as immediate:

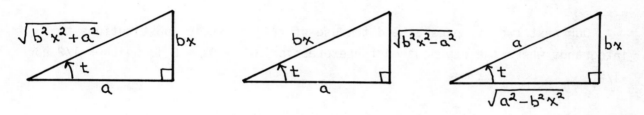

Naturally, both a and b are assumed without loss of generality to be positive.

More generally, we can consider integrands involving an expression of the form

$$\sqrt{Ax^2 + Bx + C}, \qquad A \neq 0. \tag{2}$$

Obviously, our first step is to complete the square

$$Ax^2 + Bx + C = A\left(x + \frac{B}{2A}\right)^2 + \left(C - \frac{B^2}{4A}\right).$$

First, we note that if A is negative <u>and</u> if $C - \frac{B^2}{4A}$ is negative or zero, then expression (2) is meaningless. The other three choices of sign yield one of the expressions in (1) once we identify $|A|$ as b^2 and $\left|C - \frac{B^2}{4A}\right|$ as a^2.

Example 5. Evaluate $\displaystyle\int \frac{x + 2}{\sqrt{x^2 + 2x + 5}}\, dx$.

Solution. As $x^2 + 2x + 5 = (x + 1)^2 + 4$,
we choose the substitution $x + 1 = 2 \tan t$.
Thus, $dx = 2 \sec^2 t\, dt$ and

$$\int \frac{x + 2}{\sqrt{x^2 + 2x + 5}}\, dx = \int \frac{1 + 2 \tan t}{2 \sec t} \cdot 2 \sec^2 t\, dt$$

$$= \int (\sec t + 2 \tan t \cdot \sec t)\, dt$$

$$= \log|\sec t + \tan t| + 2 \sec t + C$$

$$= \log\left| \frac{\sqrt{x^2 + 2x + 5}}{2} + \frac{x + 1}{2} \right| + \sqrt{x^2 + 2x + 5} + C.$$

Our next two examples extend the use of trigonometric substitutions to integrands where the exponent on the expression $Ax^2 + Bx + C$ is neither $1/2$ nor $-1/2$.

Example 6. Evaluate $\displaystyle\int \frac{x}{(2x - x^2)^{3/2}}\, dx$.

Solution. As $2x - x^2 = 1 - (x - 1)^2$, we choose to let $x - 1 = \sin t$.
Thus, $dx = \cos t\, dt$ and

$$\int \frac{x\, dx}{(2x - x^2)^{3/2}} = \int \frac{(1 + \sin t) \cos t\, dt}{\cos^3 t}$$

$$= \int (\sec^2 t + \sec t \tan t)\, dt$$

$$= \tan t + \sec t + C = \frac{x - 1}{\sqrt{2x - x^2}} + \frac{1}{\sqrt{2x - x^2}} + C = \frac{x}{\sqrt{2x - x^2}} + C.$$

Example 7. Evaluate $\displaystyle\int (1 - 4x^2)^{3/2}\, dx$.

Solution. Let $2x = \sin t$. Then $dx = \frac{1}{2} \cos t\, dt$ and

$$\int (1 - 4x^2)^{3/2} \, dx = \frac{1}{2} \int \cos^4 t \, dt = \frac{1}{2} \int \left(\frac{1 + \cos 2t}{2} \right)^2 dt$$

$$= \frac{1}{8} \int \left(1 + 2 \cos 2t + \frac{1 + \cos 4t}{2} \right) dt$$

$$= \frac{1}{8} \left(\frac{3t}{2} + \sin 2t + \frac{\sin 4t}{8} \right) + C$$

Now,

$$\sin 2t = 2 \sin t \cos t = 4x\sqrt{1 - 4x^2}$$

and

$$\sin 4t = 2 \sin 2t \cos 2t = 2 \sin 2t (1 - 2 \sin^2 t)$$

$$= 8x\sqrt{1 - 4x^2} \, (1 - 8x^2)$$

so that

$$\int (1 - 4x^2)^{3/2} \, dx = \frac{3}{16} \arcsin(2x) + \frac{x}{8}\sqrt{1 - x^2} \, (5 - 8x^2) + C.$$

The use of trigonometric substitutions is not restricted simply to square roots of quadratic expressions. The effect of a trigonometric substitution by virtue of identities such as: $1 + \tan^2 t = \sec^2 t$, is to replace a binomial term by a monomial. For example, $x^2 + a^2$ is replaced by $a^2 \sec^2 t$ under the substitution $x = a \tan t$. The net result is that certain algebraic rearrangements of the integrand, such as evaluation of a square root or division by the denominator, are easier to execute. Consequently, integrals of the form

$$\int \frac{Ax + B}{(Dx^2 + Ex + F)^k} \, dx \qquad \text{for } k \geq 1,$$

which arise when we apply the method of partial fractions, are easy to evaluate. We first complete the square (remember $Dx^2 + Ex + F$ is irreducible) to obtain

$$\int \frac{Ax + B}{(b^2 x^2 + a^2)^k} \, dx,$$

and then substitute $bx = a \tan t$.

Example 8. Evaluate $\int \frac{x + 3}{x^2 + 4} \, dx$.

Solution. We don't need a "trig substitution" to do this problem.

$$\int \frac{x + 3}{x^2 + 4}\, dx = \int \frac{x}{x^2 + 4}\, dx + 3 \int \frac{dx}{x^2 + 4}$$

$$= \frac{1}{2} \log|x^2 + 4| + \frac{3}{2} \arctan(\tfrac{x}{2}) + C.$$

However, we will pursue the substitution $x = 2 \tan t$ just to demonstrate how mastery of a technique can help make up for a faulty memory. Since $dx = 2 \sec^2 t\, dt$, we have

$$\int \frac{x + 3}{x^2 + 4}\, dx = \int \frac{2 \tan t + 3}{4 \sec^2 t} \cdot 2 \sec^2 t\, dt$$

$$= \int (\tan t + \tfrac{3}{2})dt = \int (\frac{\sin t}{\cos t} + \tfrac{3}{2})dt$$

$$= -\log|\cos t| + \frac{3}{2} t + C$$

$$= -\log\left|\frac{2}{\sqrt{x^2 + 4}}\right| + \frac{3}{2} \arctan(\tfrac{x}{2}) + C,$$

which you should verify to be the same as the earlier result.

<u>Example 9</u>. Evaluate $\int \dfrac{x + 2}{(x^2 - 2x + 5)^3}\, dx$.

<u>Solution</u>. Since $x^2 - 2x + 5 = (x - 1)^2 + 4$, we choose the substitution $x - 1 = 2 \tan t$. Then, $dx = 2 \sec^2 t\, dt$ and

$$\int \frac{x + 2}{(x^2 - 2x + 5)^3}\, dx = \int \frac{2 \tan t + 3}{64 \sec^6 t} \cdot 2 \sec^2 t\, dt$$

$$= \frac{1}{32} \int (\frac{2 \tan t}{\sec^4 t} + \frac{3}{\sec^4 t})dt = \frac{1}{32} \int (2 \sin t \cos^3 t + 3 \cos^4 t)dt$$

$$= -\frac{1}{64} \cos^4 t + \frac{3}{32}(\frac{3t}{8} + \frac{\sin 2t}{4} + \frac{\sin 4t}{32}) + C,$$

where we have referenced our work in Example 7. You should sketch the triangular aid for this one. Now,

$$\cos^4 t = 16/(x^2 - 2x + 5)^2,$$

$$\sin 2t = 2 \sin t \cos t = 4(x - 1)/(x^2 - 2x + 5),$$

$$\sin 4t = 2 \sin 2t(\cos^2 t - \sin^2 t) = \frac{4(x - 1)(3 + 2x - x^2)}{(x^2 - 2x + 5)^2},$$

and

$$t = \arctan(\frac{x - 1}{2}).$$

Therefore,

$$\int \frac{x + 2}{(x^2 - 2x + 5)^3} dx = -\frac{1}{4(x^2 - 2x + 5)^2} + \frac{9}{256} \arctan(\frac{x - 1}{2})$$

$$+ \frac{3}{32} \cdot \frac{x - 1}{x^2 - 2x + 5} + \frac{3}{256} \frac{(x - 1)(3 + 2x - x^2)}{(x^2 - 2x + 5)^2} + C .$$

Everything that has been done so far has a direct analog for the hyperbolic functions. With that pool of formulas at hand some results are easier to come by and look simpler. For example,

$$\text{arcsinh}(u) = \log|u + \sqrt{u^2 - 1}|.$$

Finally, we should note that the trigonometric functions are studied not simply because they are useful for substitutions or because we want to get the best view of hanging tapestries. Many physical phenomena exhibit wavelike and/or periodic behavior; for example, light, sound, and such down-to-earth quantities as gravity. Combinations of the trigonometric functions provide theoretically exact models or good empirical approximations to these physical entities. Even a grandfather clock can be measured as a sine of the times.

EXERCISES.

1. $\int \frac{x \, dx}{\sqrt{x^2 + 2x + 2}}$ 2. $\int \frac{dx}{\sqrt{3 - 2x - x^2}}$ 3. $\int x\sqrt{x^2 - 2x} \, dx$

4. $\int \frac{dx}{(x^2 + 9)^2}$ 5. $\int \frac{dx}{\sqrt{-4x - x^2}}$ 6. $\int \frac{dx}{\sqrt{4x + x^2}}$

7. $\int \frac{\sqrt{x^2 + 6x + 13}}{x + 3} \, dx$ 8. $\int \frac{x + 3}{\sqrt{x^2 + 6x + 13}} \, dx$ 9. $\int \frac{x + 4}{(9x^2 + 12x + 5)^{3/2}} \, dx$

ANSWERS.

1. $\sqrt{x^2 + 2x + 2} - \log|x + 1 + \sqrt{x^2 + 2x + 2}| + C$

2. $\arcsin(\frac{x + 1}{2}) + C$

3. $\frac{x + 1}{x^2 + 2x + 2} - \frac{1}{2} \log|x + 3| + \frac{1}{4} \log|x^2 + 2x + 2| + \frac{1}{3} \cdot \frac{(x + 1)^3}{(x^2 + 2x + 2)^{3/2}} + C$

4. $\frac{1}{54} \arctan(\frac{x}{3}) + \frac{1}{18} \cdot \frac{x}{x^2 + 9} + C$

5. $\arcsin(\frac{x + 2}{2}) + C$

6. $\log|x + 2 + \sqrt{x^2 + 4x}| + C$

7. $\log|-2 + \sqrt{x^2 + 6x + 13}| - \log|x + 3| + \frac{1}{2}\sqrt{x^2 + 6x + 13} + C$

8. $\sqrt{x^2 + 6x + 13} + C$

9. $-\frac{1}{9} \cdot \frac{30x + 19}{\sqrt{9x^2 + 12x + 5}} + C$

SECTIONS 8.6-7

There is a fundamental reason for considering techniques of integration which may seem esoteric. In practice, integrals are either numerically approximated (see Section 8.8) or found by consulting extensive tables of integrals. Sometimes only the former is possible. In either situation it is frequently useful to be able to transform the integral. One performs a substitution either to simplify the numerical work or to put the integrand in a standard form so that the tables may be consulted.

The half-angle substitution $x = 2 \arctan u$ may be easier to use if you remember that

$$\sin x = \frac{2u}{1 + u^2} \qquad \text{and} \qquad \cos x = \frac{1 - u^2}{1 + u^2}.$$

You should review the derivations on p. 351 of the text.

<u>Example</u>. Evaluate $\int \frac{1 - \cos x}{1 + \cos x}\, dx$.

From x = 2 arctan(u) we have that $dx = \frac{2}{1 + u^2}\, du$. Thus,

$$\int \frac{1 - \cos x}{1 + \cos x}\, dx = \int \frac{1 - \frac{1 - u^2}{1 + u^2}}{1 + \frac{1 - u^2}{1 + u^2}} \cdot \frac{2}{1 + u^2}\, du = 2\int \frac{u^2}{1 + u^2}\, du$$

$$= 2\int \left(1 - \frac{1}{1 + u^2}\right) du = 2(u - \arctan u) + C$$

$$= 2\left(\tan\left(\frac{x}{2}\right) - \frac{x}{2}\right) + C.$$

This may seem like just another application of an admittedly special technique. Until now the focus of your study of integration techniques has been to recognize general forms and select a method of attack like "trig substitution" or "integrate by parts." You should try, however, to remain somewhat open-minded and not stereotype all problems.

For example, at least 6 of the 12 problems for Section 8.6 on p. 353 don't require the x = 2 arctan(u) substitution. Consider the example above. We may multiply numerator and denominator by 1 - cos x to obtain

$$\int \frac{(1 - \cos x)^2}{\sin^2 x}\, dx = \int (\csc^2 x - 2 \cot x \csc x + \cot^2 x)\, dx$$

$$= -2 \cot x - x + 2 \csc x + C.$$

Or, we could use the double angle formulas.

$$\int \frac{2 \sin^2(x/2)}{2 \cos^2(x/2)}\, dx = \int \left(\sec^2\left(\frac{x}{2}\right) - 1\right) dx = 2 \tan\left(\frac{x}{2}\right) - x + C.$$

Finally, consider problem 8 on p. 353 of the text,

$$\int \frac{dx}{1 + \tan x}.$$

As a first step you might write

$$\int \frac{\cos x \, dx}{\cos x + \sin x} \, .$$

Rather than now using x = 2 arctan u (which you should work out), we will employ the substitution $x = y + \frac{\pi}{4}$. This gives us

$$\frac{1}{2} \int (1 - \tan y) dy$$

which is not too tough. Our exotic substitution was suggested by the "observation of form" that cos x + sin x looks like

$$\cos(x - a) = \cos x \cos a + \sin x \sin a$$

if cos a = sin a.

SECTION 8.8

Numerical techniques for integration and other processes is a rich and extensive subject. The methods presented in the text are the simplest, but still of great use. We shall carry the discussion a bit further to give you a better feeling for the tip of this iceberg.

Any numerical calculation is valuable only if some comment can be made on the error involved. Errors may arise in two ways. First, there is the theoretical error associated with the approximation. The interval [a,b] has been divided into n non-overlapping intervals of equal length: $\frac{b - a}{n}$. We have assumed that the function f is continuous on [a,b].

For the moment, let us also assume that f is increasing on [a,b]. It is apparent from the diagram below that the error made with the left endpoint estimate does not exceed $[f(b) - f(a)]\frac{(b - a)}{n}$. The actual error consists of the shaded regions. Each of them lies within a rectangle which has been shifted to the right to graphically exhibit their sum. In a similar way you should be able to show that the error associated with the trapezoidal estimate does not exceed $\frac{1}{2}[f(b) - f(a)] \frac{b - a}{n}$. Clearly, the trapezoidal rule does a better job.

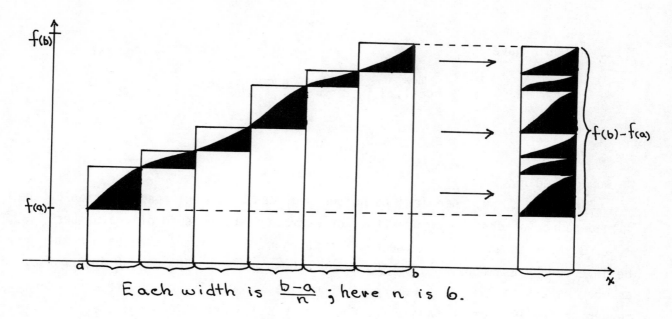

Each width is $\frac{b-a}{n}$; here n is 6.

If we drop the restriction that f is increasing, error estimates are considerably more difficult to derive. The arguments aren't tough, just a bit complicated. It may be shown that the theoretical errors for the trapezoidal rule and Simpson's rule are

$$\frac{(b-a)^3}{12n^2} f''(c) \qquad \text{and} \qquad \frac{(b-a)^5}{180n^4} f^{(4)}(c), \qquad (1)$$

where c is some point in the interval [a,b].

The text approximates (p. 359)

$$\int_1^2 \frac{1}{x} \, dx$$

with n = 5 and obtains 0.70 for the trapezoidal rule. The bound in (1) tells us this result is in error by less than

$$\left| \frac{(b-a)^3}{12n^2} f''(c) \right| = \frac{1}{300} \cdot \frac{2}{c^3} \leq \max_{c \in [1,2]} \frac{1}{150c^3} < .007$$

Had we wanted an estimate for the integral accurate to four decimal places, we would have required

$$\left| \frac{(b-a)^3}{12n^2} f''(c) \right| \leq .00005.$$

Thus, we would have

$$\max_{c\varepsilon[1,2]} \frac{1}{12n^2}\cdot\frac{2}{c^3} \leq .00005$$

or

$$\frac{1}{6n^2} \leq .00005$$

We would need $n > 57$.

For large n, any numerical technique requires a careful consideration of the second source of error: round-off. For example, suppose each of the numbers 1, 2, 3, 4, 5 is accurate to the nearest integer. This means that each number may be in error by as much as 0.5. The sum of these numbers, 15, may be in error by as much as 2.5, a 20% relative error. Sometimes cumulative round-off error can render theoretical error bounds and the entire calculation meaningless.

Let us rework the above example but assume our computer or other device can only store two significant digits. Thus,

$$\int_1^2 \frac{1}{x}\, dx = \frac{b-a}{n}[f(x_0) + 2f(x_1) + \cdots + 2f(x_4) + f(x_5)]$$

$$\simeq \frac{1}{10}[\frac{1}{1} + 2(\frac{5}{6}) + 2(\frac{5}{7}) + 2(\frac{5}{8}) + 2(\frac{5}{9}) + (\frac{1}{2})] \tag{2}$$

as in the text. Now, our narrow-minded round-off machine goes to work.

$$\int_1^2 \frac{1}{x}\, dx \simeq (0.10)[(1.0) + 2(0.83) + 2(0.71) + 2(0.62) + 2(0.44) + (0.50)]$$

$$\text{"="} \ (0.10)[(1.0) + (1.7) + (1.4) + (1.2) + (.88) + (.50)]$$

$$= (.10)[6.7] = .67 .$$

The lesson is clear. Earlier on we used (1) to show that (2) is in error by no more than .007. We found 0.70 as the approximation. Entertaining a generous round-off error, we simplified (2) in a different way and the apparent error (.70 - .67 = .03) due to round-off calculations exceeds the error of the approximating method itself.

SECTION 8.9

We shall briefly characterize the name techniques of integration. Be careful, however, not to be too anxious to categorize every integral.

The Substitution Principle may be applied to any integral. The method serves to simplify a problem to the point where you recognize the integral as a familiar formula or as something you know how to handle. Any substitution may be used, but some are more expedient than others. For the substitution $t = g(x)$, do not forget to allow for $dt = g'(x)dx$.

Integration by parts is based on the formula

$$\int u\ dv = uv - \int v\ du$$

The choice of u and dv is up to you; some choices are more expedient than others. Do not overlook the possibility that the integral appearing on the right may include some constant multiple of the original integral. Further, there are occasions when this method needs to be iterated. Though not exclusively so, the method of integration by parts often applies when the integrand involves a mixture of polynomial, trigonometric, and/or exponential functions.

The method of partial fractions is restricted to and completely resolves integrals of rational functions.

Trigonometric (and hyperbolic) substitutions are particularly useful in integrals involving quadratic expressions.

The Conic Sections

Comments on the material in this Chapter are restricted to solutions for the following exercises in the text.

Following our previous practice, some of these solutions include remarks intended to help you solve similar or related problems. Of course, you should work the exercises before consulting the given solutions.

SECTION 9.2

<u>Solution for Exercise 3a (p. 369)</u>. We first graph and label the vertices of the triangle (see diagram). The formula for the area of the triangle is $\frac{1}{2}$ hb. We arbitrarily select side DF as the base.

Thus, b = $\sqrt{(3 - (-2))^2 + ((-1) - 1)^2}$ = $\sqrt{29}$. The altitude h is the distance from the point E(2,4)(or (x_0, y_0)) to the line passing through points D and F:

5x + 2y - 1 = 0. You should notice that

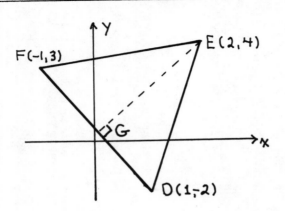

we've been careful to write the equation in this form. This is critical when we substitute in the formula for the distance from a point to a line to discover that

$$h = \frac{|Ax_0 + By_0 + C|}{\sqrt{A^2 + B^2}} = \frac{|5(2) + 2(4) - 1|}{\sqrt{29}} = \frac{17}{\sqrt{29}}.$$

Thus, the area of the triangle is 17/2 square units.

We will check this result by giving a solution that does not use this formula for the distance from a point to a line. Again choosing side DF as the base of the triangle, we have $b = \sqrt{29}$. We will determine the length of altitude EG by finding the coordinates of point G, the point of intersection of the perpendicular lines EG and DF. Line DF has slope $-\frac{5}{2}$ and its equation is given by $y - 3 = -\frac{5}{2}(x + 1)$ or $2y + 5x - 1 = 0$. The slope of line EG is 2/5 so that its equation is $y - 4 = \frac{2}{5}(x - 2)$ or $5y - 2x - 16 = 0$. The common solution of $2y + 5x - 1 = 0$ and $5y - 2x - 16 = 0$ is the point G with coordinates $(\frac{-27}{29}, \frac{82}{29})$. By the distance formula,

$$h = \sqrt{(4 - (\frac{82}{29}))^2 + (2 - (\frac{-27}{29}))^2} = \sqrt{(\frac{34}{29})^2 + (\frac{85}{29})^2}$$

$$= \frac{1}{29}\sqrt{(34)^2 + (85)^2} = \frac{17}{29}\sqrt{2^2 + 5^2} = \frac{17}{\sqrt{29}}.$$

What a delightful surprise!

You should note that the purpose of this second solution was to demonstrate that it is possible to solve problems apparently requiring the formula for the distance from a point to a line even if that formula is not used (or recalled). What in fact was done in the second solution was to follow the technique that was used to derive the formula for the distance from a point to a line. The technique is more important than the formula, but we would still recommend that you memorize the formula to save time, being careful to avoid errors in the signs of A, B, and C.

Hint for Exercise 4 (p. 369). The significance of the normal form for the equation of a straight line will be clarified in Chapter 14. For the moment, the significance of α and p can be uncovered by calculating the slope of the line and its distance from the origin.

<u>Solution for Exercise 5 (p. 369).</u> Since we are interested in the situation when the slope of ℓ is "about 3/4," we may use the diagram below and restrict our attention to $0 < \theta < \pi/2$. We are given that

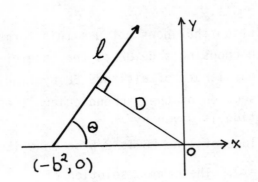

$$\frac{d\theta}{dt} = -2\pi \text{ radians/min.}$$

and

$$\tan \theta = \frac{3}{4}$$

at the moment in question. The equation of the line containing the ray ℓ is given by $(y - 0) = \tan \theta \cdot (x + b^2)$. Consequently,

$$D = \frac{|(-\tan \theta)(0) + 1(0) + (-b^2 \tan \theta)|}{\sqrt{\tan^2 \theta + 1}} = b^2 \sin \theta$$

and

$$\frac{dD}{dt} = b^2 \cos \theta \frac{d\theta}{dt} = -2\pi b^2 \cos \theta.$$

When $\tan \theta = 3/4$, $\cos \theta = 4/5$ and it follows that the distance D is decreasing at the rate of $8\pi b^2/5$ units per minute.

SECTION 9.3

<u>Solution for Exercise 1h (p. 376).</u> The given information tells us that the parabola is of the general form

$$(y - y_0)^2 = -4c(x - x_0)$$

where (x_0, y_0) are the coordinates of the vertex and c is the distance from the focus $(2,-2)$ to the vertex $(7/2,-2)$. Thus, $x_0 = 7/2$, $y_0 = -2$, $c = 3/2$ and

$(y + 2)^2 = -6(x - \frac{7}{2})$. We have taken c to be absolute rather than directed

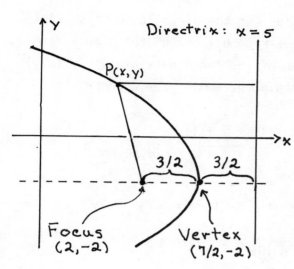

distance so that the minus sign (indicating which way the parabola opens) is more evident. If the directrix had been x = -1 instead, we would have the same values for x_0, y_0, and c, but we would have used the equation

$$(y - y_0)^2 = + 4c(x - x_0).$$

This seems simpler than trying to remember if we should write c = 3/2 or c = -3/2 and using only the equation $(y - y_0)^2 = 4c(x - x_0)$.

This problem may also be solved by appealing to the definition of a parabola. This procedure requires far more algebra but circumvents a lack of confidence in recalling the formula. Let (x,y) be the coordinates of any point P on the parabola. As the distance from P to F is equal to the distance from P to D, we have that

$$\sqrt{(x - 2)^2 + (y + 2)^2} = |x - 5|$$

or

$$(x - 2)^2 + (y + 2)^2 = (x - 5)^2$$

Thus,

$$(y + 2)^2 = x^2 - 10x + 25 - x^2 + 4x - 4$$

and, strangely enough,

$$(y + 2)^2 = -6(x - \frac{7}{2}).$$

Solution for Exercise 2e (p. 376). To solve this problem, we need to know the

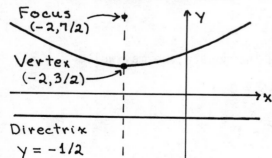

general form of the parabola. Writing

$$(x + 2)^2 = 8y - 12$$

as $(x - (-2))^2 = 4 \cdot 2(y - 3/2)$, we may read off the vertex as (-2,3/2). Since c = +2 and the x variable is squared, the parabola "opens up" so the focus is at

(-2,7/2) and y = -1/2 is the directrix.   The axis is the line x = -2.

Solution for Exercise 3a (p. 376).   The equation for the parabola in this problem
must be found from the definition and will not resemble one of the standard forms
since the directrix is not parallel to either of the coordinate axes.   If (x,y)
are the coordinates of any point on the parabola, we employ the appropriate
distance formulas to write:

$$\sqrt{(x - 1)^2 + (y - 2)^2} = \frac{|x + y + 1|}{\sqrt{1 + 1}},$$

so that by squaring both sides of the equation

$$2(x^2 - 2x + 1 + y^2 - 4y + 4) = x^2 + y^2 + 1 + 2x + 2y + 2xy$$

or

$$x^2 + y^2 - 2xy - 6x - 10y + 9 = 0.$$

SECTION 9.4

Solution for Exercise 4 (p. 383).   We first write the equation of the ellipse in
standard form

$$\frac{(x - 1)^2}{64} + \frac{(y - 0)^2}{16} = 1.$$

This ellipse has its major axis parallel to the x-axis since the number under the
x term, 64, is greater than that under the y term, 16.   As a = 8 and b = 4 the
length of the major axis is 2a = 16 and the length of the minor axis is 2b = 8.
The center of the ellipse is at (1,0) and the foci are located c units to the left
and right of this point along the major axis.   Since $c = \sqrt{a^2 - b^2} = 4\sqrt{3}$, the foci
are located at $(1 + 4\sqrt{3}, 0)$ and $(1 - 4\sqrt{3}, 0)$.

Solution for Exercise 6 (p. 383).   By completing the square, we may write

$$4x^2 + y^2 - 6y + 5 = 0$$

as

$$4x^2 + (y - 3)^2 = 4$$

so that the equation of the ellipse in standard form is

$$\frac{x^2}{1} + \frac{(y - 3)^2}{4} = 1.$$

The major axis is parallel to the y-axis (as 4 > 1) and the ellipse is centered at (0,3). We have that a = 2, b = 1, and c = $\sqrt{3}$. We compute the lengths of the major and minor axes as 4 and 2 units respectively; the foci are located at (0, 3 + $\sqrt{3}$) and (0, 3 - $\sqrt{3}$); and the vertices are located at (0,5), (0,1), (1,3), and (-1,3).

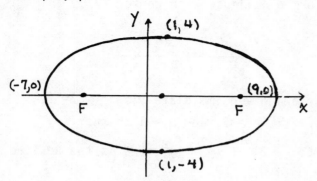

Diagram for Exercise 4

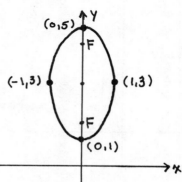

Diagram for Exercise 6

<u>Solution for Exercise 12 (p. 383)</u>. It is helpful to sketch the given information to insure that a, b, and c are properly identified. You should remember that 2a = length of major axis (the one on which foci live), 2b = length of minor axis (the shorter one), and c is the distance from the center to either focus. It is always true that $a^2 = b^2 + c^2$, which follows from the Pythagorean Theorem. Since the foci are located 6 units apart at (1,9) and (1,3), the ellipse is centered at (1,6) and we have c = 3. The length of the minor axis is

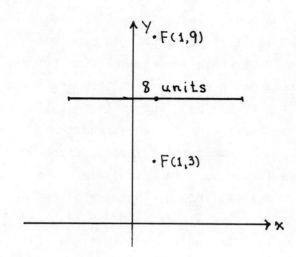

8 so b = 4.   Thus, $a = \sqrt{3^2 + 4^2} = 5$.   Since the major axis is parallel to the

y-axis,

$$\frac{(x - 1)^2}{16} + \frac{(y - 6)^2}{25} = 1$$

is the equation of the ellipse.

Solution for Exercise 14 (p. 383).   Again a graph is helpful.   Since the major axis

is parallel to the y-axis, the general

form of ellipse with center (x_0, y_0) is

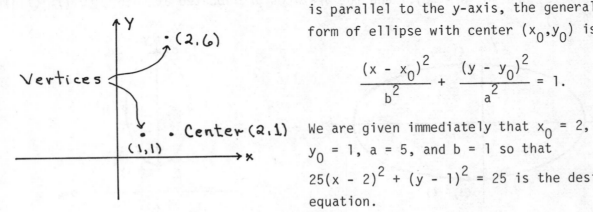

$$\frac{(x - x_0)^2}{b^2} + \frac{(y - y_0)^2}{a^2} = 1.$$

We are given immediately that $x_0 = 2$, $y_0 = 1$, a = 5, and b = 1 so that $25(x - 2)^2 + (y - 1)^2 = 25$ is the desired equation.

SECTION 9.5

Solution for Exercise 6 (p. 388).   It is always true for an ellipse that $a^2 = b^2 + c^2$.   For the hyperbola this is not the case.   Instead, $a^2 + b^2 = c^2$.   It is easy to confuse these two defining equations.   It is far better to remember what the symbols represent geometrically than to memorize the algebraic equations. The length of the major or transverse axis of each conic is 2a.   In standard form, a^2 is the larger denominator in the equation for an ellipse, but in the equation for the hyperbola a^2 appears under the positive term.   The distance from the center to a focus is c for both conics.   It may help to remember the following diagrams with C = center, F = focus, V = vertex.   These figures are representative even when the major (transverse) axis is vertical.   Notice, though, that the slopes of the asymptotes for the hyperbola are $\pm\frac{b}{a}$ if the transverse axis is horizontal and $\pm\frac{a}{b}$ if the transverse axis is vertical.   Rather than trying to keep this disparity

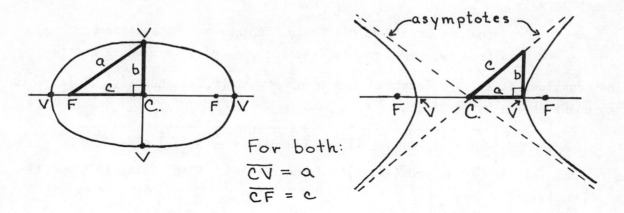

For both:
$$\overline{CV} = a$$
$$\overline{CF} = c$$

straight in your mind, we note that if you replace the "1" in the standard form by a "0" and solve, then, voilà, asymptotes. For example,

$$\frac{(x - 2)^2}{4} - \frac{(y + 3)^2}{9} = 1$$

becomes

$$\frac{(x - 2)^2}{4} - \frac{(y + 3)^2}{9} = 0$$

which may be rewritten as $y + 3 = \pm \frac{3}{2}(x - 2)$. Can you justify this apparent trickery?

Now, it is on to the solution of Exercise 6. A graph of the given information is helpful. The distance between the foci is 10 so c = 5 and the center is (2,1). a is given as 3, so b = 4. The transverse axis is parallel to the x-axis

F(-3,1)   F(7,1)

so we substitute in the form

$$\frac{(x - x_0)^2}{a^2} - \frac{(y - y_0)^2}{b^2} = 1$$

to find $16(x - 2)^2 - 9(y - 1)^2 = 144$ as the desired hyperbola.

Solution for Exercise 7 (p. 388). The transverse axis(line containing the foci) is not parallel to either of the coordinate axes, so we may not appeal to either of the standard forms. Instead, we must revert to the locus definition of the hyperbola, viz.,

> the set of all points the difference of whose distance
> from two fixed points (the foci) is plus or minus a constant.

This constant is 2a, the length of the transverse axis. Consequently, if (x,y) is any point on the hyperbola,

$$\left| \sqrt{(x - 1)^2 + (y - 1)^2} - \sqrt{(x + 1)^2 + (y + 1)^2} \right| = 2,$$

since the foci are (1,1) and (-1,-1) and hence a = 1. Simplifying this equation, we find that

$$(x - 1)^2 + (y - 1)^2 = 4 \pm 4\sqrt{(x + 1)^2 + (y + 1)^2} + (x + 1)^2 + (y + 1)^2,$$

$$- x - y - 1 = \pm \sqrt{(x + 1)^2 + (y + 1)^2},$$

$$x^2 + y^2 + 1 + 2x + 2y + 2xy = x^2 + 2x + 1 + y^2 + 2y + 1$$

or

$$2xy = 1,$$

which is indeed a mild simplification!

Solution for Exercise 14 (p. 388). From the given equation

$$\frac{(x - 1)^2}{9} - \frac{(y - 3)^2}{16} = 1,$$

we immediately have a = 3, b = 4, and (1,3) as the center. Easily, c = 5. The hyperbola "opens" in the x direction; the transverse axis lies on y = 3. Thus, the foci are located at (1 ± c, 3); i.e., (-4,3) and (6,3). The asymptotes are found as suggested above. They are given by $y - 3 = \pm \frac{4}{3}(x - 1)$.

Solution for Exercise 17 (p. 388). We first complete the square(s) to write

$$4x^2 - 8x - y^2 + 6y - 1 = 0$$

as

$$4(x - 1)^2 - (y - 3)^2 + 4 = 0$$

which is

$$\frac{(y - 3)^2}{4} - \frac{(x - 1)^2}{1} = 1$$

in standard form. Immediately then the hyperbola "opens" parallel to the y-axis with a = 2, b = 1, and c = $\sqrt{5}$. Thus, (1,3) is the center, (1, 3 ± 2) are the vertices, (1, 3 ± $\sqrt{5}$) are the foci, and y - 3 = ± 2(x - 1) are the asymptotes.

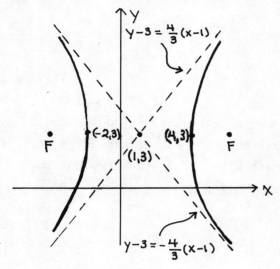

Diagram for Exercise 14

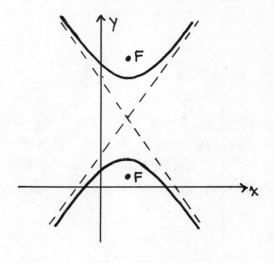

Diagram for Exercise 17

SECTION 9.6

Solution for Exercise 10 (p. 389). We begin by completing the square(s).

$$7x^2 - y^2 + 42x + 14y + 21 = 0$$

$$7(x^2 + 6x) - (y^2 - 14y) = -21$$

$$7(x^2 + 6x + 9) - (y^2 - 14y + 49) = -21 + 63 - 49$$

$$7(x + 3)^2 - (y - 7)^2 = -7$$

so

$$\frac{(y - 7)^2}{7} - \frac{(x + 3)^2}{1} = 1.$$

This is a hyperbola with vertical transverse axis (y term is "positive" in standard form). We have a = $\sqrt{7}$, b = 1, c = $2\sqrt{2}$, and (-3,7) as the center. The foci are (-3, 7 ± $2\sqrt{2}$); the vertices are (-3, 7 ± $\sqrt{7}$). The asymptotes are y - 7 = ± $\sqrt{7}$(x + 3).

<u>Hint for Exercise 13 (p. 389)</u>. The equation ab = 0 holds if and only if a = 0 or b = 0. The curve here consists of two conics.

SECTION 9.7

<u>Suggestion for Exercise 7 et al. (p. 393)</u>. Don't take the easy way out. Don't start with equation 9.7.3 or 9.7.4 from the text and then substitute the equations from 9.7.2. Instead we recommend you start fresh with 9.7.1 and carry out the calculations in detail to mimic the derivation of 9.7.3 and 9.7.4.

Volume, Work, and
Other Applications of the Integral

10

SECTIONS 10.2-3

Not all solids are solids of revolution. Further, not all volumes can be
calculated by the methods to be presented in these Sections. The more general
discussion on volume is reserved for Chapter 18. For the moment consider the
solids of revolution which are generated when the planar regions in Figure 1 are
revolved about the line L. The resulting solids are described below the
respective figures.

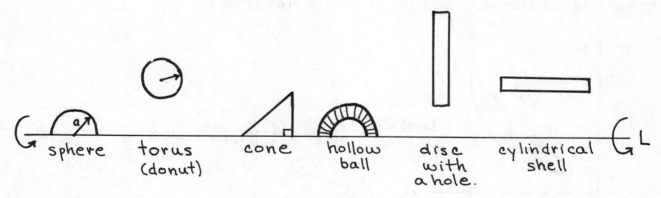

Figure 1

For our problems the line L is taken as parallel to either the x-axis or the y-axis.
An important technique for evaluating the volume of a solid of revolution is
suggested by considering the last two regions in Figure 1 as the "representative
rectangles" that proved so useful in setting up integrals to represent area. The
following comments may serve to refresh your memory. Otherwise, you should
consult Sections 5.5 and 5.10.

Consider the region R in the first quadrant bounded by the curves $y = f(x)$ and $y = g(x)$ on the interval $[a,b]$. If we use vertical representative rectangles to find the area, we obtain

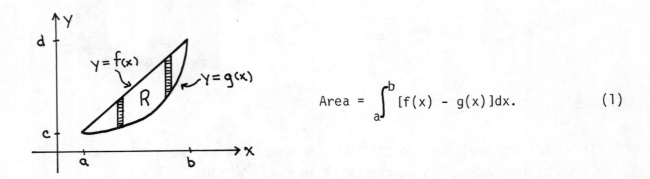

$$\text{Area} = \int_a^b [f(x) - g(x)]dx. \qquad (1)$$

If we use horizontal representative rectangles to find the area, we need to solve both $y = f(x)$ and $y = g(x)$ for x in terms of y. We obtain $x = F(y)$ and $x = G(y)$, respectively. Although $F = f^{-1}$ and $G = g^{-1}$, this other notation has been selected to simplify appearances. The integral for the area is then

$$\text{Area} = \int_c^d [G(y) - F(y)]dy. \qquad (2)$$

Suppose that the region R is revolved about the x-axis. The vertical representative rectangles generate discs. Just as we may think of the rectangles as lined up side by side in the x direction to comprise the area of R, we may consider the discs to be laminated together to form the volume of the solid. Their outer radius is $f(x)$; their inner radius is $g(x)$. The variable of integration is x. If we informally consider the thickness of each disc to be dx, we then obtain

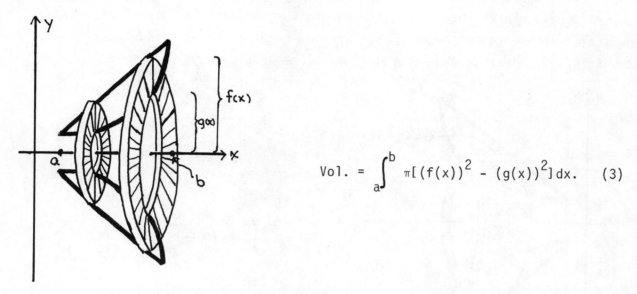

$$\text{Vol.} = \int_a^b \pi[(f(x))^2 - (g(x))^2]\,dx. \quad (3)$$

On the other hand, if the region R is revolved about the y-axis, the vertical representative rectangles generate shells. These shells may be laminated together the way a roll of toilet paper is to form the volume of the solid. The radius of these cylindrical shells is x; their height is f(x) - g(x). The variable of integration is x. If we informally consider the thickness of the shell's wall to be dx, then

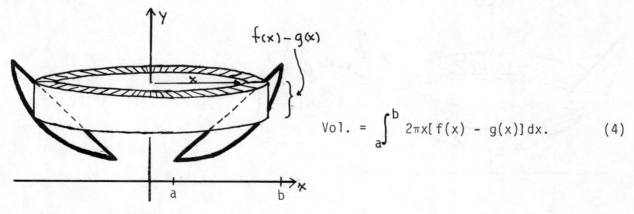

$$\text{Vol.} = \int_a^b 2\pi x[f(x) - g(x)]\,dx. \quad (4)$$

This solid is different from that obtained by revolving region R about the x-axis. In general, the volumes of two such solids are not equal.

We consider next what happens if we use the horizontal, rather than the vertical, representative rectangles. When we revolve region R about the x-axis, we obtain shells rather than discs. The volume is given by

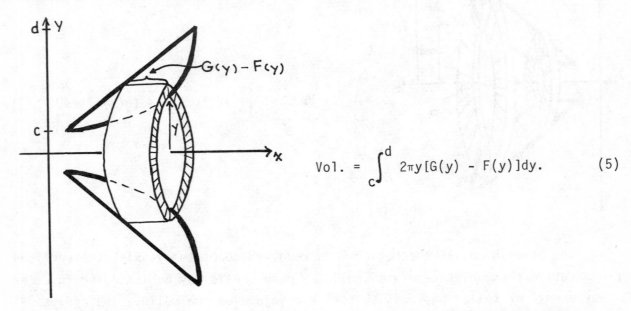

$$\text{Vol.} = \int_c^d 2\pi y[G(y) - F(y)]dy. \qquad (5)$$

When we revolve region R about the y-axis, we obtain discs rather than shells. The volume is given by

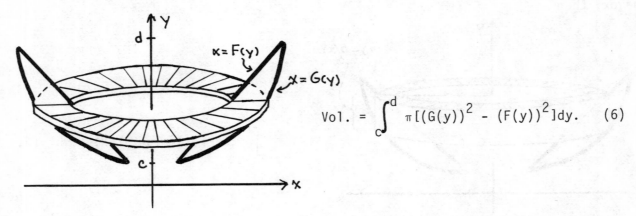

$$\text{Vol.} = \int_c^d \pi[(G(y))^2 - (F(y))^2]dy. \qquad (6)$$

Two observations are now possible. Equations (3) and (6) represent volumes calculated using the <u>disc method</u>. Both equations have the form

$$\int \pi \left[(\quad)^2 - (\quad)^2 \right] d\underline{\quad} . \tag{7}$$

outer
radius   inner
radius   "thickness"

Equations (4) and (5) represent volumes calculated using the <u>shell</u> <u>method</u>. Both
equations have the form

$$\int 2\pi (\quad) \left[(\quad) - (\quad) \right] d\underline{\quad} . \tag{8}$$

radius   height   "thickness"

Next, compare equation (1) to equations (3) and (4). If we ignore the form of the
integrals that pertains to volume (say, the π and the squares in (3)), all three
integrals appear the same. This makes sense. All three integrals are based upon
the same representative rectangle. That is, the same distances or lengths are
used, but differently. Similarly, compare equation (2) to equations (5) and (6).

   Unfortunately, many students have a propensity for confusing formulas (3) -
(6). This happens because they are more intent on memorizing results than
mastering techniques. All that we have are two ways of calculating the volumes of
each of two solids. Whether the solid results from revolution about the x-axis or
about the y-axis and whether you employ the disc or shell method to represent the
volume, the following <u>single</u> procedure always applies to solids of revolution.

1. Sketch a graph of the planar region to be revolved.
2. Clearly recognize the line (axis) about which the region will be
   revolved.
3. Proceed as if you were going to evaluate the area of the region.
   Select the direction for your representative rectangles. Determine
   the "upper" and "lower" boundaries and the intervals where these
   boundaries are in effect.
4. The selected axis and selected representative rectangle <u>dictate</u>
   whether you are using the shell or disc method.
5. Record the form for this dictated method.
6. Read the necessary lengths and dimensions from your diagrams.

<u>Example 1</u>. Using both the shell and disc methods represent with integrals the
volumes of the solids obtained by revolving region R about the x-axis and the
y-axis, respectively. The region R is bounded by the curves $y = x^2$ and $y = 2x$.
<u>Solution</u>. We first determine the points of intersection of the curves and sketch
the region. If we use vertical representative rectangles, then for each part of
the problem we will have one or more integrals of the form $\int \cdots$ dx. If we now
revolve the region about the x-axis, we are calculating the volume using the disc
method so the form of the integral is

$$\int \pi [(\quad)^2 - (\quad)^2] dx.$$

From the figure it is apparent that we should write

$$V = \int_0^2 \pi [(2x)^2 - (x^2)^2] dx.$$

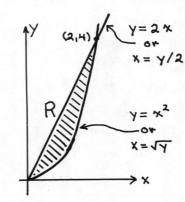 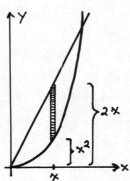 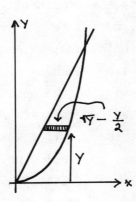

Next, designating horizontal representative rectangles, we have integral(s)
of the form $\int \cdots$ dy. Revolving the region about the x-axis, we can be more
specific, i.e.,

$$\int 2\pi (\quad) [(\quad) - (\quad)] dy.$$

We can determine the details from the figure.

$$V = \int_0^4 2\pi y [(\sqrt{y}) - (\tfrac{y}{2})] dy.$$

For the solid generated by revolving the region about the y-axis, the volume
is

$$\int_0^2 2\pi x[(2x) - (x^2)]dx \qquad\qquad \text{by the shell method}$$

and

$$\int_0^4 \pi[(\sqrt{y})^2 - (\tfrac{y}{2})^2]dy \qquad\qquad \text{by the disc method.}$$

The shell and disc methods are subject to the same anomalies that arise with area problems. First, one method may give rise to a simpler integral to evaluate. Second, if more than one curve constitutes the upper (and/or lower) boundary of the region, then more than one integral will be needed to represent the volume.

Example 2. Calculate the volume of the solid generated by revolving region R about the x-axis, where region R is bounded by the curves $x = 0$, $y = 2$, and $y = e^x$.

Solution. By the disc method, we have a simple integral to evaluate.

$$\text{Volume} = \int_0^{\log 2} \pi[(2)^2 - (e^x)^2]dx$$

$$= \pi(4x - \frac{e^{2x}}{2})\Big|_0^{\log 2}$$

$$= \pi(4\log 2 - 2) - \pi(0 - \tfrac{1}{2})$$

$$= \pi(\log 16 - \tfrac{3}{2}).$$

Using the shell method, we obtain

$$\text{Volume} = \int_1^2 2\pi y(\log y - 0)dy = \pi(\frac{-y^2}{2} + y^2\log y)\Big|_1^2$$

$$= \pi(-2 + 4\log 2) - \pi(-\tfrac{1}{2} + 0) = \pi(\log 16 - \tfrac{3}{2}),$$

where it was necessary for us to integrate by parts. You should supply the details.

Example 3. Determine the volume of the solid obtained by revolving the triangle with vertices $(1,2)$, $(4,5)$, and $(3,6)$ about the line $x = -2$.

Solution. This is not an easy problem. If we use vertical representative rectangles, the shell method is dictated. We will have one or more integrals of the form

$$\int 2\pi(\quad)[(\quad) - (\quad)]dx.$$

We need the equations for the lines determining the sides of the triangle. We solve these equations for y in terms of x since the height of the rectangle is measured vertically (direction of y-axis). The distance from the line x = -2 to a rectangle is the radius of a typical shell. This distance is x + 2. Two integrals are needed just as two integrals would be needed to find the area of the triangle.

$$\text{Volume} = \int_1^3 2\pi(x + 2)[(2x) - (x + 1)]dx$$

$$+ \int_3^4 2\pi(x + 2)[(-x + 9) - (x + 1)]dx$$

If we designate horizontal representative rectangles, then the disc method is dictated by the rotation about x = -2. Again, we need a carefully labeled diagram. The volume is given by

$$\text{Volume} = \int_2^5 \pi[(y - 1 + 2)^2 - (\tfrac{y}{2} + 2)^2]dy$$

$$+ \int_5^6 \pi[(9 - y + 2)^2 - (\tfrac{y}{2} + 2)^2]dy$$

Notice how the integrals are changed by using x = -2 rather than the y-axis (x = 0) as the axis of rotation. Both calculations of volume (fortunately) yield 27π cubic units.

There is another type of solid whose volume we may calculate at this juncture. These solids may be characterized as follows. A line ℓ is passed through the solid and all planes perpendicular to this line cut similar cross sections in the solid. For example, the similar cross sections cut by all planes perpendicular to a diameter of a sphere are circles. Again, the similar cross sections cut by all planes perpendicular to the vertical axis of an Egyptian pyramid are squares. Sometimes these solids are also solids of revolution, but not very often.

There are two ways in which volume problems concerning these solids may be posed. First, you may be given the solid and it is up to you to select and represent the similar cross sections. Second, you may be given a base upon which similar cross sections of a specified type (semi-circular, triangular, square, et al.) are to be built. The integral to represent the volume is found in the same way for both cases. We will example the second type of problem.

Example 4. The <u>base</u> of a solid is the parabolic region in the plane bounded by the curves $y = 4 - x^2$ and $y = 1$. Cross sections of this solid perpendicular to the x-axis are <u>squares</u>. What is the volume of this solid?

Solution. Once again our representative rectangles are useful. The approximate volume for a cross section of width dx is

$$[(4 - x^2) - (1)]^2 \cdot dx .$$

Therefore, the volume of the solid is (relying on symmetry)

$$2 \int_0^{\sqrt{3}} (3 - x^2)^2 dx = 2(9x - 2x^3 + \frac{x^5}{5}) \Big|_0^{\sqrt{3}}$$

$$= \frac{48\sqrt{3}}{5} .$$

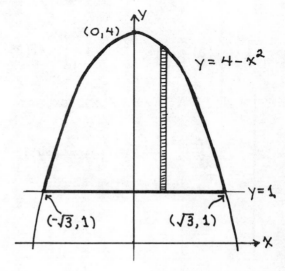

You should verify that if the cross sections had been squares which were perpendicular to the y-axis instead of the x-axis, then the volume V would have been

$$V = \int_{1}^{4} [2\sqrt{4 - y}]^2 dy = 18,$$

which is 96% as large as the previous volume.

EXERCISES. Use one or more integrals to represent the volume of the solid obtained by revolving the specified region about the specified line first by the disc method and then by the shell method.

1. About the x-axis; R bounded by $y = x^2$, $x = y^3$.

2. About the line x = -2; R bounded by $y = x^2$, $x = y^3$.
3. About the y-axis; R bounded by $x = |y|$, $x = 2$.
4. About the line y = 3; R bounded by $x = |y|$, $x = 2$.

ANSWERS.

1. $\displaystyle\int_{0}^{1} \pi[(x^{1/3})^2 - (x^2)^2]dx;$ $\qquad\qquad \displaystyle\int_{0}^{1} 2\pi y[(\sqrt{y}) - (y^3)]dy.$

2. $\displaystyle\int_{0}^{1} \pi[(2 + \sqrt{y})^2 - (2 + y^3)^2]dy;$ $\qquad \displaystyle\int_{0}^{1} 2\pi(x + 2)[x^{1/3} - x^2]dx.$

3. $\displaystyle\int_{-2}^{0} \pi[(2)^2 - (-y)^2]dy + \int_{0}^{2} \pi[(2)^2 - (y)^2]dy;$ $\quad \displaystyle\int_{0}^{2} 2\pi x[(x) - (-x)]dx.$

4. $\displaystyle\int_{0}^{2} \pi[(3 + x)^2 - (3 - x)^2]dx;$

$\qquad\qquad \displaystyle\int_{-2}^{0} 2\pi(3 - y)[(2) - (-y)]dy + \int_{0}^{2} 2\pi(3 - y)[(2) - (y)]dy.$

SECTIONS 10.4-6

Comments on the material in these Sections is restricted to solutions for the
following exercises in the text.

    p. 418 (Section 10.4): 2c, 7b, 12, 14.
    p. 422 (Section 10.5): 2, 7b.
    p. 425 (Section 10.6): 6a.

Following our previous practice, some of these solutions include remarks intended
to help you solve similar or related problems. Of course, you should work the
exercises before consulting the given solutions.

Solution for Exercise 2c (p. 418). The force function for this problem is
f(x) = kx. The work done in stretching a spring from A feet beyond its natural
length to B feet beyond its natural length (so that the corresponding lengths
of the spring are L + A and L + B feet, respectively) is given by

$$\text{work} = \int_A^B kx \, dx.$$

We are given that

$$W = \int_0^a kx \, dx$$

which enables us to solve for k. Evaluating the integral, we calculate that
$k = 2W/a^2$. Thus, the work done in stretching the spring from A = (ℓ + a) - ℓ = a
to B = (ℓ + 2a) - ℓ = 2a feet beyond its natural length is

$$\int_a^{2a} \frac{2W}{a^2} x \, dx = \frac{2W}{a^2} \cdot \frac{x^2}{2}\Big|_a^{2a} = 3W \text{ ft.-lbs.}$$

Notice how this result reinforces your notion of spring stretching inasmuch as the
longer it gets the harder it gets.

Solution for Exercise 7b (p. 418). Consider a cross section of the container x
feet from its top. The cross section is a circle whose radius, R, may be found
by similar triangles. Using these triangles we obtain

$$\frac{r}{h} = \frac{R}{h - x}$$

so that the cross sectional area is $\pi R^2 = \pi r^2 (h - x)^2 / h^2$. This section is to be lifted $k + x$ feet; the liquid weighs σ lbs/cu.ft. The work done in pumping out the upper $\frac{1}{2} h$ feet of liquid is given by

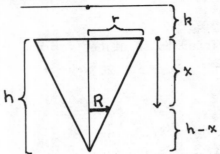

$$\int_0^{h/2} \sigma(k + x) \pi \frac{r^2(h - x)^2}{h^2} \, dx.$$

We can avoid a considerable amount of algebra by writing this integral as

$$\frac{\sigma \pi r^2}{h^2} \int_0^{h/2} [\underbrace{k(h - x)^2 - (h - x)^3 + h(h - x)^2}_{= x(h - x)^2}] dx,$$

which we evaluate as

$$\frac{\sigma \pi r^2}{h^2} \left(-\frac{k}{3}(h - x)^3 + \frac{(h - x)^4}{4} - \frac{h}{3}(h - x)^3 \right) \Bigg|_0^{h/2} = \sigma \pi r^2 h \left(\frac{56k + 11h}{192}\right) \text{ ft.-lbs.}$$

In a problem such as this it is important to select and adhere to a fixed frame of reference. In the solution above we measured from the top of the container. If instead we measure from the bottom of the container, then R is rx/h and the work is given by

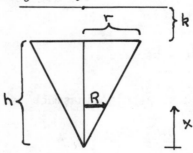

$$\int_h^{h/2} \sigma(k + h - x) \pi \frac{r^2 x^2}{h^2} \, dx.$$

On the other hand, if we measure from the level to which the liquid is being pumped, then R is $\frac{r}{h}(k + h - x)$ and the work is given by

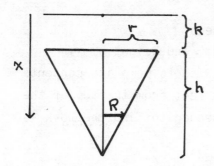

$$\int_{k}^{k + \frac{h}{2}} \sigma(x)\pi\frac{r^2}{h^2}(k + h - x)^2dx .$$

<u>Solution for Exercise 12 (p. 418)</u>. The dimensions of the container are not relevant. Suppose the container (viewed as a point mass at its center of gravity) has been raised x feet. This takes x/n seconds. During this time p · (x/n) gallons have leaked out of the container. Water weighs σ lbs./gal., so the container and its contents weigh w - σxp/n lbs. when the container is x feet off the ground. The required work is given by

$$\int_{0}^{m} (w - \frac{\sigma px}{n})dx \qquad\qquad (\sigma = 8.3)$$

where presumably the integrand is never negative.

<u>Solution for Exercise 14 (p. 418)</u>. Suppose the beam has been raised x feet. The force needed at this instant to lift the beam and the remaining 50-x feet of cable is

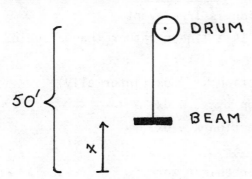

$$F(x) = (50 - x)6 + 800.$$

The work involved in raising the beam 20 feet by winding up the cable is

$$\int_{0}^{20} [(50 - x)6 + 800]dx = 20,800 \text{ ft.-lbs.}$$

For some reason, even those students who consistently can solve pumping problems have difficulty solving lifting problems. For the latter type of problem there is a tendency to include an "extra x" for "distance." Some careful reflection may lead you to realize that pumping and lifting problems are very similar. If not, perhaps the following analysis will help.

Assume we have a right circular cylinder of radius r and height h. The cylinder is full of a fluid weighing w lbs./cu.ft. We wish to pump its contents to the top of the cylinder. Consider a cross section x feet from the top of the cylinder. A step by step development of the integral needed to find the work is as follows:

Area of cross section is πr^2

Volume of cross section is $\pi r^2 dx$ (informally)

Weight of this section is $w \pi r^2 dx$

Work to raise this section is $x w \pi r^2 dx$

Thus,

$$W = \int_0^h x w \pi r^2 dx \qquad \text{is the required work.}$$

Now, consider a cable of radius r and length h suspended from a winch. The cable weighs w lbs./ft. (Ponder this.) We wish to wind up the entire cable. Consider a cross section of the cable x feet from the winch. A step by step development of the integral needed to find the work is as follows:

Area of cross section is <u>irrelevant</u>.

Volume of cross section is <u>irrelevant</u>.

(Weight of cable is a linear rather than a cubic measurement.)

"Length" of cross section   dx (informally)

Weight of this section   w dx

Work to raise this section x w dx

Thus,

$$W = \int_0^h x w \, dx \qquad \text{is the required work.}$$

<u>Solution for Exercise 2 (p. 422)</u>. If we place the vertex of the parabola at the origin of the uv-plane and use the "v-axis" as the axis of symmetry, then the parabola must be given by $v = u^2$. (Why?) Hence, for $u = \frac{w(x)}{2}$ we find that

$(\frac{w(x)}{2})^2 = 16 - x$, or $w(x) = 2\sqrt{16 - x}$. From Formula 10.5.2 in the text, the fluid pressure is given by

$$\int_0^{16} 70x(2\sqrt{16 - x})dx.$$

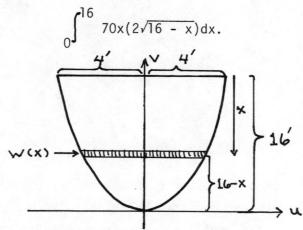

<u>Solution for Exercise 7b (p. 422)</u>. We need to solve the following for A (the extra depth).

$$\int_8^{14} (62.5)x\,(10)dx = \frac{1}{2}\int_{8 + A}^{14 + A} (62.5)x\,(10)dx.$$

The integral on the left is the answer to part (a).

<u>Solution for Exercise 6a (p. 425)</u>. Would you believe that

$$P.V. = \int_3^4 (1000 + 80t)e^{-0.06t}dt$$

is the outcome for this income problem?

Polar Coordinates; Parametric Equations

11

We have already noted that trigonometric functions are useful for modeling many natural phenomena which exhibit periodic behavior. The use of polar coordinates is also natural. For example, spirals occur frequently in nature as in the shell of the nautilus (a mollusk) and in the arrangement of petals on flowers. The polar coordinate form for a logarithmic spiral: $r = e^{\theta}$, is far simpler than its counterpart in rectangular coordinates: $\log|x^2 + y^2| = \arctan(y/x)$. (Diagram 1).

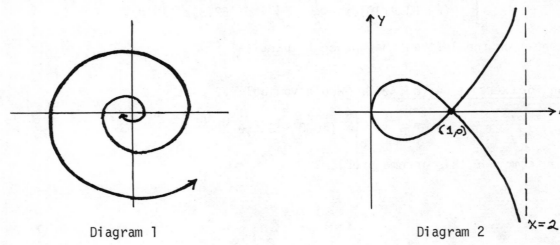

Diagram 1   Diagram 2

There are a number of things about polar coordinates which at first would make their study appear difficult. First, virtually all of your previous work has been set in a rectangular coordinate system. You will be thinking "left or right followed by up or down" as on a map of city streets rather than "direction followed by forwards or backwards" as on a radar screen. This difficulty in using

216

a different frame of reference will be most apparent in Section 11.4. Experience
will help to dissipate the problem.

Second, some complications occur because points in the polar plane do not
have a unique representation. However, it is these very complications which make
parametric equations (for example, polar coordinates) so useful. The curve in
Diagram 2 is a strophoid. Its rectangular equation is $y^2 = x(x - 1)^2/(2 - x)$.
Notice that the curve passes through the point (1,0) twice. This is hard to
specify in rectangular coordinates. However, the polar form of this strophoid is
$r = \sec \theta + \tan \theta$, and we can note that the curve passes through the same point
when $\theta = 0$ and when $\theta = \pi$. That is, the polar points [1,0] and [-1,π] are
geometrically the same point.

The complications that accompany this multiple representation of points arise
when we test for symmetry. In rectangular coordinates, the curve $y = f(x)$ is
symmetric about the y-axis if and only if $f(x) = f(-x)$. Such a simple necessary
and sufficient condition cannot be written for polar curves. For example, if
[r,θ] and [$r, \pi - \theta$] both lie on the curve, we may conclude the curve is symmetric
about the y-axis. However, we may not conclude that symmetry is lacking if this
condition fails. The curve $r^2 = 4 \cos \theta$ is symmetric about the y-axis, but not
by means of the above test. The symmetry may be shown by rewriting the test point
[$r, \pi - \theta$] as [$-r,-\theta$]. The points [r,θ] and [$-r,-\theta$] both satisfy the equation.

SECTION 11.2

Like anything else, the graphing of polar curves is very easy once you get the
knack of it. We have three suggestions. First, it is inefficient and often
frustrating to simply plot a lot of points and then hope to connect them in the
proper way; thus, don't! Second, try to start to recognize certain types of
curves. After all it is only because you know the shape of $y = x^2$ and have a
penchant for simple looking curves that the plodding point plotting technique
gives you

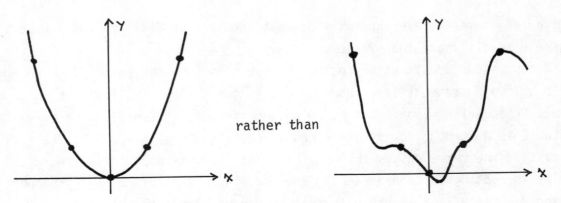

rather than

Shortly, we shall categorize some of the standard polar curves.

Our third point is a suggested procedure. Try to graph the polar curve an interval at a time rather than point by point. That is to say, in between carefully selected and plotted points determine the behavior of r: what is its range of values?, is r increasing?, is r decreasing? To make it easier to answer these questions, you should record the graphs of whatever trigonometric functions are involved. Two examples will now be executed in detail to illustrate this procedure.

Example 1. Sketch the polar graph of $r = 1 - 2 \cos \theta$.
Solution. We begin by recording

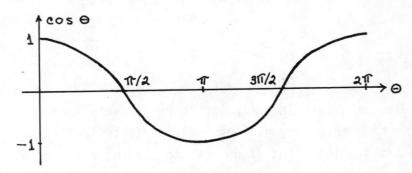

We shall graph $r = 1 - 2 \cos \theta$ for θ from 0 to 2π. We notice from this graph of $\cos \theta$ and some quick algebra that $r = 0$ when $\theta = \pi/3$ or $5\pi/3$ and that $|r|$ is greatest (= 3) when $\theta = \pi$. Therefore, we will graph $r = 1 - 2 \cos \theta$ on each of the following intervals: $[0,\pi/3]$, $[\pi/3,\pi/2]$, $[\pi/2,\pi]$, $[\pi,3\pi/2]$, $[3\pi/2,5\pi/3]$, $[5\pi/3,2\pi]$.

As θ increases from 0 to $\pi/3$, $\cos \theta$ decreases from 1 to 1/2. Therefore, r increases from -1 to 0. (Check this claim.) The result is shown in Diagram A.

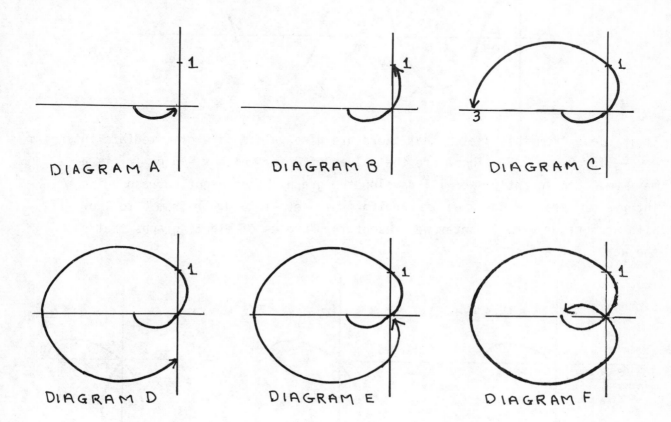

As θ increases from π/3 to π/2, cos θ decreases from 1/2 to 0 so that r increases from 0 to 1. The result is exhibited in Diagram B.

As θ increases from π/2 to π, cos θ decreases from 0 to -1 so that r increases from 1 to 3. The result is shown in Diagram C.

Continuing this "graphing by intervals" for the next three selected intervals, we progressively complete the graph in Diagrams D, E, and F.

Example 2. Sketch the polar graph of r = cos 2 θ.
Solution. We begin with the following sketch.

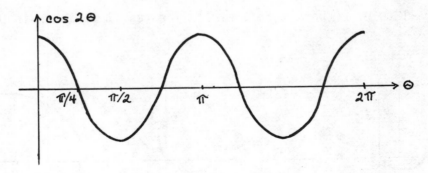

It is clear from this sketch that there are nine values of θ of immediate interest: θ = $\frac{n\pi}{4}$ for $0 \le n \le 8$. These are the values of θ for which r = 0 or |r| is a maximum. Consequently, we will develop the graph on the eight corresponding intervals where r(≑ cos 2 θ) is ranging from 1 or -1 to 0, or from 0 to 1 or -1. With no further verbal banter we present the "pieces of eight" graphs that pave the way for r = cos 2 θ.

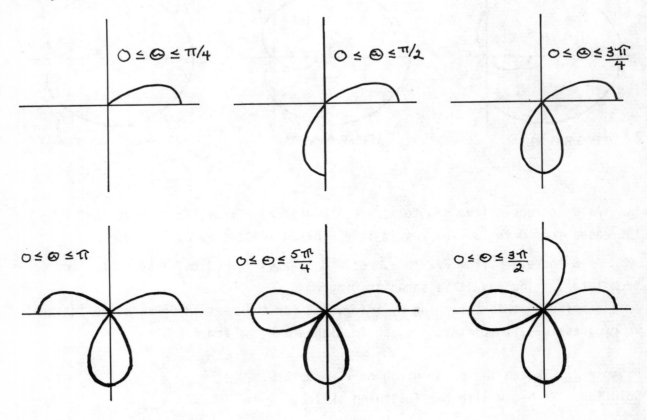

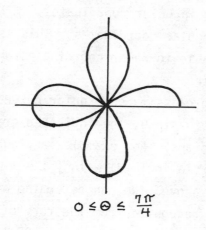

$$0 \le \Theta \le \frac{7\pi}{4}$$

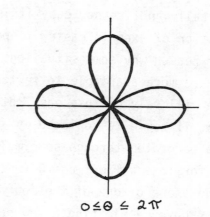

$$0 \le \Theta \le 2\pi$$

It is time to characterize the forms of some standard polar curves. Consider the four cardioids sketched below.

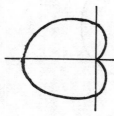 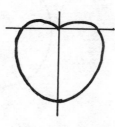 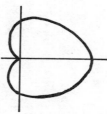 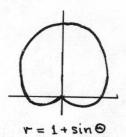

$r = 1 - \cos \Theta$ $r = 1 - \sin \Theta$ $r = 1 + \cos \Theta$ $r = 1 + \sin \Theta$

These four cardioids not only look alike, but they are very closely related algebraically.

$$r = 1 - \cos \theta = 1 - \cos(\theta - 0)$$
$$r = 1 - \sin \theta = 1 - \cos(\theta - \pi/2)$$
$$r = 1 + \cos \theta = 1 - \cos(\theta - \pi)$$
$$r = 1 + \sin \theta = 1 - \cos(\theta - 3\pi/2)$$

To assure that you appreciate this <u>rotation of axes</u>, see if you can sketch the graph of

$$r = 1 - \frac{\sqrt{2}}{2}(\cos \theta - \sin \theta) = 1 - \cos(\theta - \pi/4).$$

Incidentally, notice how easy it is to rotate axes in polar coordinates. A
translation of axes is easier to perform in rectangular coordinates. This is just
another reason why some situations are more adaptable to polar coordinates while
others are more adaptable to rectangular coordinates.

We will rely on this easy rotation of axes to classify some polar curves.
That is, if cos θ is replaced by -cos θ or sin θ or -sin θ, the indicated graph
is simply rotated through some multiple of π/2 radians. For example, any equation
of the form r = a + b · cos θ (where cos could be sin instead) with $|\frac{a}{b}| = 1$ will
have the shape of one of the four cardioids drawn above. We can determine which
curve we have by plotting a few points. Thus, we recognize from the <u>form</u> of the
equation that the graph of r = -2 + 2 sin θ is a cardioid with a cusp. Plotting
the points corresponding to θ = 0, π/2, π, and 3π/2 as shown on the left, we

correctly position the cardioid as shown on the right.

The <u>constants</u> a and b used below are <u>not</u> zero.

<u>CARDIOIDS</u>: r = a + b cos θ.
    If r is <u>never</u> zero on [0,2π) the cardioid has a "dimple" if $1 < |\frac{a}{b}| < 2$ and
        is convex for $|\frac{a}{b}| \geq 2$.
    If r is zero <u>once</u> on [0,2π), the cardioid has a cusp.
    If r is zero <u>twice</u> on [0,2π), the cardioid has an inner loop. Sometimes this
        curve is called a limaçon.

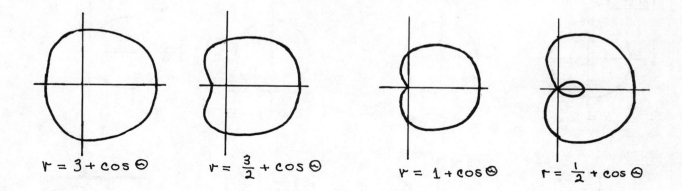

$r = 3 + \cos \Theta$ $r = \frac{3}{2} + \cos \Theta$ $r = 1 + \cos \Theta$ $r = \frac{1}{2} + \cos \Theta$

<u>CONICS:</u> $r = \dfrac{1}{a + b \cos \theta}$.

If $a + b \cos \theta$ is <u>never</u> zero on $[0, 2\pi)$, we have an ellipse. One of the foci is at the pole (origin).

If $a + b \cos \theta$ is zero <u>once</u> on $[0, 2\pi)$, we have a parabola. The focus is at the pole (origin).

If $a + b \cos \theta$ is zero <u>twice</u> on $[0, 2\pi)$, we have a hyperbola. One of the foci is at the pole (origin).

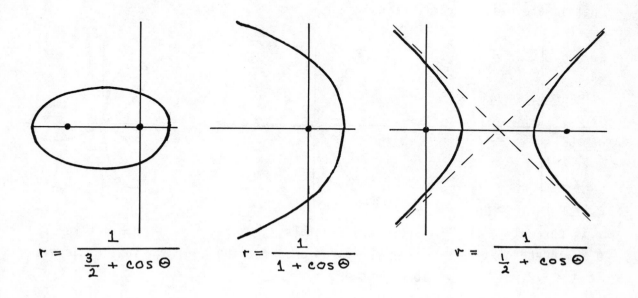

$r = \dfrac{1}{\frac{3}{2} + \cos \Theta}$ $r = \dfrac{1}{1 + \cos \Theta}$ $r = \dfrac{1}{\frac{1}{2} + \cos \Theta}$

CIRCLES: r = a <u>or</u> r = a cos θ.

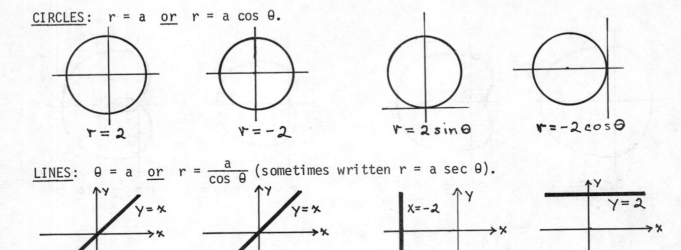

LINES: θ = a <u>or</u> r = $\frac{a}{\cos \theta}$ (sometimes written r = a sec θ).

PETAL CURVES: r = a·cos(nθ) <u>or</u> r = a·sin(nθ), with n an integer.

　　　If n is <u>odd</u>, there are n petals.

　　　If n is <u>even</u>, there are 2n petals.

You should notice that the petals are evenly distributed, so once you position
one petal by finding two successive values of θ at which r = 0 the remaining
petals are easy to plot.   Consider r = cos 3θ.

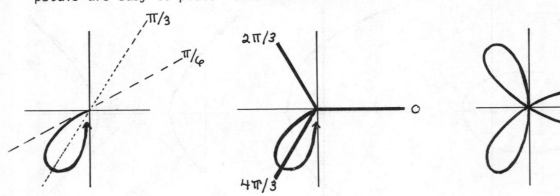

r is zero at
θ = π/6 and next
at θ = π/2.

There are 3 petals, arranged
every 2π/3 radians from the
one just drawn.

Now you're
in the clover.

LEMNISCATES (of Bernoulli): $r^2 = a \cos 2\theta$.

Notice that because the r term is squared, <u>two</u> points will be plotted for some values of θ, whereas for other values of θ (when a·cos 2θ is negative) <u>no</u> points are plotted.

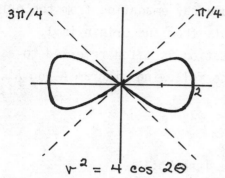

$r^2 = 4 \cos 2\theta$

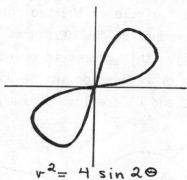

$r^2 = 4 \sin 2\theta$

Knowing these forms will not entirely suffice, but it should greatly simplify your work. Sometimes, though rarely, you may come across a curve requiring the graphing by intervals technique.

$r = 2 + \csc \theta$

(a conchoid)

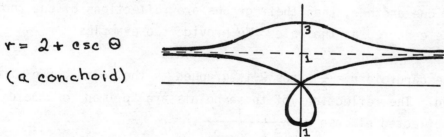

On other occasions you may have the polar form of a familiar rectangular curve in hand. Then, the conversions x = r cos θ and y = r sin θ are needed.

$r = \sin \theta - 4 \cos \theta$ becomes $(x + 2)^2 + (y - \frac{1}{2})^2 = \frac{17}{4}$.

$r(\sin \theta - 4 \cos \theta) = 3$ becomes $y - 4x = 3$.

$r = \sec \theta \tan \theta$ becomes $y = x^2$.

It may not be unusual for you to be asked to graph only <u>part</u> of a familiar polar curve. Several such problems appear in the exercises at the end of this Section. But, first, an aside which you should find very interesting.

It was observed above that the polar equations for conics and cardioids are essentially reciprocals for one another. This algebraic connection is both useful and fascinating. First, we need a definition.

Two points in the polar plane are said to be inverse images of one another in the unit circle if the two points lie on the same ray emanating from the origin and the product of the distances of these two points from the origin is 1. Technically, this geometric operation or transformation is often referred to as an inversion or reflection in the unit circle. In the following diagram A and A', B and B', C and C' are inverse images of one another.

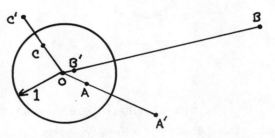

lengths:

$\overline{OA}$ = 1/2, $\overline{OA'}$ = 2, $\overline{OA} \cdot \overline{OA'}$ = 1.

$\overline{OB}$ = 4 , $\overline{OB'}$ = 1/4, $\overline{OB} \cdot \overline{OB'}$ = 1.

$\overline{OC}$ = 3/4, $\overline{OC'}$ = 4/3, $\overline{OC} \cdot \overline{OC'}$ = 1.

Now, if the polar equations for a cardioid and an ellipse are exact reciprocals of one another, then their graphs are reflections of one another in the unit circle centered at the pole. We provide two examples.

<u>Example 3.</u> The cardioid $r = \frac{3}{2} - \cos\theta$ is graphed as the solid curve with certain points lettered. The reflections of these points are "primed" on the dotted line graph of the reflected ellipse $r = \dfrac{1}{\frac{3}{2} - \cos\theta}$.

One of the foci for the ellipse is located at the pole. This is always true for polar conics in standard form. As a review of earlier work, use this information to derive the rectangular form for this conic. You should find that

$$5x^2 - 8x + 9y^2 = 4.$$

You can check this result by converting $r = \dfrac{1}{\frac{3}{2} - \cos\theta}$ to rectangular coordinates.

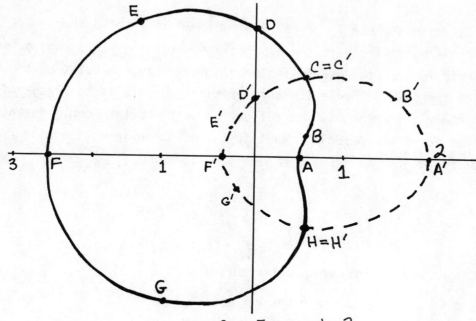

Sketch for Example 3.

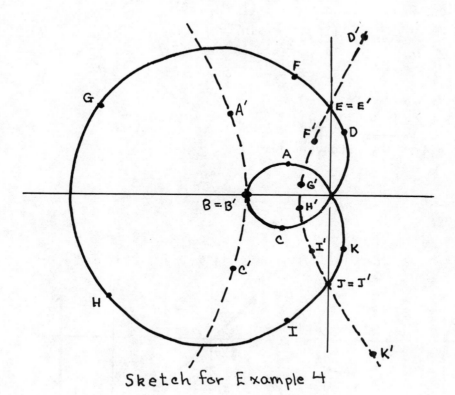

Sketch for Example 4

__Example 4.__   The cardioid $r = 1 - 2 \cos \theta$ is graphed as a solid line. Its reflection in the unit circle is the hyperbola $r = \dfrac{1}{1 - 2 \cos \theta}$. Again, selected points and their inverse images are lettered. Naturally, no point on the hyperbola corresponds to the pole. The left branch of the hyperbola is the image of the inner loop; the right branch is the reflection of the outer loop. Again, the pole is a focus for the hyperbola. What is the equation in rectangular coordinates for this hyperbola?

EXERCISES

Sketch the graphs of the following polar curves.

1. $r = -3 \cos \theta$ 2. $r + \cos \theta = 1$ 3. $\theta = -\pi/4$

4. $r = 2 \sec \theta$ 5. $r = 2 \sin \theta$ 6. $r \csc \theta = 2$

7. $r = 2 - \cos \theta$ 8. $r = 2/(1 - \cos \theta)$ 9. $r = 1 - 3 \sin \theta$

10. $r = \sin 2\theta$ 11. $r^2 = \sin 2\theta$ 12. $r = 6/(2 + \sin \theta)$

13. $r = 3 + 2 \sin \theta$ 14. $r^2 = 4, \ 0 \le \theta < 3\pi/4$

15. $\theta = \pi/4, \ |r| < 2$ 16. $r = -3 - \cos \theta, \ 0 < \theta \le 5\pi/4$

17. $r = \sin \theta, \ 0 \leqq \theta < \pi/2$ 18. $r = \cos 3\theta, \ \pi/3 < \theta \le \pi$

19. $r = 1/(1 + 2 \sin \theta), \ 0 \le \theta < 3\pi/2$

20. $x^2 + y^2 = 4\sqrt{x^2 + y^2}$ (Convert first)

ANSWERS

1. circle   2. cardioid with   3. line   4. line   5. circle
                    cusp

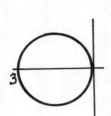   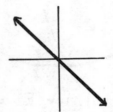   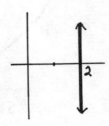   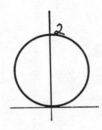

6. circle (with hole, why?)

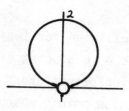

7. cardioid (convex)

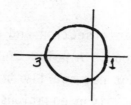

8. parabola

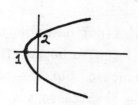

9. cardioid (with loop)

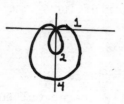

10. petals (4)

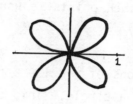

11. lemniscate

12. ellipse

13. cardioid (dimple)

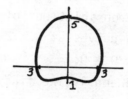

14. parts of a circle

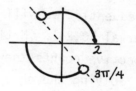

15. line segment

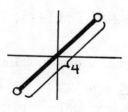

16. part of convex cardioid

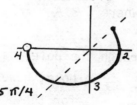

17. semicircle

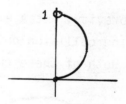

18. part of 3-petal

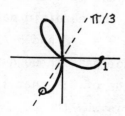

19. part of hyperbola

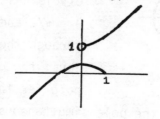

20. circle and its center. Hmm! Why?

SECTION 11.3

The distinction between points of intersection and points of collision for two curves should be carefully understood. A point of collision is also a point of intersection but the converse is not necessarily true. Collision points are found by the simultaneous solution of the given equations. Intersection points which are not also collision points cannot be found algebraically by simultaneously solving the equations. The discovery of such points is usually a matter of luck sometimes abetted by accurate graphs.

Example 1. Determine the collisions of $r = 1 - 2 \cos \theta$ and $r = \cos \theta$.

Solution. The curve $r = \cos \theta$ is a circle. This curve is fully traced out on the interval $[0,\pi]$ or for that matter on any interval of length π. However, only half of the graph of $r = 1 - 2 \cos \theta$ is generated for θ in the interval $[0,\pi]$. Consequently, we will find all common solutions of the two equations for θ in the interval $[0,2\pi]$. Notice that both curves at $\theta = 2\pi$ have returned to the respective points they occupied at $\theta = 0$.

Solving the equations simultaneously, we find that

$$1 - 2 \cos \theta = r = \cos \theta,$$
$$\cos \theta = 1/3.$$

For brevity, let $A = \arccos 1/3$ where $0 < A < \pi/2$. Thus, $[1/3, A]$ and $[1/3, 2\pi - A]$ are the collision points.

Both of these curves pass through the pole, as their graphs suggest. The

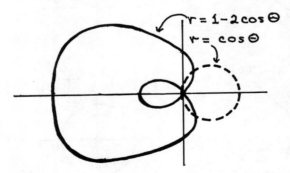

r = 1 - 2 cos Θ
r = cos Θ

pole, however, is not a point of collision for these two curves, but only a point of intersection. For θ between 0 and 2π, the circle passes through the pole when $\theta = \pi/2$ and $3\pi/2$, but the cardioid passes through the pole when $\theta = \pi/3$ and $5\pi/3$. Thus, the curves do not pass through the pole for the same value of θ; that is, the curves do not pass through the pole simultaneously. This is why the pole is only an intersection point. Solving the equations simultaneously reveals only those intersections that occur for the same values of θ. We might then characterize collisions as simultaneous intersections. Despite repeated efforts of the American motorist, there is a difference between an intersection and a collision.

Example 2. Determine the collisions of $r = \cos 3\theta$ and $r = \cos \theta$.

Solution. You should graph these curves before continuing. As both of these curves are fully traced out (i.e., the complete graph is determined) for values of θ in the interval $[0,\pi]$, we need only find solutions of

$$\cos 3\theta = r = \cos \theta \qquad \text{for } \theta \text{ in the interval } [0,\pi]. \tag{1}$$

Since

$$\cos 3\theta = \cos \theta \cos 2\theta - \sin \theta \sin 2\theta$$

$$= \cos \theta (\cos 2\theta - 2 \sin^2\theta)$$

$$= \cos \theta (4 \cos^2\theta - 3),$$

we may restate (1) as

$$4 \cos \theta(\cos^2\theta - 1) = 0.$$

Thus, $\theta = 0, \pi/2, \pi$. The points of collision are $[1,0]$ and $[0,\pi/2]$. The point $[-1,\pi]$ is the same as $[1,0]$. The petal curve intersects but does not collide with the circle at the pole when $\theta = \pi/6$ and $5\pi/6$. What this means is that we have a point (the pole) which under one representation, $[0,\pi/2]$, is a collision point but under another representation, $[0,\pi/6]$, is just a point of intersection.

As noted earlier, collision points require the specification of when as well as where. Here we might identify θ as the parameter, i.e., the variable in terms of which the position of a point is determined. Equations in polar coordinates are a specific, but highly practical example of parametric equations.

SECTION 11.4

Application of the area formula in polar coordinates requires that we determine the upper and lower boundaries of the region and the intervals over which those curves are indeed the upper and lower boundaries. (Actually, it may be more appropriate to think in terms of "outer" and "inner" curves.) Because of the non-unique representation of points in the polar plane, one must be especially careful in determining the interval for which a curve is a boundary. That is, one should be certain the given curve is indeed traced out for the given values of θ. Our second example examines this point. Since a given polar equation may generate two points for a given value of θ (e.g.: the lemniscate $r^2 = 4 \cos 2\theta$), there may well be a question as to what region is represented by the integral. We resolve this question as follows.

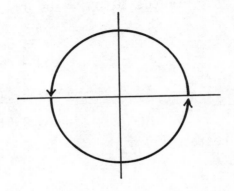

Example 1. For θ in the interval $[0,\pi]$ the polar equation $r^2 = 4$ generates the upper <u>and</u> lower semicircles centered at the origin with radius 2. Since the circle is traced out for $0 \leq \theta \leq \pi$, does

$$\int_0^\pi \frac{1}{2}[\pm\, 2]^2 \; d\theta \; \text{ for } r = f(\theta) = \pm\, 2 \text{ give}$$

us the area of the circle?

Solution. No! The value of the integral is 2π rather than 4π. The reason why the integral only gives us the area for a semicircle is that <u>$r = f(\theta)$ is a function</u>. Thus, we must select $r = f(\theta) = 2$ (the upper semicircle) or $r = f(\theta) = -2$. Because of the symmetry involved and because $f(\theta)$ is squared in the formula

$$\int_{\theta_1}^{\theta_2} \frac{1}{2}[f(\theta)]^2 \; d\theta,$$

it doesn't matter whether we choose the plus sign or the minus sign.

This same comment applies in general. For example,

$$\int_{-\pi/4}^{\pi/4} \frac{1}{2}[4 \cos 2\theta]^2 \; d\theta$$

will give us the area of one half of the lemniscate, even though the entire curve (both loops) is traced out for $-\pi/4 \leq \theta \leq \pi/4$. Further, if $f(\theta) < 0$ for $\theta_1 \leq \theta \leq \theta_2$, then

$$\int_{\theta_1}^{\theta_2} \frac{1}{2}[f(\theta)]^2 \, d\theta$$

will still represent the area of the region bounded by $\theta = \theta_1$, $\theta = \theta_2$, and the negative quantity $f(\theta)$ <u>since</u> the effect of squaring $f(\theta)$ in the integral is to ignore the sign of $f(\theta)$. Here we are finding the area of the region which is symmetric through the pole to the original region. In fact, as long as $\theta_1 < \theta_2$, the integral for polar area will always be non-negative. Thus, there is no counterpart in polar area problems to the idea of "negative area" as discussed for area problems in rectangular coordinates. There is, though, a problem of equivalent difficulty as exampled next.

<u>Example 2.</u> Find the area of the region which lies inside the cardioid $r = 1 - 2 \cos \theta$ but not inside its inner loop.
<u>Solution.</u> The most common error made in this and many similar problems is to fail to clearly determine those intervals on which a given curve is traced out. It is certainly correct to think that if $r = f(\theta)$ and $r = g(\theta)$ are traced out as shown on the right for $\theta_1 \leq \theta \leq \theta_2$, then the area of the shaded region is given by

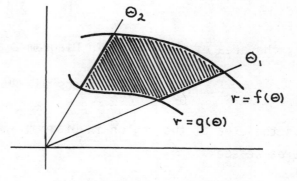

$$\int_{\theta_1}^{\theta_2} \frac{1}{2}[(f(\theta))^2 - (g(\theta))^2] \, d\theta .$$

Problems arise when both the inner and outer curves are generated by the same equation but for different values of θ.

The region whose area we want is shaded in Diagram 1. It is not unusual for some to hurriedly state that the upper and lower boundaries are traced out as θ goes from $\pi/3$ to $5\pi/3$. Such is true for the outer curve but not the inner curve. It is traced out for $-\pi/3 \leq \theta \leq \pi/3$ or, if you prefer to stay in $[0, 2\pi]$, for

$0 \le \theta \le \pi/3$ and $5\pi/3 \le \theta \le 2\pi$. The common error that leads us to write

$$\text{Area} = \int_{\pi/3}^{5\pi/3} \frac{1}{2}[(1 - 2 \cos \theta)^2 - (1 - 2 \cos \theta)^2] \, d\theta$$

as the area of the shaded region is often missed due to algebraic folly. Clearly, the value of this last integral is zero, so it can't be right.

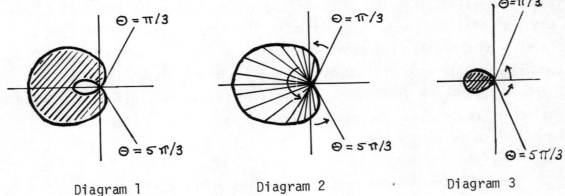

Diagram 1   Diagram 2   Diagram 3

We need to note that

$$\int_{\pi/3}^{5\pi/3} \frac{1}{2}[1 - 2 \cos \theta]^2 \, d\theta$$

is the area of the region in Diagram 2 and that

$$\int_{0}^{\pi/3} \frac{1}{2}[1 - 2 \cos \theta]^2 \, d\theta \; + \; \int_{5\pi/3}^{2\pi} \frac{1}{2}[1 - 2 \cos \theta]^2 \, d\theta$$

is the area of the region in Diagram 3. The difference of these values is the area we seek.

Our final example is directed towards more subtle versions of the common error just noted.

Example 3. Using integrals, represent the areas of the labeled regions in the following diagram with boundaries $r = 3 \cos \theta$ and $r = 2 - \cos \theta$.

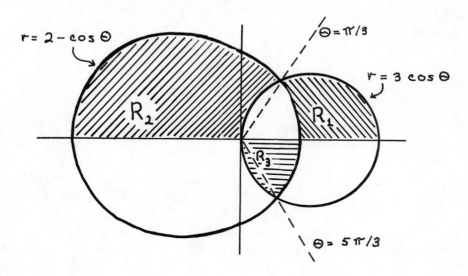

Solution.

$$\text{Area of } R_1 = \int_0^{\pi/3} \frac{1}{2}[(3 \cos \theta)^2 - (2 - \cos \theta)^2]\, d\theta$$

$$\text{Area of } R_2 = \int_{\pi/3}^{\pi} \frac{1}{2}[2 - \cos \theta]^2\, d\theta - \int_{\pi/3}^{\pi/2} \frac{1}{2}[3 \cos \theta]^2\, d\theta$$

$$\text{Area of } R_3 = \int_{\pi/2}^{2\pi/3} \frac{1}{2}[3 \cos \theta]^2\, d\theta + \int_{5\pi/3}^{2\pi} \frac{1}{2}[2 - \cos \theta]^2\, d\theta$$

You should carefully verify each of these representations.

EXERCISES

Using one or more integrals, represent the area of each of the following polar regions.

1. Inside $r = 4$ and to the right of $r = 2 \sec \theta$.

2. Inside $r = 2$ but outside $r = 4 \cos \theta$.

3. Inside the inner loop of $r = 1 - 2 \sin \theta$.

4. Inside $r = 4$ and between the lines $\theta = \pi/2$ and $r = 2 \sec \theta$.

5. Interior to both $r = 1 - \sin \theta$ and $r = \sin \theta$.

6. Inside one petal of $r = 5 \cos 6\theta$.

7. The region bounded by $r = 1/(1 - \cos \theta)$ and $r = \sec \theta$.

8. Inside $r = -4 \cos \theta$.

9. Inside $r^2 = 4$ and below $r(\sin \theta + \cos \theta) = -1 + \sqrt{3}$.

ANSWERS

1. $2 \int_0^{\pi/3} \frac{1}{2}[(4)^2 - (2 \sec \theta)^2]d\theta$

2. $\int_{\pi/3}^{5\pi/3} \frac{1}{2}(2)^2 d\theta - 2 \int_{\pi/3}^{\pi/2} \frac{1}{2}(4 \cos \theta)^2 d\theta$

3. $2 \int_{\pi/6}^{\pi/2} \frac{1}{2}(1 - 2 \sin \theta)^2 d\theta$

4. $2\left\{ \int_0^{\pi/3} \frac{1}{2}(2 \sec \theta)^2 d\theta + \int_{\pi/3}^{\pi/2} \frac{1}{2}(4)^2 d\theta \right\}$

5. $2\left\{ \int_0^{\pi/6} \frac{1}{2}(\sin \theta)^2 d\theta + \int_{\pi/6}^{\pi/2} \frac{1}{2}(1 - \sin \theta)^2 d\theta \right\}$

6. $24 \int_0^{\pi/12} \frac{1}{2}(5 \cos 6\theta)^2 d\theta$

7. $\int_{\pi/3}^{5\pi/3} \frac{1}{2}(\frac{1}{1 - \cos \theta})^2 d\theta + \int_{-\pi/3}^{\pi/3} \frac{1}{2}(\sec \theta)^2 d\theta$

8. $\int_{\pi}^{2\pi} \frac{1}{2}(-4 \cos \theta)^2 d\theta$

9. $\int_{2\pi/3}^{11\pi/6} \frac{1}{2}(2)^2 d\theta + \int_{-\pi/6}^{2\pi/3} \frac{1}{2}(\frac{-1 + \sqrt{3}}{\sin \theta + \cos \theta})^2 d\theta$

SECTION 11.5

Parameters are most often thought of in terms of time, temperature, or some other controlling (physical) variable(s). We generally regard a parameter to be a "behind the scenes" variable. Curves which are expressed parametrically may cross themselves giving rise to more than one tangent line at a point. There may be a disparity between intersection points and collision points, as previously discussed in the case of polar coordinates. These ambiguities can be resolved as long as we remember that a parameter tells us in a general sense <u>when</u>, as well as <u>where</u>.

We appeal to the chain rule to calculate the slope of tangent lines to curves expressed parametrically.

Example 1. Find the equation of the tangent line to the curve given parametrically
by

$$x(t) = t^3 - t \qquad\qquad y(t) = 2^t$$

at $t = 1$.

Solution. Since $x = 0$ and $y = 2$ when $t = 1$, the point of tangency is $(0,2)$. As
$\frac{dy}{dx} = \frac{dy}{dt} \cdot \frac{dt}{dx} = \frac{dy}{dt}/\frac{dx}{dt}$ by the chain rule, we have

$$\frac{dy}{dx} = \frac{2^t \log 2}{3t^2 - 1}$$

so that the slope of the tangent line when $t = 1$ is $\log 2$. Thus, $y - 2 = x \log 2$
is the equation of the desired tangent line.

To insure the existence of a tangent line at a point, we assume that $x'(t)$
and $y'(t)$ are continuous and that $x'(t)$ and $y'(t)$ are not both zero. A
parametrically expressed curve has a horizontal tangent at $t = t_0$ iff $x'(t_0) \neq 0$
and $y'(t) = 0$. It has a vertical tangent at $t = t_0$ iff $x'(t) = 0$ and $y'(t) \neq 0$.
You may well wonder what happens if $x'(t) = 0$ and $y'(t) = 0$. Might not

$$\frac{dy}{dx} = \frac{y'(t)}{x'(t)}$$

give us something other than $\frac{0}{0}$ if we first made some algebraic maneuvers as when
we were evaluating limits?

Consider the curve expressed by $x(t) = \cos^3 t$ and $y(t) = \sin^3 t$ for all real
numbers t.

$$\frac{dy}{dx} = \frac{dy/dt}{dx/dt} = \frac{3 \sin^2 t \cos t}{-3 \cos^2 t \sin t} = -\tan t,$$

which seems to make sense for $t = 0$ ($\frac{dy}{dx}$ apparently equals zero) even though
$x'(0) = y'(0) = 0$. Our algebra has misled us, as you will see in a moment. We
can eliminate the parameter in this problem to obtain

$$x^{2/3} + y^{2/3} = 1.$$

Therefore,

$$\frac{2}{3} x^{-\frac{1}{3}} + \frac{2}{3} y^{-\frac{1}{3}} \frac{dy}{dx} = 0, \tag{1}$$

which defines $\frac{dy}{dx}$ as long as neither x nor y is zero. Notice that when t = 0, we have x = 1 and y = 0. The error of rewriting equation (1) as

$$\frac{dy}{dx} = -\sqrt[3]{y/x} \tag{2}$$

without noting that equation (1) requires x ≠ 0 and y ≠ 0 is the analog of the error in our parametric approach above. Equations (1) and (2) are not equivalent. Similarly,

$$\frac{3 \sin^2 t \cos t}{-3 \cos^2 t \sin t} \neq -\tan t \qquad \text{if } t = 0.$$

The lesson of this example is that if x'(t) = y'(t) = 0 so $\frac{dy}{dx} = \frac{"0"}{0}$, we must accept that result. The graph for the curve under discussion explains this misinterpretation at t = 0. As the curve approaches (1,0), it approaches horizontal and levels off but does not have a tangent line. Notice that at each of the four points where x'(t) = y'(t) = 0 this curve has a corner or cusp.

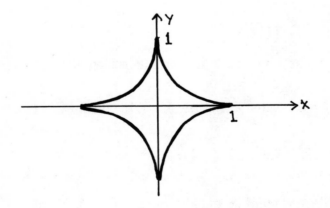

0 ≤ t < 2π will give one complete copy of the curve (an astroid).

One also needs to be careful when determining tangent lines for polar curves. It is not simply enough to specify where you want a tangent line; you must also specify when. For example, the cardioid: r = 1 - 2 cos θ, intersects the origin twice during one complete tracing of the curve. To say that we want the equation of the tangent line to the cardioid at the point [0,0] is misleading on two counts. First, the point $r = \theta = 0$ does not satisfy the equation of the cardioid, so it would be incorrect to substitute these values into any calculations of the tangent line. Second, there are in fact two tangent lines to the cardioid at the origin, one when $\theta = \pi/3$ and one when $\theta = 5\pi/3$.

Whether we intend to represent the tangent line(s) in polar or rectangular coordinates, we need to be able to find the slope of the tangent line(s). There are two ways to find the slope. One could use

$$y = r \sin \theta$$
$$x = r \cos \theta$$
$$r^2 = x^2 + y^2$$

to write the polar equation in rectangular coordinates and then find dy/dx.

Or, one could differentiate the above equations to find that

$$\frac{dy}{d\theta} = \sin \theta \frac{dr}{d\theta} + r \cos \theta$$

and

$$\frac{dx}{d\theta} = \cos \theta \frac{dr}{d\theta} - r \sin \theta.$$

By the chain rule, we have that

$$\frac{dy}{dx} = \frac{dy}{d\theta} \cdot \frac{d\theta}{dx} = \frac{dy/d\theta}{dx/d\theta}.$$

Example 2. Represent each of the tangent lines to the curve $r = 1 - 2 \cos \theta$ at the pole in polar and rectangular form.

Solution. The given cardioid passes through the pole when $\theta = \pi/3$ and $\theta = 5\pi/3$. To find the slope of the tangent line, we begin by writing

$$\frac{dy}{dx} = \frac{dy}{d\theta} \cdot \frac{d\theta}{dx} = \frac{dy/d\theta}{dx/d\theta},$$

$$y = r \sin \theta = \sin \theta - 2 \sin \theta \cos \theta,$$
$$x = r \cos \theta = \cos \theta - 2 \cos^2\theta.$$

Thus,

$$\frac{dy}{d\theta} = \cos \theta - 2 \cos^2\theta + 2 \sin^2\theta$$

and

$$\frac{dx}{d\theta} = -\sin \theta + 4 \cos \theta \sin \theta,$$

so that

$$\frac{dy}{dx} = \frac{\cos \theta - 2 \cos^2\theta + 2 \sin^2\theta}{-\sin \theta + 4 \cos \theta \sin \theta} \ .$$

When $\theta = \pi/3$, $\frac{dy}{dx} = \sqrt{3}$ and the equation of the tangent line is $y = \sqrt{3}\ x$ in rectangular form. To convert this to polar form, we substitute $y = r \sin \theta$ and $x = r \cos \theta$ to arrive at $\tan \theta = \sqrt{3}$ or $\theta = \pi/3$.

When $\theta = 5\pi/3$, $\frac{dy}{dx} = -\sqrt{3}$ and the tangent line equation is $y = -\sqrt{3}\ x$ or $\theta = 5\pi/3$.

You should be suspicious at this point, if not curious, about the polar form for these tangent lines. In general we have

$$\frac{dy}{dx} = \frac{dy/d\theta}{dx/d\theta} = \frac{\sin \theta \dfrac{dr}{d\theta} + r \cos \theta}{\cos \theta \dfrac{dr}{d\theta} - r \sin \theta} \ .$$

When $r = 0$

$$\frac{dy}{dx} = \tan \theta,$$

so that the slope of the tangent line (at the pole) for $\theta = A$ is simply the inclination of the line $\theta = A$. This is no surprise at all; it's the definition of slope. This is true, however, only at the origin.

Example 3. Give the polar and rectangular representations for the tangent line to the curve $r = 1 - 2 \cos \theta$ at the point where $\theta = 2\pi/3$.

Solution. As above, we will use

$$\frac{dy}{dx} = \frac{dy/d\theta}{dx/d\theta} \ , \ y = r \sin \theta, \text{ and } x = r \cos \theta.$$

Thus,

$$\frac{dy}{d\theta} = \frac{d}{d\theta}(\sin \theta(1 - 2 \cos \theta)) = \cos \theta - 2 \cos^2\theta + 2 \sin^2\theta$$

and

$$\frac{dx}{d\theta} = -\sin\theta + 4\cos\theta\sin\theta.$$

When $\theta = 2\pi/3$, $dy/d\theta = 1/2$ and $dx/d\theta = -3\sqrt{3}/2$. Therefore, $dy/dx = -1/3\sqrt{3}$. Further, $r = 2$ when $\theta = 2\pi/3$ so that $y = \sqrt{3}$ and $x = -1$. The equation of the tangent line is

$$y - \sqrt{3} = -\frac{1}{3\sqrt{3}}(x + 1)$$

or
$$3\sqrt{3}\,y + x - 8 = 0,$$

which can be written as

$$r(3\sqrt{3}\sin\theta + \cos\theta) = 8$$

in polar coordinates.

A common trap to which students often fall prey is to evaluate $\frac{dr}{d\theta}$ as the slope of the tangent line. In the last example, the value of $\frac{dr}{d\theta}$ is $\sqrt{3}$, but the value of $\frac{dy}{dx} = -1/3\sqrt{3}$. In general, dy/dx and $dr/d\theta$ are different. They represent totally different things.

In Section 11.2 of the Supplement the shape of a cardioid, $r = a + b\cos\theta$, was characterized in terms of how many times r is zero on $[0, 2\pi]$. It was stated that the curve has a dimple if $1 < |\frac{a}{b}| < 2$. Can you verify this claim? As a start, notice that such curves have four vertical tangents.

Here is another problem for you to consider. The curve C is given by the parametric equations $y = y(t)$ and $x = x(t)$, where t is the parameter. The quantities dy/dx, $x'(t)$, and $y'(t)$ help us discuss the behavior of C. In fact,

$$\frac{dy}{dx} = \frac{y'(t)}{x'(t)}.$$

What about d^2y/dx^2, $y''(t)$ and $x''(t)$? What do they mean? Further, can you show that

$$\frac{d^2y}{dx^2} \text{ is not } \frac{y''(t)}{x''(t)},$$

which may run contrary to your intuition?

SECTION 11.6

We begin by providing a solution for Exercise 1 on p. 454. We shall reset the
scene. Suppose a wheel of radius r is rolling without slipping along a level
road. The motion of a fixed point on the outer edge of the wheel is suggested in
Diagram 1.

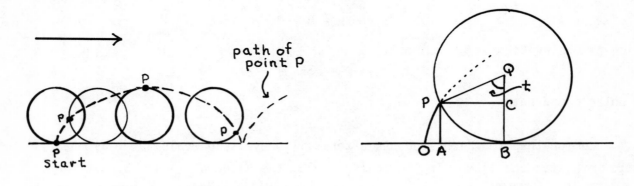

Diagram 1   Diagram 2

The resulting curve is called a <u>cycloid</u>. Diagram 2 represents the action of
point P during the first quarter turn of the wheel, $0 \leq t \leq \pi/2$. Angle BQP is t
and O is the starting position of P. We will take O as the origin and the road
as the x-axis. A little geometrical thought shows the coordinates (x,y) of the
point P may be found as follows:

$$x = \overline{OA} = \overline{OB} - \overline{AB} = \overparen{PB} - \overline{AB} = rt - r \sin t$$

$$y = \overline{AP} = \overline{BC} = \overline{BQ} - \overline{CQ} = r - r \cos t.$$

This same set of parametric equations holds for t outside the initial restriction
$[0,\pi/2]$. For example if $\pi/2 \leq t \leq \pi$ (see Diagram 3), we could repeat the above
analysis to find that

$$x = rt - r \sin t.$$

However, the figure may lead us to think that we should write

$$y = r + r \cos t$$

instead of $y = r - r \cos t$. However we should remember that in this case cos t is

negative.   Therefore, y = r - r cos t will still be the correct representation.
We may apply a similar argument in
the other quadrants so that the cycloid
is given by

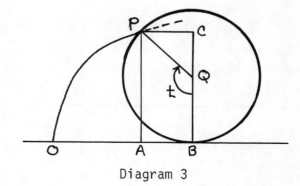

$$x = r(t - \sin t)$$
$$y = r(1 - \cos t)$$

for all t. Clearly, one would be
hard-pressed to derive a direct
relationship between x and y <u>without</u>
the use of parameters.

Diagram 3

    There are a number of related problems of interest. We shall merely state
them. Suppose our wheel is rolling again.

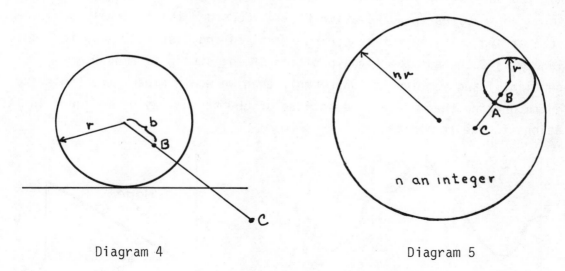

Diagram 4   Diagram 5

In Diagram 4, what is the path followed by point B? (answer: x = rt - b sin t,
y = r - b cos t); by point C? (answer: ?). Suppose now the wheel is rolling
around inside another wheel. What is the path followed by point A?, by point B?,
by point C?

    Finally, the cycloid and its friends are not just mathematical recreations.
Suppose you are the manufacturer of frictionless(!) playground slides and want to
market the fastest ride in the West. Your janitor, the mathematician, will

quickly inform you that a piece of a cycloid positioned as shown rather than a

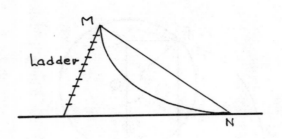

straight line will provide the quickest
descent from point M to point N. Your
local bartender, the physicist, may point
out an even better solution using some
Newtonian results and suggest you place
point N below point M and let the kids
jump.

SECTION 11.7

It is not too surprising that the formula for the length of the arc y = f(x) from
x = a to x = b involves the derivative f'(x). Consider the interval [a,b] to be
an elastic string. If the entire string is raised and distorted in some fashion
(see Diagram 1) such that the final position of the string is smooth
(differentiable) and continuous (unbroken), then we would expect the string to
have length greater than b - a. Deviations of the string may be measured in its
inclinations to the horizontal.

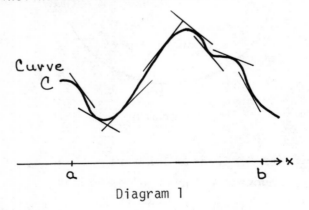

Diagram 1

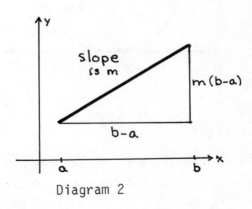

Diagram 2

The derivative as the slope of these inclinations is then a good measure of the
twisting and turning employed to arrive at the final position. This intuitive
view does not tell us that the integrand is $\sqrt{1 + (f'(x))^2}$, but notice in
particular that if the final position of the string is a straight line segment of

slope m (Diagram 2), then the length is $(b - a)\sqrt{1 + m^2}$.

It really is only necessary to remember the arc length formula

$$\ell(C) = \int_{t_1}^{t_2} \sqrt{[x'(t)]^2 + [y'(t)]^2}\, dt \qquad (1)$$

for a curve C given parametrically. This result easily applies to a curve given as $y = f(x)$ on $[a,b]$ if we let $t_1 = a$, $t_2 = b$, $x(t) = t$, and $y(t) = f(t)$.

$$\ell(C) = \int_a^b \sqrt{1 + [y'(t)]^2}\, dt = \int_a^b \sqrt{1 + [f'(x)]^2}\, dx$$

For a curve given in polar coordinates as $r = f(\theta)$, we simply need to remember that

$$x = r \cos \theta \qquad \text{and} \qquad y = r \sin \theta$$

and apply equation (1) rather than remembering

$$\ell(C) = \int_{\theta_1}^{\theta_2} \sqrt{[f(\theta)]^2 + [f'(\theta)]^2}\, d\theta.$$

Example 1. Find the length of one arch of the cycloid (see Section 11.6) given by the equations

$$x(t) = a(t - \sin t), \quad y(t) = a(1 - \cos t), \quad 0 \le t \le 2\pi.$$

Solution. Since $x'(t) = a(1 - \cos t)$ and $y'(t) = a \sin t$ the length $\ell(C)$ of the arch is given by

$$\ell(C) = \int_0^{2\pi} \sqrt{[x'(t)]^2 + [y'(t)]^2}\, dt = \int_0^{2\pi} \sqrt{a^2(1 - \cos t)^2 + a^2 \sin^2 t}\, dt$$

$$= a \int_0^{2\pi} \sqrt{2(1 - \cos t)}\, dt$$

From the identity $\sin x = +\sqrt{(1 - \cos 2x)/2}$ for $0 \le x \le \pi$, we have

$$\ell(C) = 2a \int_0^{2\pi} \sin(\tfrac{t}{2})\, dt \doteq -4a \cos(\tfrac{t}{2})\Big|_0^{2\pi} = 8a.$$

<u>Example 2.</u> Find the length of the polar curve r = sec θ for 0 ≤ θ ≤ π/3.

<u>Solution</u>. First we note that

$$x(\theta) = \sec \theta \cdot \cos \theta = 1 \text{ and } y(\theta) = \sec \theta \cdot \sin \theta = \tan \theta$$

so that

$$x'(\theta) = 0 \quad \text{and} \quad y'(\theta) = \sec^2\theta.$$

Thus, the length ℓ(C) of the arc is given by

$$\ell(C) = \int_0^{\pi/3} \sqrt{(x'(\theta))^2 + (y'(\theta))^2} \, d\theta = \int_0^{\pi/3} \sec^2 \theta \, d\theta = \tan \theta \Big|_0^{\pi/3} = \sqrt{3}$$

We remark that the length of an arc is never less than the distance between the points it connects. (Why?) Here the endpoints are (1,0) for θ = 0 and (1,√3) for θ = π/3. The distance between these points is √3 units. Perhaps now you remember that r = sec θ is the polar form for the straight line x = 1.

SECTION 11.8

We remarked at the start of the last section that the involvement of f'(x) in a formula for the arc length ℓ(C) of y = f(x) on [a,b] should be no surprise. The formula for the area of the associated surface of revolution

$$S = \int_a^b 2\pi f(x) \sqrt{1 + [f'(x)]^2} \, dx \qquad (1)$$

should similarly appeal to your common sense. If a horizontal line segment is spun about the x-axis, the surface area of the resulting cylinder is easily

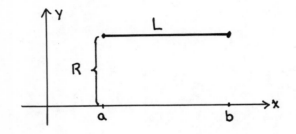
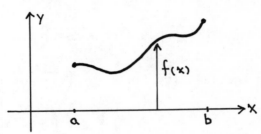

2π R $\cdot$ L. We have just seen how $\int_a^b \sqrt{1 + [f'(x)]^2}\, dx$ represents L in the more

general case of arc length. Considering $f(x)$ to be an approximation or generalization of R, we might consider formula (1) to be "intuitively obvious." Similarly, these comments may help you memorize the formula for the parametric case

$$S = \int_a^b 2\pi\, y(t)\sqrt{[y'(t)]^2 + [x'(t)]^2}\, dt.$$

FURTHER COMMENTS ON INTEGRATION

Two theoretical views of the integral are presented in the text. The Darboux method utilizes upper and lower sums. The integral and its applications have been presented to you in this context. There is an appealing ease in viewing, both physically and theoretically, the integral as a limit resulting from some pinching process by these upper and lower approximations.

The other theoretical view presented by the text involves Riemann Sums. Here the integral is viewed as the limit of some approximate or representative quantity. You may recall our use of the term "representative rectangle" in the discussions of areas and volume.

There are several other ways to define an "integral." Such a discussion is best left to advanced courses. It is not our purpose here to forecast that work. Nor is it our purpose to indicate a preference between the two definitions you have seen. Rather, it is our aim to strengthen your feeling for what an integral is. We shall do so by quickly revisiting several of the applications you may have studied. In order to avoid paralleling the text's developments of these topics, we shall follow the Riemannian approach at an intuitive level.

Suppose $g(x)$ is a continuous function on the set (interval) of real numbers [a,b]. We may consider the integral

$$\int_a^b g(x)\, dx \tag{1}$$

to represent the limit of a special sum involving the function g on the interval
[a,b]. To simply say that an integral is the limit of a sum is too imprecise.
We must specify what exactly is being summed. In particular, we partition the
interval [a,b] into a number of subintervals: $[x_i, x_{i+1}]$, $i = 0, \cdots, n - 1$ with
x_0 = a and x_n = b, evaluate the function g at any point in each subinterval:
$c_i \in [x_{i-1}, x_i]$, and compute the sum

$$g(c_1) \cdot (x_1 - x_0) + g(c_2) \cdot (x_2 - x_1) + \cdots + g(c_n) \cdot (x_n - x_{n-1}). \qquad (2)$$

We then ask what happens if more and more subintervals are formed so that the
longest of them is made arbitrarily short. The answer is that the continuity of
g on [a,b] is sufficient to guarantee that the sum (2) has a limit. This limit is
called the integral, expression (1).

If the function g is positive on the interval [a,b], then an easy interpretation of expression (2) is that of the sum of the areas of n rectangles having
$(x_i - x_{i-1})$ as the width of their bases and $g(c_i)$ as their respective heights.
Just how well expression (2) approximates the area of the region bounded by
x = a, x = b, y = 0, and y = g(x) is dependent upon how well $g(c_i)$ approximates or
represents the average value of g on $[x_{i-1}, x_i]$. Without repeating earlier
justifications, let us simply note that since g is continuous on [a,b], $g(c_i)$ is
representative of g on $[x_{i-1}, x_i]$ if we take a "sufficient" number of intervals
each of which is "sufficiently" small. That is, we evaluate the known-to-exist
limit of expression (2).

Suppose now that the interval of real numbers represents a range of polar
angles: a to b, in the polar plane. Suppose further that r = f(θ) is a positive
and continuous function defined on this "polar interval." The formula for the
area of a sector of a circle of radius R and central angle A (in radians) is given
by $\frac{AR^2}{2}$. The following diagram suggests that $\frac{1}{2}[f(c_i)]^2(\theta_i - \theta_{i-1})$ is representative of the polar area subtended by the arc r = f(θ) and the rays $\theta = \theta_{i-1}$ and
$\theta = \theta_i$. Thus expression (2) is (sufficiently) representative of the (polar) area
bounded by r = f(θ) and the rays θ = a and θ = b, and

$$\int_a^b g(x)dx = \int_a^b \frac{[f(\theta)]^2}{2} d\theta$$

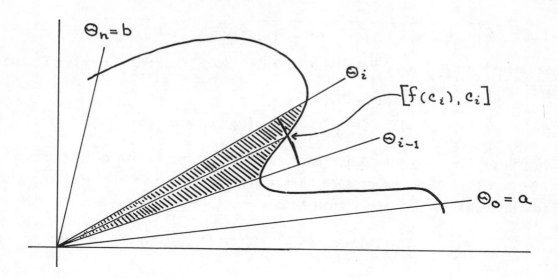

represents the limit of the sum of the representative approximations:

$$\frac{1}{2}[f(c_i)]^2(\theta_i - \theta_{i-1}).$$

Here we are making the identification $g(x) = \frac{1}{2}[f(x)]^2$.

   Let us recall now our earlier work with the differential. We discovered that the value of a function and its derivative at one point, x, could be used to give a sufficiently representative value for the function at another point, x + h. That is, we employed the formula

$$f(x + h) \simeq f(x) + h \cdot f'(x),$$

even though no discussion was given on just how to choose x and h to obtain "good" approximations. Nevertheless, let us presume we may so choose h, since as h approaches zero so does the error in our approximation of f(x + h). Consulting the following diagram we can compute the length of line segment AB as:

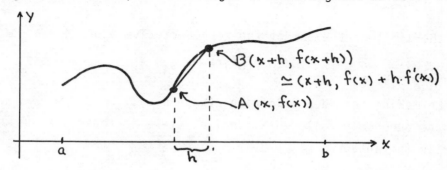

$$\sqrt{(x + h - x)^2 + (f(x + h) - f(x))^2} \simeq h \cdot \sqrt{1 + (f'(x))^2}.$$

If we now identify x_{i-1} as x, x_i as $x + h$, and x as c_i, then the length of line segment AB is approximately

$$\sqrt{1 + (f'(c_i))^2} \cdot (x_i - x_{i-1}) \simeq g(c_i)(x_i - x_{i-1}) \qquad (3)$$

If the length of line segment AB is representative of the length of the arc $y = f(x)$ from A to B, then expression (3) is representative of the arc length over this short interval and the arc length from $x = a$ to $x = b$ is given by:

$$\int_a^b g(x)dx = \int_a^b \sqrt{1 + (f'(x))^2} \, dx,$$

which is the limit of the sum of the representative approximations:

$$\sqrt{1 + f'(c_i)^2} \cdot (x_i - x_{i-1}).$$

Here we are making the identification $g(x) = \sqrt{1 + f'(x)^2}$. It should be noted that by the rules for composition of functions, g is continuous if f' is and we have assumed that f' is continuous.

Finally, suppose that $[a,b]$ is an interval of straight line distances measured from some fixed reference point (the origin) and that $f(s)$ is a function describing the force acting along this line upon some object at position s. Then, if $f(c_i)$ is representative of the force during the short space $[s_{i-1}, s_i]$, then $f(c_i) \cdot (s_i - s_{i-1}) = g(c_i) \cdot (x_i - x_{i-1})$ is representative of the work expended during this space $[s_{i-1}, s_i]$. The total work expended from $s = a$ to $s = b$ is given by:

$$\int_a^b f(s)ds$$

which is the limit of the sum of the representative approximations:

$$f(c_i)(s_i - s_{i-1}).$$

Here we are making the identification $g(x) = f(x)$.

We have purposely followed a parallel style in these discussions. We may now describe the integral

$$\int_a^b g(x)dx$$

as the limit of the sum of representative approximations of the form

$$g(c_i)(x_i - x_{i-1}),$$

where the symbols x_i and c_i are as defined earlier and the limit is taken as the length of the longest subinterval approaches zero. The continuity of g is a sufficient but not necessary condition for the existence of this limit, the integral. Consequently, any situation which submits to these conditions will give rise to an integral.

For example, let us consider an important notion from Probability and Statistics. Suppose we want to study the life of light bulbs. For the sake of the study we will ignite N light bulbs at the same time and assume that no bulb will continuously burn for more than a year; $0 \le t \le 365$ in days. We evenly partition this interval and record how many bulbs burn out during each subinterval. We then construct a frequency histogram:

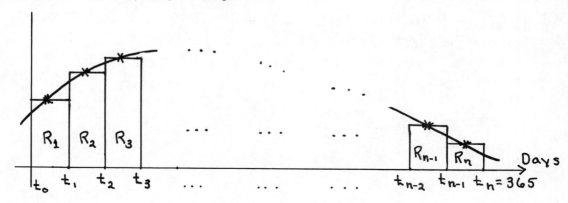

To represent our results the height of each rectangle is chosen so that the area of the rectangle is the number of bulbs that burned out during the interval. Thus, the quantity

$$(\text{area } R_1 + \text{area } R_2 + \text{area } R_3)/N$$

is the _probability_ that a light bulb will burn out in t_3 or fewer days. For purposes of interpolation we may approximate our discrete results (the histogram) by passing a continuous curve through the midpoint of the top of each rectangle.

Naturally, as we take finer partitions the approximating curve will be more accurate. Call the curve $y = g(x)$. Notice then that the sum of the areas of the rectangles is of the form

$$g(c_1)(t_1 - t_0) + g(c_2)(t_2 - t_1) + \cdots + g(c_n)(t_n - t_{n-1})$$

which has

$$\int_0^{365} g(t)dt$$

as a limiting value for some function g. Though the function g may well be only an approximation based on empirical results, we can now make statements like:

$B = \frac{1}{A} \int_0^x g(t)dt$ is the probability that a light bulb will burn out during the first x days, where $A = \int_0^{365} g(t)dt$ with t as time in days;

$1 - B$ is the probability that a light bulb will last at least x days;

and
$\frac{1}{A} \int_a^b g(t)dt$ is the probability that the life of a light bulb is between a and b days. (Is N = A?)

Sequences;
Indeterminate Forms; Improper Integrals

12

PRELIMINARY COMMENTS

The role of approximations in mathematics is too frequently obscured by the need
for precision of expression and a clear logical foundation in the development of
mathematical ideas. A significant number of applications of mathematics require
approximations. Normally, these approximations are the result of some iterative
procedure that produces a succession of hopefully better and better approximations.
The collection of these successive approximations is called a sequence because the
order in which the approximations occur in the collection is important. We are
concerned with the order of the iterative values since we need to know if further
approximations will improve the estimate and, if so, how many iterations are
needed to find an approximation within some predetermined allowance for error. As
this language suggests, we will be interested in the limit, if it exists, of a
sequence and how many terms or values in the sequence are required before we
obtain an acceptable approximation of this limit. The ϵ and δ notation for limits
of functions of real numbers has a direct analogy for sequences. Sequences are
simply defined as functions whose domains are restricted to the set of positive
integers.

Suppose we wish to approximate the square root of a given positive number x.
The process we will use is sometimes called divide and average. Let a > 0 be our
first guess or estimate of $\sqrt{x}$. (Normally, a is taken equal to 1). Our second
approximation, b, is calculated as $\frac{1}{2}(a + \frac{x}{a})$. Our third approximation, c, is
calculated as $\frac{1}{2}(b + \frac{x}{b})$. It is common to use subscripts in recording successive
approximations and elements of a sequence. We have $s_1 = a$, $s_2 = b$, $s_3 = c$, and

253

so on. The basic iterative step in our algorithm for calculating square roots may be written as

$$s_{n + 1} = \frac{1}{2}(s_n + \frac{x}{s_n}) \qquad \text{for } n \geq 1. \qquad (1)$$

For example, let us look at the first few approximations for the square root of $x = 3$. We shall take the first term, s_1, in this sequence of approximations as 1. Then, using the iterative formula (1), we have

$$s_2 = \frac{1}{2}(s_1 + \frac{3}{s_1}) = \frac{1}{2}(1 + \frac{3}{1}) = 2.0000,$$

$$s_3 = \frac{1}{2}(s_2 + \frac{3}{s_2}) = \frac{1}{2}(2 + \frac{3}{2}) = 1.7500,$$

and

$$s_4 = \frac{1}{2}(s_3 + \frac{3}{s_3}) = 1.7321,$$

which is fairly close.

To see why $\lim_{n \to \infty} s_n = \sqrt{x}$, we begin by noting that if one of the numbers s_n or x/s_n is smaller than $\sqrt{x}$, then the other is greater than $\sqrt{x}$. (Why?) Next, $s_{n + 1}$ is the average of s_n and x/s_n, so it must lie between s_n and x/s_n. It is also true that $x/s_{n + 1}$ lies between s_n and x/s_n. (Why?) Thus,

$$\left| s_{n + 1} - \sqrt{x} \right| < \left| s_{n + 1} - \frac{x}{s_{n + 1}} \right| < \left| s_n - \frac{x}{s_n} \right|.$$

In fact,

$$\left| s_{n + 1} - \frac{x}{s_{n + 1}} \right| < \frac{1}{2}\left| s_n - \frac{x}{s_n} \right| \qquad (2)$$

as you may verify by substituting equation (1) in the left hand side of inequality (2). The inequality in (2) tells us that the error at each step of the sequence of approximations is less than one-half the error of the previous approximation. Using induction and (2), we may show that

$$\left| s_{n + 1} - \frac{x}{s_{n + 1}} \right| < \frac{1}{2^n}\left| s_1 - \frac{x}{s_1} \right|. \qquad (3)$$

What does the inequality in (3) tell us? Well, since the product of the two numbers $s_{n + 1}$ and $x/s_{n + 1}$ is x and

$$\frac{1}{2^n}\left|s_1 - \frac{x}{s_1}\right|$$

can be made arbitrarily small for arbitrarily large n, it must be that s_{n+1} (and x/s_{n+1}) can be made arbitrarily close to $\sqrt{x}$.

As an example, we consider $x = 4$ with $s_1 = 1$ and $n = 5$. From (3) we can anticipate the error in s_6.

$$\left|s_6 - \sqrt{4}\right| < \frac{1}{2^5}\left|1 - \frac{4}{1}\right| = \frac{3}{2^5}$$

so that

$$1.90625 < s_6 < 2.09375.$$

Some 'quick' calculations carried out with accuracy to four decimal places actually show that

$$s_1 = 1.00000, \quad s_2 = 2.50000, \quad s_3 = 2.05,$$

$$s_4 = 2.00061, \quad s_5 = 2.00000, \quad s_6 = 2.00000,$$

which is a significantly better approximation than expected!

The discussion that preceded this example could be dressed up to constitute a proof of our divide and average algorithm to compute square roots. The actual process is relatively simple. The numerical algorithms that may be developed to evaluate derivatives, integrals, and roots of equations as might arise in max-min problems are also relatively simple. The proofs of these techniques are sometimes not so simple. Nonetheless, the proofs are critical for they tell us when we may apply the procedure and what kind of accuracy we can obtain.

Now we consider another approximation by sequences. We shall need the following result, which you should verify. If f and g are integrable functions and $f(x) \le g(x)$ on the interval [a,b], then

$$\int_a^x f(x)dx \le \int_a^x g(x)dx \qquad \text{for } a \le x \le b.$$

As $\cos x \le 1$ for all real numbers x, we have

$$\int_0^x \cos x \, dx \le \int_0^x 1 \cdot dx \qquad x > 0,$$

or

$$\sin x \le x.$$

Next, we have

$$\int_0^x \sin x \, dx \le \int_0^x x \, dx,$$

$$-\cos x + 1 \le x^2/2,$$

or

$$1 - \frac{x^2}{2} \le \cos x.$$

Again,

$$\int_0^x (1 - \frac{x^2}{2}) dx \le \int_0^x \cos x \, dx,$$

so

$$x - \frac{x^3}{6} \le \sin x.$$

Continuing this process we have

$$\cos x \le 1 - \frac{x^2}{2} + \frac{x^4}{24}$$

and

$$\sin x \le x - \frac{x^3}{6} + \frac{x^5}{120}.$$

Consider the set of inequalities

$$x - \frac{x^3}{6} \le \sin x \le x - \frac{x^3}{6} + \frac{x^5}{120},$$

which we may write as

$$\left| \sin x - (x - \frac{x^3}{6}) \right| \le \frac{|x^5|}{120}$$

if we drop the restriction that x is positive. What we have is a cubic polynomial

that approximates sin x with an error not in excess of $\frac{|x^5|}{120}$. Notice that if

$0 \le x \le \pi/4$, $x - x^3/6$ gives us an estimate of sin x with error not exceeding

$(\pi/4)^5/120 < .003.$

This approximation is not satisfactory for most practical purposes. However, it is important to note that the <u>sequence</u> of polynomials:

$$s_1(x) = x$$

$$s_2(x) = x - x^3/3! \qquad\qquad n! = n(n - 1) \cdots 3 \cdot 2 \cdot 1$$

$$s_3(x) = x - \frac{x^3}{3!} + \frac{x^5}{5!}$$

$$s_4(x) = x - \frac{x^3}{3!} + \frac{x^5}{5!} - \frac{x^7}{7!}$$

$$s_n(x) = x - \frac{x^3}{3!} + \cdots + (-1)^{n + 1} \frac{x^{2n - 1}}{(2n - 1)!},$$

approximates sin x to any desired accuracy. That is $\lim\limits_{n \to \infty} s_n(x) = \sin x$. Since this limit may be shown to exist and equal sin x, we may claim that the infinite sum $x - \frac{x^3}{3} + \cdots + (-1)^{n + 1} \frac{x^{2n - 1}}{(2n - 1)!} + \cdots$ is equal to sin x. This infinite sum is called an infinite series.

Quite naturally it turns out that polynomial approximations such as those exampled above are of immense practical value because of their simplicity. The purpose of this and the next Chapter is to provide an introduction to this topic of infinite series.

SECTIONS 12.1-2

It is one thing to show that a sequence converges. It is usually a separate question altogether to actually determine the value of that limit. This distinction may not strike you at first because the exercises are selected as ones you should be able to do.

There are a variety of ways in which we will be examining functions whose domains are the set of positive integers. We will consider sequences

$$a_1, \ a_2, \ a_3, \ \cdots, \ a_n, \ \cdots$$

where we may make the identification $f(n) = a_n$ and ask about $\lim\limits_{n \to \infty} f(n)$. We will consider infinite series

$$a_1 + a_2 + \cdots + a_n + \cdots,$$

where we will not only ask about $\lim\limits_{n \to \infty} a_n$, but also $\lim\limits_{n \to \infty} s_n$, where

$s_n = a_1 + a_2 + \cdots + a_n$. We will even be concerned with

$$\lim_{n \to \infty} g(n) = \lim_{n \to \infty} \int_1^n F(t)dt.$$

We need to cautiously observe that these limits specify the domain variable as an integer. This fact can make a difference. For example,

$$\lim_{n \to \infty} \sin(n\pi) = \lim_{n \to \infty} 0 = 0,$$

but $\lim\limits_{x \to \infty} \sin(x\pi)$ does not exist.

If $\lim\limits_{x \to \infty} f(x) = L$, then $\lim\limits_{n \to \infty} f(n) = L$. However, we may not conclude that

$\lim\limits_{n \to \infty} f(n)$ does not exist if $\lim\limits_{x \to \infty} f(x)$ does not exist. To solidify your under-

standing of this point, answer the following question. If $\lim\limits_{x \to \infty} f(x)$ and

$\lim\limits_{n \to \infty} f(n)$ both exist, are the limit values equal? The answer is yes, but why?

Suppose we wanted to know if the sequence $f(n) = a_n = n^2$ was increasing. By definition we are asking if $a_n < a_{n + 1}$ for all $n \geq 1$. We could appeal to the derivative of f, but for this appeal the domain of f cannot be just the positive integers. We would have to consider $f(x) = x^2$ for all $x > 0$. Since $f'(x) = 2x > 0$ for $x > 0$, we may say that the sequence $f(n) = a_n = n^2$ is increasing for $n \geq 1$. Here as well, though, we must be careful in making the transition from n to x, from integers to reals. The derivative of $f(x) = x \sin \pi x$ is not zero for all $x \geq 1$ even though the sequence $f(n) = n \sin n\pi = 0$ is constant.

We re-emphasize the need for caution by noting that the sequence

$f(n) = \dfrac{1}{n - 11.5}$ is bounded (by -2 and 2) even though the function $f(x) = 1/(x-11.5)$ is not bounded.

A number of limit theorems on sequences are stated and proven in the text. You should carefully study the details and observe how the results and their verification are analogous to the material developed for functions of a real variable. You should further recall the admonitions stated at that time. For example, if a_n and b_n are sequences, then we must know that $\displaystyle\lim_{n \to \infty} a_n$ and $\displaystyle\lim_{n \to \infty} b_n$ exist before we may say that $\displaystyle\lim_{n \to \infty} (a_n + b_n) = \lim_{n \to \infty} a_n + \lim_{n \to \infty} b_n$.

There is nothing wrong with examining the "first few" terms of a sequence if its behavior has you stymied, but you should be aware that this practice could be misleading. Further, to prove any of your observations you must consider and manipulate the general term. Our first example illustrates these points.

Example 1. Discuss the boundedness, monotonicity, and convergence of the sequence $a_n = \dfrac{n^2 + 100}{n}$.

Solution. The first few terms are $a_1 = 101$, $a_2 = 52$, $a_3 = 36.\overline{3}$, $a_4 = 29$ and $a_5 = 25$. We might conjecture that we have a decreasing sequence which must converge as it is bounded below by zero. Let us test these statements. For a_n to be decreasing we need $a_n > a_{n+1}$. That is, can we show that

$$\frac{n^2 + 100}{n} > \frac{(n + 1)^2 + 100}{n + 1} \ ?$$

This inequality is equivalent to

$$n^2(n + 1) + 100 > n(n + 1)^2,$$

or $\qquad n(n + 1)(n - n - 1) + 100 > 0,$

or $\qquad\qquad\qquad\qquad 100 > n(n + 1).$

This last inequality is only true for $n \leq 9$ so a_n is not decreasing. We have shown that a_n is increasing for $n > 9$. For all practical purposes, we may state

that the sequence a_n is increasing. (But, we should note the restriction n > 9).
The same steps that led us to conclude that a_n is not decreasing may be used as
the proof that a_n is increasing for n > 9.

The fact that a_n is increasing doesn't tell us whether or not the sequence is
bounded above. However, $a_n = n + \frac{100}{n} > n$ for $n \geq 1$ and the sequence {n} is <u>known</u>
to be unbounded. We conclude that a_n is also unbounded. Necessarily, a_n does not
converge.

Each of these conclusions could have been reached had we considered instead
the function $f(x) = \frac{x^2 + 100}{x}$ and its derivative $f'(x) = 1 - 100/x^2$. You should
supply the details.

<u>Example 2</u>. Discuss the boundedness, monotonicity, and convergence of the sequence
$a_n = \dfrac{4n}{\sqrt{4n^2 + 1}}$.

<u>Solution</u>. Certainly, the sequence is bounded below by zero. To determine if
there is an upper bound, we could assume one exists and call it B without
attempting to initially assign B a value. We then test our assumption.

$$\frac{4n}{\sqrt{4n^2 + 1}} \leq B$$

$$4n \leq B\sqrt{4n^2 + 1}$$

$$16n^2 \leq 4B^2n^2 + B^2$$

$$n^2(16 - 4B^2) \leq B^2$$

Now it is easier to select a value for B. The right hand side of the last
inequality is never negative. The left hand side is negative for B > 2. The
sequence is bounded above by any number greater than or equal to 2. (In a moment
we shall show the sequence to be decreasing so that the <u>least</u> upper bound is
$a_1 = 4/\sqrt{5} < 2$.)

We could have approached the question of boundedness in a different way. We
could note that $\sqrt{4n^2 + 1}$ is approximately 2n. We should be careful with such

approximations (see Example 3 below) and only use them as a first guess. Our first guess here leads us to an upper bound of 2 but to prove this we need to put into symbols the implicit thought behind this approximation. We could write

$$\frac{4n}{\sqrt{4n^2 + 1}} < \frac{4n}{\sqrt{4n^2}} = \frac{4n}{2n} = 2.$$

If our sequence were $b_n = \dfrac{4n}{\sqrt{4n^2 - 1}}$ instead, our inequalities would be different.

$$\frac{4n}{\sqrt{4n^2 - 1}} > \frac{4n}{\sqrt{4n^2}} = 2$$

We need to write something like

$$\frac{4n}{\sqrt{4n^2 - 1}} < \frac{4n}{\sqrt{3n^2}} = \frac{4}{\sqrt{3}}$$

to obtain an upper bound.

To determine the monotonicity of a_n we could test an inequality ($a_n < a_{n+1}$) or examine a derivative as in Example 1. Instead, let us rewrite a_n as

$$\frac{4n}{\sqrt{4n^2 + 1}} = \frac{4}{\frac{1}{n}\sqrt{4n^2 + 1}} = \frac{4}{\sqrt{4 + \frac{1}{n^2}}} \, .$$

Now, $1/n^2$ is known to be decreasing so $4 + 1/n^2$ is decreasing. The square root function is an increasing function, so $\sqrt{4 + 1/n^2}$ is still a decreasing sequence. Thus, a_n must be increasing (fixed numerator and decreasing denominator results in increasing quotient). We have then another technique for handling monotonicity. Can you apply this technique to show that the sequence $b_n = \dfrac{4n}{\sqrt{4n^2 - 1}}$ is decreasing?

Since a_n is bounded above and increasing we know that $\lim\limits_{n \to \infty} a_n$ exists. With a_n rewritten as $4/\sqrt{4 + 1/n^2}$ we may apply various limit theorems to conclude that the value of this limit is 2.

Example 3. Evaluate $\displaystyle\lim_{n \to \infty} \frac{1}{\sqrt{n^2 + n} - n}$.

Solution. It is not unusual to think that $\sqrt{n^2 + n}$ is approximately equal to n and

conclude that $\dfrac{1}{\sqrt{n^2 + n} - n}$ is not bounded above. In fact, writing

$$\frac{1}{\sqrt{n^2 + n} - n} \cdot \frac{\left(\frac{1}{n}\right)}{\left(\frac{1}{n}\right)} = \frac{1/n}{\sqrt{1 + \frac{1}{n}} - 1}$$

may lead one to the same conclusion if he looks at the denominator only. Notice
that in its new form the numerator of a_n also approaches zero. Remember that in
the statement on limits of quotients: a_n/b_n, if $b_n \to 0$ we cannot conclude that
$\displaystyle\lim_{n \to \infty} a_n/b_n$ fails to exist unless it is known that $\displaystyle\lim_{n \to \infty} a_n \neq 0$. To evaluate this
limit we need to use a technique learned some time back involving the quadratic
surd. That is,

$$\lim_{n \to \infty} \frac{1}{\sqrt{n^2 + n} - n} = \lim_{n \to \infty} \frac{1}{\sqrt{n^2 + n} - n} \cdot \frac{\sqrt{n^2 + n} + n}{\sqrt{n^2 + n} + n}$$

$$= \lim_{n \to \infty} \frac{\sqrt{n^2 + n} + n}{n}$$

$$= \lim_{n \to \infty} \left(\sqrt{1 + \frac{1}{n}} + 1\right) = 2,$$

which may be a slight surprise.

Finally, limits of the form $\displaystyle\lim_{x \to \infty} f(x)$ may be rewritten as $\displaystyle\lim_{t \downarrow 0} f(\frac{1}{t})$ by
making the substitution $x = 1/t$. For the current example we find that

$$\lim_{x \to \infty} \frac{1}{\sqrt{x^2 + x} - x} = \lim_{t \downarrow 0} \frac{1}{\sqrt{\frac{1}{t^2} + \frac{1}{t}} - \frac{1}{t}} = \lim_{t \downarrow 0} \frac{t}{\sqrt{t + 1} - 1} .$$

Can you show that this limit is related to the derivative of $f(x) = \sqrt{x}$ evaluated at
$x = 1$?

SECTION 12.3

Comments on this Section consist of solutions for Exercises 3, 8, 10, 12, 18, 20, 23, and 24 on p. 491 of the text. Following our previous practice the solutions include comments that may help you solve other problems.

Solution for Exercise 3. We observe that $0 < (2/n)^n < (2/3)^n$ for $n \geq 3$. Since $(2/3)^n \to 0$ as $n \to \infty$, we conclude from the Pinching Theorem that the limit is zero.

Solution for Exercise 8. First we write $n^{\alpha/n}$ as $(n^{1/n})^\alpha$. We know that $\lim_{n \to \infty} n^{1/n} = 1$. Since $f(x) = x^\alpha$ is continuous at $x = 1$, $\lim_{n \to \infty} n^{\alpha/n} = f(1) = 1$. (Note the use of Theorem 12.2.12.)

Solution for Exercise 10. We use Theorem 12.2.7 and then result 12.3.2, which you should look up.

$$\lim_{n \to \infty} \frac{3^{n+1}}{4^{n-1}} = 12 \lim_{n \to \infty} \left(\frac{3}{4}\right)^n = 0.$$

Solution for Exercise 12. We begin by writing:

$$(n + 2)^{1/n} = \left(\frac{n+2}{n} \cdot n\right)^{1/n} = \left(1 + \frac{2}{n}\right)^{1/n} \cdot n^{1/n}.$$

We know that $\lim_{n \to \infty} n^{1/n} = 1$. Next, we have

$$1 < \left(1 + \frac{2}{n}\right)^{1/n} \leq 1 + \frac{2}{n},$$

so the Pinching Theorem applies. Thus,

$$\lim_{n \to \infty} (n + 2)^{1/n} = \lim_{n \to \infty} \left(1 + \frac{2}{n}\right)^{1/n} \cdot \lim_{n \to \infty} n^{1/n} = 1.$$

Solution for Exercise 18. Things aren't always what they may seem to be. For example, this problem may seem quite hard, since we cannot evaluate

$$\int \sin(x^2)\,dx.$$

We recall that

$$\left| \int_a^b f(x)\,dx \right| \le \int_a^b |f(x)|\,dx$$

and that

$$\int_a^b f(x)\,dx \le \int_a^b g(x)\,dx,$$

if $f(x) \le g(x)$ for all x in $[a,b]$. Thus,

$$0 \le \left| \int_{-1/n}^{1/n} \sin(x^2)\,dx \right| \le \int_{-1/n}^{1/n} |\sin(x^2)|\,dx \le \int_{-1/n}^{1/n} 1\,dx = \frac{2}{n}.$$

It follows from the Pinching Theorem that the limit is zero.

<u>Solution for Exercise 20</u>. We have yet to learn how to handle

$$\int e^{-x^2}\,dx,$$

so we need to circumvent the need to evaluate the integral. We shall appeal to
the statement

$$\int_a^b f(x)\,dx \le M(b - a),$$

where M is the maximum of f on $[a,b]$. Notice that this inequality may be viewed
as an "upper sum" approximation. A diagram for positive f:

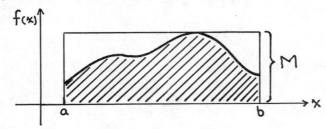

reveals the motivation. This type of approximation for integrals using rectangles
is a standard technique.

   For the problem at hand, we observe that e^{-x^2} is positive and decreasing.
This permits us to write

$$0 \leq \int_{n}^{n+1} e^{-x^2} dx \leq e^{-n^2} [(n+1) - n] = e^{-n^2}.$$

Since $e^{-n^2} \to 0$ as $n \to \infty$, we conclude from the Pinching Theorem that the sequence converges to zero.

Solution for Exercise 23. It helps to notice that $\sin(n\pi) = 0$ for all $n \geq 1$. Our sequence: $0, 0, 0, \cdots$, converges to 0.

Solution for Exercise 24. The sequence $\pi/n \to 0$ as $n \to \infty$. However, n^2 is unbounded. We need to combine the expressions in a different way. We shall appeal to

$$\lim_{x \downarrow 0} \frac{\sin x}{x} = \lim_{t \to \infty} \frac{\sin(1/t)}{1/t} = 1.$$

This suggests that we write

$$n^2 \cdot \sin(\pi/n) = n\pi \cdot \frac{\sin(\pi/n)}{\pi/n}.$$

The sequence $n\pi$ is still unbounded, but since $\frac{\sin(\pi/n)}{\pi/n} \to 1$ (rather than zero) a conclusion is possible. Namely, the sequence diverges.

SECTIONS 12.4-6

L'Hospital's rule is often remembered strictly as the conclusion of a theorem; namely, the limit of a quotient is the limit of the quotient of the derivatives of the numerator and denominator. Often overlooked is the hypothesis in this theorem which tells us when this quotient of derivatives wizardry is legitimate. For example, it is clear that

$$\lim_{x \to 0} \frac{1 + \sin 2x}{x}$$

does not exist. But if we misapply (since $\lim_{x \to 0} x = 0$ but $\lim_{x \to 0} (1 + \sin 2x) \neq 0$) L'Hospital's rule we come up with

$$\lim_{x \to 0} \frac{1 + \sin 2x}{x} = \lim_{x \to 0} \frac{2 \cos 2x}{1} = 2.$$

Suppose that a and b are non-zero numbers. If we try to evaluate the limit of a quotient by substitution, there are nine possible outcomes:

$$\frac{a}{b}, \frac{0}{a}, \frac{0}{\infty}, \frac{a}{\infty}, \frac{a}{0}, \frac{\infty}{0}, \frac{\infty}{a}, \frac{0}{0}, \frac{\infty}{\infty} .$$

The first four outcomes are well-defined limits. In the next three outcomes there is no limit. Whenever we encounter one of the last two cases, some additional algebra is called for. L'Hospital's Rule extends our capabilities so that we can handle any quotient limit. We can also handle any product now since we may write

$$f(x) \cdot g(x) = \frac{f(x)}{1/g(x)} = \frac{g(x)}{1/f(x)} .$$

Finally, exponentials may be converted into products and thence into quotients. If

$$y = [f(x)]^{g(x)},$$

then

$$\log y = g(x) \cdot \log f(x) = \frac{1}{1/g(x)} \cdot \log(f(x))$$

and the limit of y may be found using the continuity of the log function.

<u>Example 1</u>. Evaluate $\lim_{x \downarrow 0} (\tan x)^{\sin x}$.

<u>Solution</u>. Let $y = (\tan x)^{\sin x}$. Then

$$\log y = \sin x \log(\tan x) = \frac{\log(\tan x)}{\csc x} .$$

Since $\lim_{x \downarrow 0} \log(\tan x)$ is $-\infty$ and $\lim_{x \downarrow 0} \csc x$ is ∞, we may apply L'Hospital's Rule.

$$\lim_{x \downarrow 0} \frac{\log(\tan x)}{\csc x} = \lim_{x \downarrow 0} \frac{\sec^2 x/\tan x}{-\csc x \cot x} .$$

With a little algebra we find that

$$\lim_{x \downarrow 0} \frac{\sec^2 x}{-\csc x} = \lim_{x \downarrow 0} \left(- \frac{\sin x}{\cos^2 x}\right) = 0.$$

Thus

$$0 = \lim_{x \downarrow 0} \log y = \log\left(\lim_{x \downarrow 0} y\right)$$

or

$$e^0 = \lim_{x \to 0} y.$$

So,

$$\lim_{x \to 0} (\tan x)^{\sin x} = 1.$$

Example 2. Evaluate $\lim\limits_{x \to \infty} (x^2 \sin \frac{1}{x} - x).$

Solution. We may rewrite this problem as

$$\lim_{x \to \infty} \frac{\sin(\frac{1}{x}) - \frac{1}{x}}{1/x^2} .$$

The limits of the numerator and denominator are both zero as $x \to \infty$. (Verify this.)
We may apply L'Hospital's Rule to find that

$$\lim_{x \to \infty} \frac{\sin(\frac{1}{x}) - \frac{1}{x}}{1/x^2} = \lim_{x \to \infty} \frac{-\frac{1}{x^2} \cos(\frac{1}{x}) + \frac{1}{x^2}}{-2/x^3}$$

which simplifies to

$$\lim_{x \to \infty} \frac{\cos(\frac{1}{x}) - 1}{2/x} .$$

Once again the limits of the numerator and denominator are zero as $x \to \infty$, so we
are justified in applying L'Hospital's Rule a second time to obtain:

$$\lim_{x \to \infty} \frac{\frac{1}{x^2} \sin(\frac{1}{x})}{-2/x^2} = -\frac{1}{2} \lim_{x \to \infty} \sin(\frac{1}{x}) = 0.$$

We could have originally used the substitution $x = \frac{1}{t}$ to restate the problem
as

$$\lim_{t \to 0} \frac{\sin t - t}{t^2} .$$

The same work as above is called for, but perhaps the algebra would have been
simpler.

Example 3. Evaluate $\lim\limits_{x \to 1} \log x \cdot e^{\frac{1}{x - 1}} .$

Solution. In its present form the product approaches a limit of the form $0 \cdot \infty$ so we may rewrite the product as a quotient:

$$\lim_{x \downarrow 1} \log x \cdot e^{\frac{1}{x - 1}} = \lim_{x \downarrow 1} \frac{\log x}{e^{-1/(x - 1)}} \, ,$$

and apply L'Hospital's Rule:

$$\lim_{x \downarrow 1} \frac{1/x}{\frac{1}{(x - 1)^2} e^{-1(x - 1)}} = \lim_{x \downarrow 1} \frac{(x - 1)^2}{xe^{-1/(x - 1)}} \, .$$

To simplify the work, we note that $\lim_{x \downarrow 1} \frac{1}{x} = 1$ and examine

$$\lim_{x \downarrow 1} \frac{(x - 1)^2}{e^{-1/(x - 1)}} = \lim_{x \downarrow 1} \frac{e^{1/(x - 1)}}{1/(x - 1)^2}$$

Two successive applications of L'Hospital's Rule yield $+ \infty$ for this limit and, consequently, the value of the original limit is $+ \infty$.

Example 4. Evaluate $\lim_{x \uparrow 1} \log x \cdot e^{\frac{1}{x - 1}}$.

Solution. The brave may proceed as in Example 3 but since we've tried that (on scratch paper) let us begin with the substitution $t = x - 1$. Our limit becomes

$$\lim_{t \uparrow 0} \log(t + 1) \, e^{\frac{1}{t}} = 0, \quad \text{without using L'Hospital's Rule.}$$

Right!

Example 5. Evaluate $\lim_{x \to 0} \dfrac{\int_0^x \dfrac{dt}{2 + t}}{x}$.

Solution.

$$\lim_{x \to 0} \frac{\int_0^x \frac{dt}{2 + t}}{x} = \lim_{x \to 0} \frac{\log(x + 2) - \log 2}{x} = \lim_{x \to 0} \frac{\frac{1}{x + 2} - 0}{1} = \frac{1}{2} \, .$$

Since $\lim_{x \to 0} \int_0^x f(t)dt = 0$, can you use the Fundamental Theorem of Calculus to show that

$$\lim_{x \to 0} \frac{\int_0^x f(t)dt}{x} = f(0)?$$

<u>Example 6</u>. Evaluate $\lim\limits_{x \to 3} \dfrac{x^2 - 3}{x - 5}$.

<u>Solution</u>. If you apply L'Hospital's Rule, you obtain 6 as the value of the limit.
However, L'Hospital's Rule does <u>not</u> apply in this instance. Why? What is the
actual value of this limit? (Hint: the answer is -3.)

    The text presents tests (p. 493 and p. 502) for horizontal and vertical
asymptotes. In general, the curve y = f(x) has the line y = mx + b as an
asymptote if and only if at least one of the following limits is zero:

$$\lim_{x \to -\infty} [f(x) - (mx + b)] \qquad \text{or} \qquad \lim_{x \to \infty} [f(x) - (mx + b)].$$

As an example we will show that y = 3x is an asymptote for the hyperbola

$$x^2 - \frac{y^2}{9} = 1.$$

We shall consider the upper portion of the right hand branch of the hyperbola:

$$y = \sqrt{9x^2 - 9} \ .$$

Thus,

$$\lim_{x \to \infty} [\sqrt{9x^2 - 9} - 3x] = \lim_{x \to \infty} [\sqrt{9x^2 - 9} - 3x] \cdot \frac{\sqrt{9x^2 - 9} + 3x}{\sqrt{9x^2 - 9} + 3x}$$

$$= \lim_{x \to \infty} \frac{-9}{\sqrt{9x^2 - 9} + 3x} = 0$$

and we are done.

SECTION 12.7

There is nothing improper about improper integrals. Our primary need for improper

integrals is in the next Chapter with the Integral Test for infinite series. We
repeat the introductory comment that

$$\lim_{x \to \infty} \int_a^x f(t)dt \text{ and } \lim_{n \to \infty} \int_a^n f(t)dt$$

may not agree. As an example consider $f(t) = \cos(\pi t)$. A heavier use of improper
integrals arises in a Junior level course in complex analysis under the topic of
contour integration (of functions of a complex variable).

    You should keep in mind that improper integrals don't arise only when one
of the limits of integration is ∞. We must also stay alert for any peculiarities
in the domain of the integrand. That is, to evaluate

$$\int_0^4 \frac{dx}{x(x - 2)}$$

we need to consider

$$\lim_{a \downarrow 0} \int_a^1 \frac{dx}{x(x - 2)} + \lim_{b \uparrow 2} \int_1^b \frac{dx}{x(x - 2)} + \lim_{c \downarrow 2} \int_c^4 \frac{dx}{x(x - 2)}$$

and each of these three limits must exist if the original (improper) integral is
going to exist (converge).

Infinite Series

13

Comments on this Section consist of solutions for Exercises 2i, 6b, and 8b on p. 517 of the text.

Solution for Exercise 2i.

$$\sum_{k=1}^{3} (-1)^{k+1} \left(\frac{1}{2}\right)^{2k-1} = (-1)^2 \left(\frac{1}{2}\right) + (-1)^3 \left(\frac{1}{2}\right)^3 + (-1)^4 \left(\frac{1}{2}\right)^5 = \frac{1}{2} - \frac{1}{8} + \frac{1}{32} = \frac{13}{32} .$$

Solution for Exercise 6b. There are 6 terms so the index should run, for example, from 1 to 6 or 0 to 5. Since the exponents on the b's go from 0 to 5, we shall let the index k do the same. Thus far we have each term as plus or minus some power of a times b^k. We notice that the sum of exponents on a and b in each term is 5. So a^{5-k} is matched with b^k. The signs alternate so we would have $(-1)^k$ or $(-1)^{k+1}$ depending on which works. It turns out that we have

$$\sum_{k=0}^{5} (-1)^k a^{5-k} b^k .$$

Solution for Exercise 8b. To verify

$$\sum_{k=1}^{n} (2k - 1) = n^2 \tag{1}$$

by induction, we observe that statement (1) is true for n = 1. Namely,

$$\sum_{k=1}^{1} (2k - 1) = 1 = 1^2.$$

Now we assume that statement (1) is true for some $n = N \geq 1$. Thus,

$$\sum_{k=1}^{N+1} (2k - 1) = \sum_{k=1}^{N} (2k - 1) + (2N + 1)$$

$$= N^2 + (2N + 1)$$

by assumption. But, $N^2 + 2N + 1 = (N + 1)^2$. Since the truth (assumed) of (1) for $n = N$ implies (1) is also true for $n = N + 1$, we are done.

We can also verify (1) by using problem 8a,

$$\sum_{k=1}^{n} k = \frac{1}{2}n(n + 1).$$

We shall separately sum the odds and evens as follows.

$$\sum_{k=1}^{2n} k = \sum_{k=1}^{n} (2k - 1) + \sum_{k=1}^{n} 2k = \sum_{k=1}^{n} (2k - 1) + 2 \sum_{k=1}^{n} k$$

Thus,

$$\frac{1}{2}(2n)(2n + 1) = \sum_{k=1}^{n} (2k - 1) + 2 \cdot \frac{1}{2}(n)(n + 1),$$

which we may rearrange to yield (1).

Now that you've seen the "slick" solution, perhaps the following solution will seem easy.

$$\sum_{k=1}^{n} (2k - 1) = \sum_{k=1}^{n} 2k + \sum_{k=1}^{n} (-1) = 2 \sum_{k=1}^{n} k - n = n(n + 1) - n = n^2.$$

SECTION 13.2

In this Chapter we will be concerned with whether or not a given infinite series

$$a_1 + a_2 + \cdots + a_n + \cdots \tag{1}$$

converges. To determine the convergence or divergence of this infinite series
we must focus our attention on the sequence of partial sums

$$a_1, \; a_1 + a_2, \; a_1 + a_2 + a_3, \; \cdots, \; a_1 + a_2 + \cdots + a_n, \; \cdots \qquad (2)$$

for the series in expression (1). You should clearly distinguish at the outset
between the sequence in display (2) and the sequence

$$a_1, \; a_2, \; a_3, \; \cdots, \; a_n, \; \cdots \qquad (3)$$

The tendency to confuse series and sequences is fostered in part by the fact that
series are defined in terms of sequences. There is a big difference between
sequences (2) and (3). Consider the case of $a_n = 1$ for all $n \geq 1$. Sequence (2)
diverges whereas sequence (3) converges to 1. Again, if $a_n = 1/2^n$, sequence (2)
is the familiar geometric series that converges to 2 whereas sequence (3)
converges to zero. You should verify these claims.

In determining the convergence of series (1) we will have occasion to examine
both sequences (2) and (3). If sequence (2) converges then the limit is the sum
of series (1) and the limit of sequence (3) must be zero. On the other hand, if
sequence (3) converges to say L, then sequence (2) and, thus, series (1) by
definition diverge if $L \neq 0$ and might converge if $L = 0$.

The two sequences:

$$\tfrac{1}{3}, \; 5, \; 2, \; \pi, \; a_5, \; a_6, \; \cdots, \; a_n, \; \cdots$$

and

$$a_5, \; a_6, \; a_7, \; \cdots, \; a_n, \; \cdots \; ,$$

either both converge or both diverge. If they converge, they do so to the same
limit. The associated series:

$$\tfrac{1}{3} + 5 + 2 + \pi + a_5 + a_6 + \cdots + a_n + \cdots$$

and

$$a_5 + a_6 + \cdots + a_n + \cdots \; ,$$

either both converge or both diverge. However, if they converge, they do so to
different values. The lesson is clear. The first few terms of a sequence may be
ignored when answering the existence question on convergence, but these same terms

may not be overlooked in evaluating the actual sum of a convergent series.

It is frequently useful to alter the indexing of an infinite series to put it in a more manageable form. For example, we may write

$$\sum_{k=1}^{\infty} \frac{1}{(k+1)^2} \quad \text{as} \quad \sum_{n=2}^{\infty} \frac{1}{n^2} .$$

This type of change is very much like the Substitution Principle used in integration. In our example we used $n = k + 1$ and noted that if the first value of k is 1 then the first value of n is 2.

The number of available techniques for evaluating the sum of a series is rather limited. The two types of sums you can handle at this stage are exampled below. The sum of the geometric series

$$\sum_{k=0}^{\infty} ar^k$$

exists when $|r| < 1$ and is equal to $\frac{a}{1-r}$, where a is the first term and r is the ratio of successive terms.

Example 1. Evaluate $\displaystyle\sum_{n=3}^{\infty} \frac{3 + 2^{2n+1}}{5^{n-1}}$.

Solution. We rewrite the series as the sum of two geometric series:

$$\sum_{n=3}^{\infty} \frac{3}{5^{n-1}} \quad \text{and} \quad \sum_{n=3}^{\infty} \frac{2^{2n+1}}{5^{n-1}} = 10 \sum_{n=3}^{\infty} \left(\frac{4}{5}\right)^n .$$

For the first series, the first term is 3/25 and the ratio is 1/5. The sum is

$$\frac{3/25}{1 - 1/5} = \frac{3}{20} .$$

The sum of the second series is

$$10 \frac{(4/5)^3}{1 - 4/5} = \frac{128}{5} .$$

The sum of the original series is

$$\frac{3}{20} + \frac{128}{5} = \frac{103}{4} .$$

Example 2. Evaluate $\displaystyle\sum_{k=1}^{\infty} \frac{1}{k(k+3)}$.

Solution. We take our cue from the technique of partial fractions.

$$\frac{1}{k(k+3)} = \frac{1}{3}[\frac{1}{k} - \frac{1}{k+3}]$$

This gives us

$$\sum_{k=1}^{\infty} \frac{1}{k(k+3)} = \frac{1}{3} \sum_{k=1}^{\infty} (\frac{1}{k} - \frac{1}{k+3})$$

$$= \frac{1}{3}[(1 - \frac{1}{4}) + (\frac{1}{2} - \frac{1}{5}) + (\frac{1}{3} - \frac{1}{6}) + (\frac{1}{4} - \frac{1}{7}) + (\frac{1}{5} - \frac{1}{8}) + \ldots].$$

The sum "telescopes" in the sense that almost all of the terms are matched by a term with the opposite sign. Our telescoping sum collapses (whence the phrase "telescopes") to leave us

$$\frac{1}{3}[1 + \frac{1}{2} + \frac{1}{3}] = \frac{11}{18}$$

as the sum of the series.

SECTIONS 13.3-4

The tests which may be applied to determine the convergence or divergence of a given infinite series are reviewed below along with some suggestions as to how one might decide which tests to use for a given series.

The theorem which states:

If the series $\sum a_k$ converges, then $a_k \to 0$,

is the simplest test to apply. Some divergent series may be shown to diverge by this test but no convergent series may be shown to converge by this test. Frequently, this test is overlooked and, if not overlooked, misused. That is, sometimes the converse, which is false, is used. If $a_k \not\to 0$, the series must diverge. However, if $a_k \to 0$, no conclusion is possible and some other test is needed. Perhaps you can avoid this pitfall by noticing that $1/k$ and $1/k^2$ both approach zero, but $\sum 1/k$ diverges whereas $\sum 1/k^2$ converges.

The Integral Test is also frequently overlooked because other tests are seemingly simpler to use. Naturally, one should only use the Integral Test where one can evaluate the integral involved! Unlike the other tests, the Integral Test is completely conclusive; that is, if you can evaluate the integral, then you can conclude whether or not the series converges. Since the Integral Test essentially involves an improper integral, you should be careful and write

$$\lim_{n \to \infty} \int_c^n f(x)dx \qquad \text{rather than} \qquad \int_c^\infty f(x)dx$$

More importantly you should remember that there may be a difference between

$$\lim_{n \to \infty} \int_c^n f(t)dt \qquad \text{and} \qquad \lim_{x \to \infty} \int_c^x f(t)dt.$$

It is the first integral that is of interest to us here.

The Basic Comparison Test requires that you be familiar with the behavior of a good number of series. In many situations a p-series is a good choice. Unless the behavior of the series you are using for the comparison is readily known, as in the case of a p-series, you should verify the convergence or divergence of the series used. Further, you should remember that the comparison need only hold for some sufficiently large value of the subscript and <u>all</u> larger values. (See Example 1 below.) The decision of when to try the Basic Comparison Test as your first test is quite dependent on the number of series you are familiar with and, therefore, on how many exercises you have tried. Implicit in this familiarity is the ability to recognize that the behavior of the series

$$\sum \frac{3n - 2}{n^2 + 1} \qquad \text{and} \qquad \sum \frac{1}{n}$$

is very similar. (See Example 1.)

The Root Test and the Ratio Test are established using a geometric series and the Basic Comparison Test. If the limit examined in either test is 1, the test is inconclusive. The Root Test is normally used only in those series involving some kind of exponential. The Ratio Test is also useful for such series as well as those involving factorials. The Root Test cannot normally be used on series where the general term is a rational function (quotient of two polynomials); and, for such series, the Ratio Test is always inconclusive.

In the statement of the Limit Comparison Test:

If Σa_n and Σb_n are series of positive terms and

$\lim\limits_{n \to \infty} \dfrac{a_n}{b_n} = L$ with $0 < L < \infty$, then Σa_n converges if

and only if Σb_n converges,

you should take special note of the fact that $L \neq 0$. As with the Basic Comparison Test, you need to be familiar with a fair number of series in order to be able to use this test. You will probably have to use L'Hospital's Rule to evaluate the limit. Suppose that Σa_n is the series you are investigating and Σb_n is the series you select for the comparison. You know that $a_n \to 0$ or else the series is known to diverge. Also, you want $b_n \to 0$ or else you can't satisfy the hypothesis of $0 < L < \infty$. Thus, L'Hospital's Rule may be used. Finally, recall again that there may be a difference between

$$\lim_{n \to \infty} \frac{f(n)}{g(n)} \qquad \text{and} \qquad \lim_{x \to \infty} \frac{f(x)}{g(x)} .$$

In each of the following examples we determine whether or not the given series converges. The solutions are constructed to example as many of the tests as possible, so a given solution may not be the most expedient solution or the way you would do the problem. In some instances a number of solutions are given.

Example 1. $\displaystyle\sum_{n=1}^{\infty} \frac{3n - 2}{n^2 + 1}$.

Solution. Since $\lim\limits_{n \to \infty} \dfrac{3n - 2}{n^2 + 1} = 0$, this series may either diverge or converge and

we must use other tests. The general term in the series is a rational function, so the Ratio Test is inconclusive and the Root Test may be too hard to use. But, the Basic Comparison Test used with a p-series may be just what we want. Since the "order of magnitude" of $(3n - 2)/(n^2 + 1)$ is $1/n$ and $\Sigma 1/n$ diverges, we will try to show the series is divergent. For our comparison we will use $\Sigma \frac{c}{n}$ and let the inequalities suggest a good choice for the constant $c > 0$ (even though $c = 3$ appears at this point to be a good bet).

$$\frac{c}{n} < \frac{3n - 2}{n^2 + 1}$$

$$cn^2 + c < 3n^2 - 2n$$

$$0 < (3 - c)n^2 - 2n - c$$

This last inequality is never true for $n \geq 1$ if $c \geq 3$. (Verify this claim.) Let us try $c = 2$. Continuing, we have:

$$0 < n^2 - 2n - 2$$
$$3 < (n - 1)^2$$
$$\sqrt{3} < (n - 1)$$
$$1 + \sqrt{3} < n$$

Since these steps are reversible, we know that if $c = 2$, then

$$\frac{2}{n} < \frac{3n - 2}{n^2 + 1}$$

for $n \geq 3$ and the original series diverges.

The same reasoning that led us to the choice of $\Sigma \frac{c}{n}$ above also applies for the Limit Comparison Test. Here, however, the choice of c is not critical.

$$\lim_{n \to \infty} \frac{(3n - 2)/(n^2 + 1)}{c/n} = \lim_{n \to \infty} \frac{3n^2 - 2n}{c(n^2 + 1)} = \lim_{n \to \infty} \frac{3 - \frac{2}{n}}{c(1 + \frac{1}{n^2})} = \frac{3}{c} .$$

Since $0 < \frac{3}{c} < \infty$, any value of c (positive of course) will do and again the original series diverges.

Not many would consider using the integral test here but notice how easily it works.

$$\lim_{n \to \infty} \int_1^n \frac{3x - 2}{x^2 + 1} \, dx = \lim_{n \to \infty} \left. \left(\frac{3}{2} \log|x^2 + 1| - 2 \arctan x \right) \right|_1^n$$

$$= \lim_{n \to \infty} \left(\frac{3}{2} \log|n^2 + 1| - 2 \arctan n \right) - \frac{3}{2} \log 2 + 2 \arctan 1$$

$$= \lim_{n \to \infty} \frac{3}{2} \log|n^2 + 1| - \pi - \frac{3}{2} \log 2 + \frac{\pi}{2} = \infty.$$

Example 2. $\displaystyle\sum_{n = 1}^{\infty} \frac{n!}{(2n + 3)!}$.

Solution. This series cries out for the Ratio Test which is fortunate as none of the other tests seem to apply.

$$\lim_{k \to \infty} \frac{(k + 1)! \big/ (2k + 5)!}{k! \big/ (2k + 3)!} = \lim_{k \to \infty} \frac{(k + 1)!}{k!} \cdot \frac{(2k + 3)!}{(2k + 5)!}$$

$$= \lim_{k \to \infty} \frac{k + 1}{(2k + 4)(2k + 5)} = 0 .$$

Thus the series converges.

To insure that you are at ease with the factorial notation, see if you can use

$$2 \cdot 4 \cdot 6 \cdot \ldots (2n - 2)(2n) = 2^n \cdot n!$$

to show that

$$\frac{n!}{(2n + 3)!} = \frac{1}{1 \cdot 3 \cdot 5 \cdot \ldots \cdot (2n - 1) \cdot 2^n \cdot (2n + 1)(2n + 2)(2n + 3)} .$$

Hmm? Now the Basic Comparison Test doesn't look so useless here. Namely,

$$\frac{n!}{(2n + 3)!} < \frac{1}{(2n + 1)(2n + 2)(2n + 3)} < \frac{1}{(2n + 1)^3}$$

for $n \geq 1$ and we can show that

$$\sum_{n = 1}^{\infty} \frac{1}{(2n + 1)^3}$$

converges by the Integral Test or the Limit Comparison Test.

Example 3. $\displaystyle\sum_{k=2}^{\infty} \frac{\log k}{2^k}$.

Solution. We begin this time with the Root Test because of the presence of 2^k.

$$\lim_{k \to \infty} \left(\frac{\log k}{2^k}\right)^{1/k} = \frac{1}{2} \lim_{k \to \infty} (\log k)^{1/k}.$$

There are a variety of ways to evaluate this limit. As a matter of review, let us consider them.

First, $1 \leq \log k \leq k$ for $k \geq 3$ so that

$$1 \leq (\log k)^{1/k} \leq k^{1/k}$$

and the Pinching Theorem gives us

$$\lim_{k \to \infty} (\log k)^{1/k} = 1.$$

Or, we could invoke L'Hospital's Rule as follows.

Let $y = (\log x)^{1/x}$.

Then, $\log y = \frac{1}{x} \log(\log x),$

$$\lim_{x \to \infty} \frac{\log(\log x)}{x} = \lim_{x \to \infty} \frac{1}{x \log x} = 0,$$

and $\displaystyle\lim_{x \to \infty} \log y = 0$. Thus, $\displaystyle\lim_{x \to \infty} y = e^0 = 1$.

In either event we have

$$\lim_{k \to \infty} \left(\frac{\log k}{2^k}\right)^{1/k} = \frac{1}{2}.$$

The series converges.

Had we chosen to try the Ratio Test we would have found

$$\lim_{k \to \infty} \frac{\log(k + 1)\big/2^{k + 1}}{\log(k)\big/2^k} = \frac{1}{2} \lim_{k \to \infty} \frac{\log(k + 1)}{\log k} = \frac{1}{2}.$$

Here we appealed to L'Hospital's Rule to show that

$$\lim_{x \to \infty} \frac{\log(x + 1)}{\log x} = \lim_{x \to \infty} \frac{x}{x + 1} = 1.$$

An interesting question may occur to you at this point. Do the Ratio Test and Root Test always agree? In particular, if one test is inconclusive (limit is one), will the other test be inconclusive? The answer is no.

Example 4. $\displaystyle\sum_{k = 1}^{\infty} \frac{2}{4 + \sqrt{k}}$.

Solution. The form of the general term in the series suggests that neither the Root Test nor the Ratio Test will prove conclusive. For the Basic Comparison Test we will try $\Sigma c/\sqrt{k}$ for some $c > 0$ and try to show our series divergent.

$$\frac{c}{\sqrt{k}} < \frac{2}{4 + \sqrt{k}}$$

$$4c + c\sqrt{k} < 2\sqrt{k}$$

$$4c \qquad < (2 - c)\sqrt{k}$$

It appears now that any choice of $0 < c < 2$ will work. For example, if we select $c = 3/2$

$$\frac{3}{2\sqrt{k}} < \frac{2}{4 + \sqrt{k}}$$

for $k > 144$ and the series diverges.

As in Example 1, the choice for $c > 0$ is not critical when we apply the Limit Comparison Test.

$$\lim_{k \to \infty} \frac{c/\sqrt{k}}{2/(4 + \sqrt{k})} = \lim_{k \to \infty} \frac{c(4 + \sqrt{k})}{2\sqrt{k}} = \frac{c}{2} \lim_{k \to \infty} \left(1 + \frac{4}{\sqrt{k}}\right) = \frac{c}{2} .$$

Again, it appears that the Limit Comparison Test is easier to use than the Basic Comparison Test.

Example 5. $\displaystyle\sum_{n = 2}^{\infty} \frac{\log n}{n\sqrt{n}}$.

Solution. The Ratio Test and Root Test are inconclusive. Let us try the Basic Comparison Test with a p-series. Specifically, let us try to show the series is convergent by comparing the series to $\Sigma 1/n^p$ for some $p > 1$ to be determined.

$$\frac{\log n}{n\sqrt{n}} < \frac{1}{n^p}$$

$$\log n < n^{\frac{3}{2} - p} \qquad\qquad (1)$$

We appeal to the following fact: If $f(a) > g(a)$ for some number a and $f'(x) \geq g'(x)$ for $x > a$, then $f(x) > g(x)$ for $x \geq a$. Thus, we need to determine a (and p) such that

$$\frac{1}{x} < (\frac{3}{2} - p)x^{\frac{1}{2} - p},$$

or

$$1 < (\frac{3}{2} - p)x^{\frac{3}{2} - p} \qquad\qquad \text{for } 0 < a < x. \qquad (2)$$

For this last inequality to be true we must have $0 < p < \frac{3}{2}$. Since we began with $p > 1$, we may only select a value for p such that $1 < p < 3/2$. For simplicity we select $p = 5/4$. Now we need a value for a where inequalities (1) and (2) hold. Looking at (2) first,

$$1 < \frac{1}{4}a^{1/4},$$

a must be greater than 256. Unfortunately inequality (1) is not true for $n = 257$. However, $e^{12} > 256 = 2^8$ and inequality (1) holds for $n > e^{12}$, as you should verify. We conclude that the original series converges since

$$\frac{\log n}{n\sqrt{n}} < \frac{1}{n^{5/4}}$$

for $n > e^{12}$.

This is all fine and dandy, but there must be a simpler way to do this problem. Once again let us see if the Limit Comparison Test is easier. Assume $p > 1$.

$$\lim_{n \to \infty} \frac{\log n / n\sqrt{n}}{1/n^p} = \lim_{n \to \infty} \frac{\log n}{n^{3/2 - p}}$$

This limit is infinite unless $p < 3/2$. For $1 < p < 3/2$ we may use L'Hospital's Rule.

$$\lim_{x \to \infty} \frac{\log x}{x^{3/2 - p}} = \lim_{x \to \infty} \frac{1/x}{(\frac{3}{2} - p)x^{1/2 - p}}$$

$$= \frac{1}{\frac{3}{2} - p} \lim_{x \to \infty} \frac{1}{x^{3/2 - p}} = 0,$$

no matter what p between 1 and 3/2 we may select. Since the limit is zero, the Limit Comparison Test does not apply. You should take heed of this instance where the Basic Comparison Test works but the Limit Comparison Test (for the same series of comparison) does not work.

Finally, let us consider the Integral Test. We integrate by parts with $u = \log x$ and $dv = x^{-3/2} dx$ to obtain

$$\int_2^n x^{-3/2} \log x \, dx = -2x^{-1/2} \log x \Big|_2^n + \int_2^n 2x^{-3/2} dx$$

$$= (-2x^{-1/2} \log x - 4x^{-1/2}) \Big|_2^n$$

$$= -2 \frac{\log n + 2}{\sqrt{n}} + \sqrt{2}(2 + \log 2).$$

L'Hospital's Rule gives us

$$\lim_{x \to \infty} \frac{\log x + 2}{\sqrt{x}} = \lim_{x \to \infty} \frac{1/x}{1/(2\sqrt{x})} = \lim_{x \to \infty} \frac{2}{\sqrt{x}} = 0.$$

Therefore,

$$\lim_{n \to \infty} \int_2^n \frac{\log x}{x\sqrt{x}} \, dx = \sqrt{2}(2 + \log 2)$$

and the series converges.

SECTION 13.5

A series in which successive terms have opposite signs is called an alternating series.

$$a_1 - a_2 + a_3 + \cdots + (-1)^{n-1} a_n + \cdots, \qquad a_i > 0 \text{ for } i \geq 1. \qquad (1)$$

Associated with the series in (1) is the series with the signs all the same.

$$a_1 + a_2 + a_3 + \cdots + a_n + \cdots . \qquad (2)$$

If series (2) converges, then series (1) converges and we say that series (1) converges absolutely. If series (2) diverges, then series (1) may yet converge or diverge. If series (2) diverges and series (1) converges, we say that series (1) converges conditionally (conditional on the presence of the alternating signs). If series (1) converges, then series (2) may either converge or diverge. And, accordingly, we say that series (1) converges absolutely or it converges conditionally.

It is rather simple to test the convergence of an alternating series. Since we normally want to know if this convergence is absolute or conditional, we first test the series with the signs the same as in (2). If we obtain convergence by the tests of the previous Sections we are done, for then series (1) converges absolutely. Otherwise, our series (1) will either converge (conditionally) or diverge. At this level of study we have but one test for convergence at our disposal - the Alternating Series Theorem:

If a_k is a decreasing sequence of positive numbers and

$a_k \to 0$, then $\Sigma(-1)^k a_k$ converges.

If the hypotheses of this theorem are met and we have previously shown that Σa_k is divergent, then the series converges conditionally. If the hypotheses of the theorem are not satisfied and Σa_k diverges, there are 2 possibilities.

1. a_k does not approach zero.

2. a_k approaches zero but a_k is not a decreasing sequence.

Naturally, the first possibility tells you the series is divergent. The second possibility is more disquieting. There are three possibilities.

(a) You need to use the definition of a series and examine the limit of the partial sums of the series.

(b) You have made an error.

(c) Your instructor has not given you the question (s)he meant to or (s)he
    doesn't want you to write a perfect exam.

Unfortunately, too many students begin with option (c).

In each of the following examples we determine if the given series converges
absolutely, converges conditionally, or diverges.

<u>Example 1.</u> $\displaystyle\sum_{k=2}^{\infty} (-1)^k \frac{2}{4 + \sqrt{k}}$.

<u>Solution.</u> In Example 4 of the previous Section we showed that

$$\sum_{k=1}^{\infty} \frac{2}{4 + \sqrt{k}}$$

diverges. Therefore, we turn to the Alternating Series Theorem. Since

$$a_k = \frac{2}{4 + \sqrt{k}}$$

approaches 0, we need to show that

$$\frac{2}{4 + \sqrt{k + 1}} < \frac{2}{4 + \sqrt{k}} \qquad (3)$$

or

$$2(4 + \sqrt{k}) < 2(4 + \sqrt{k + 1})$$

or

$$\sqrt{k} < \sqrt{k + 1} .$$

This last inequality is true and the steps taken to derive it are reversible, so
inequality (3) is true. Therefore, the series converges conditionally.

You should remember that had inequality (3) seemed too formidable, you could
have considered the derivative of $f(x) = 2/(4 + \sqrt{x})$ even though this technique is
not always easier.

<u>Example 2.</u> $\displaystyle\sum_{n=1}^{\infty} (-1)^{n+1} \frac{n}{n^2 + 100}$.

<u>Solution.</u> Since $\displaystyle\int_{1}^{n} \frac{x}{x^2 + 100} dx = \frac{1}{2} \log|n^2 + 100| - \frac{1}{2} \log|101|$

and $\displaystyle\lim_{n \to \infty} \log \sqrt{\frac{n^2 + 100}{101}} = \infty$, the series

$$\sum_{n=1}^{\infty} \frac{n}{n^2 + 100}$$

diverges. Rather easily, we have $\displaystyle\lim_{n \to \infty} \frac{n}{n^2 + 100} = 0$. Thus, let us try to show

that $a_n > a_{n+1}$ for $a_n = n/(n^2 + 100)$.

$$\frac{n}{n^2 + 100} > \frac{n + 1}{(n + 1)^2 + 100}$$

$$n(n + 1)^2 > (n + 1)n^2 + 100$$

$$0 > n(n + 1)(-1) + 100$$

$$n(n + 1) > 100$$

$$n > 9 .$$

It is false then that $a_n > a_{n+1}$ for $n \geq 1$. However, since the sequence a_n is
decreasing for $n \geq 10$ we may yet use the Alternating Series Theorem to conclude
that the series converges (conditionally). Can you justify this slight bending
of the hypothesis of the theorem? What changes are needed in the proof of the
theorem if we reform the theorem as follows?

> If a_n is a sequence of positive numbers which approaches zero,
> and, for some integer N, $a_{n+1} < a_n$ for all $n \geq N$, then the series
>
> $$\sum_{n=1}^{\infty} (-1)^n a_n \text{ converges.}$$

SECTIONS 13.6-7

The series expansion

$$f(x) = \sum_{n=0}^{\infty} f^{(n)}(a) \frac{(x - a)^n}{n!}$$

is usually referred to in one of two ways: the Taylor series in powers of x-a, or the Taylor series about _a_. When a = 0 we call this series the Maclaurin series expansion, viz.,

$$f(x) = \sum_{n=0}^{\infty} f^{(n)}(0) \frac{x^n}{n!} .$$

This expansion may be referred to in a number of ways: the Maclaurin series, the Taylor series in powers of x, or the Taylor series about zero.

The region of convergence for a series involving some variable x consists of those values of x for which the series converges. This topic is discussed in Section 13.9. As you would expect, the region of convergence for a Taylor series consists of those values of x for which the remainder $R_n(x)$ approaches zero. However, a power series represents a function only when the series converges. For example, the function

$$f(x) = \frac{1}{1 - x}$$

is defined for all $x \neq 1$. However, the Maclaurin series expansion for f:

$$\sum_{n=0}^{\infty} x^n, \qquad \text{(Remember your geometric series.)}$$

converges only for $-1 < x < 1$ rather than the entire domain of f.

The only source of possible difficulty in finding the Taylor series expansion about a given point for a given function lies in the determination of the general form for $f^{(n)}(a)$. As demonstrated by the examples that follow, there are three helpful guides to keep in mind:

(1) Sometimes it is easier to find the general form of $f^{(n)}(x)$ and then substitute in _a_ rather than trying to determine the general form of $f^{(n)}(a)$ from the first few values. This is especially true if you have made some arithmetical simplifications. We suggest that you don't.

(2) Since successive terms are generated by differentiation, there may be a build up of coefficients very much like factorials. For example, $\frac{d^4}{dx^4}(x^n) = n(n-1)(n-2)(n-3)x^{n-4} = n!x^{n-4}/(n-4)!$. Alternating signs come about in a similar way. Be careful that the exponent you use

is appropriate. That is, for $n \geq 1$, $(-1)^n$ gives you minus, plus, minus, $\cdots$, whereas $(-1)^{n+1}$ gives you plus, minus, plus, $\cdots$.

(3) It is quite possible that the first few values of $f^{(n)}(a)$ do not fit into the general form. In this case you simply write these terms separate from the general sum.

One must use mathematical induction on the form of $f^{(n)}(x)$ to prove that the expression for $f^{(n)}(a)$ is correct. We shall not do so, nor is it expected of you.

Example 1. Determine the Taylor series expansion about $a = 1$ for the function $f(x) = x^2 + e^{-x}$.

Solution. We organize our work in a table, a practice you may find useful to mimic. For more complicated functions it is sometimes necessary to calculate the derivatives on the side.

n	$f^{(n)}(x)$	$f^{(n)}(1)$
0	$x^2 + e^{-x}$	$1 + e^{-1}$
1	$2x - e^{-x}$	$2 - e^{-1}$
2	$2 + e^{-x}$	$2 + e^{-1}$
3	$- e^{-x}$	$- e^{-1}$
4	e^{-x}	e^{-1}
5	$-e^{-x}$	$- e^{-1}$

It certainly appears that $f^{(n)}(x) = (-1)^n e^{-x}$ for $n \geq 3$.

Thus, $f(x) = f(1) + f'(1)(x - 1) + f''(1)\dfrac{(x - 1)^2}{2} + \displaystyle\sum_{n = 3}^{\infty} f^{(n)}(1)\dfrac{(x - 1)^n}{n!}$

$= 1 + e^{-1} + (2 - e^{-1})(x - 1) + (2 + e^{-1})\dfrac{(x - 1)^2}{2} + \displaystyle\sum_{n = 3}^{\infty} \dfrac{(-1)^n}{e}\dfrac{(x - 1)^n}{n!}.$

There is a simpler way in which we could have done this example. Remembering the Maclaurin expansion for e^x:

$$e^x = \sum_{n = 0}^{\infty} \frac{x^n}{n!} ,$$

we could have written

$$e^{-x} = e^{-(x - 1) - 1} = e^{-1} \cdot e^{-(x - 1)} = e^{-1} \sum_{n = 0}^{\infty} \frac{(-1)^n (x - 1)^n}{n!}. \tag{1}$$

Further, as it is a fairly quick calculation to show that

$$x^2 = 1 + 2(x - 1) + (x - 1)^2, \tag{2}$$

we could have simply added the series in (1) and (2) to obtain the desired expansion.

This approach to the problem raises two points. You would do well to memorize the Maclaurin series expansions for certain functions. A minimal list includes $\sin x$, $\cos x$, $1/(1 - x)$, $\log(1 + x)$, and $\arctan x$. (Section 13.8 discusses these last two.) Next, the Taylor series expansion about $\underline{a}$ of the polynomial $p(x)$ of degree N is always a polynomial of the same degree. Can you prove this? (Big hint! Show that the Nth derivative of $p(x)$ evaluated at $x = a$ is never zero and all higher order derivatives are always zero.)

Example 2. Determine the Maclaurin series expansion for $f(x) = x/(1 + x^2)^2$.
Solution.

n	$f^{(n)}(x)$	$f^{(n)}(0)$
0	$\dfrac{x}{(1 + x^2)^2}$	0
1	$\dfrac{1 - 3x^2}{(1 + x^2)^3}$	1
2	$\dfrac{-12x(1 - x^2)}{(1 + x^2)^4}$	0
3	$\dfrac{-12(5x^4 - 10x^2 + 1)}{(1 + x^2)^5}$	-12

Fearful that $f^{(4)}(0) = 0$ and that our patience is ebbing, let us ask if there is another way to do the problem. We notice that

$$\int \frac{x}{(1 + x^2)^2} \, dx = -\frac{1}{2} \cdot \frac{1}{1 + x^2} + C.$$

Also,

$$\int \frac{1}{1 + x^2} \, dx = \arctan x + C.$$

Thus,

$$\frac{d^2}{dx^2} (\arctan x) = \frac{-2x}{(1 + x^2)^2}.$$

The Maclaurin series expansion for arctan x is

$$x - \frac{x^3}{3} + \frac{x^5}{5} - \cdots = \sum_{n = 0}^{\infty} (-1)^n \frac{x^{2n + 1}}{2n + 1}. \qquad (3)$$

Therefore, the Maclaurin series expansion for $f(x)$ is minus one-half times the second derivative of this expression.

$$f(x) = -\frac{1}{2} \sum_{n = 1}^{\infty} (-1)^n (2n + 1) \frac{(2n)x^{2n - 1}}{2n + 1} \qquad (4)$$

$$= \sum_{n = 0}^{\infty} (-1)^n (n + 1) x^{2n + 1}. \qquad (5)$$

You should carefully check the adjustments made in the index in (3), (4), and (5). Finally, notice that we could have treated $1/(1 + x^2)$ as a geometric series

$$\sum_{n = 0}^{\infty} (-1)^n (x^2)^n$$

and differentiated but once.

The lesson of this last example is that although there are but a few series we need to memorize, we can have many more at our fingertips by bearing in mind the derivatives and anti-derivatives of those functions. This technique is discussed in Section 13.10.

The Lagrange form of the remainder for the Taylor series for $f(x)$ about _a_ is

$$R_n = \frac{f^{(n)}(c)(x - a)^n}{n!} \qquad \text{with c between a and x.}$$

Normally we need the general form of $f^{(n)}(x)$ when using Taylor series to make approximations. The approximations can only be made in the region of convergence of the series to be assured that $R_n \to 0$. However, we do not know if R_n is strictly decreasing. To be on the safe side when computing how many terms of the series to use for the approximation, we are usually a bit generous in determining a bound for $f^{(n)}(c)$.

Example 3. Use a Taylor series to approximate $\sqrt{2}$ to an accuracy of .001.
Solution. We will use $f(x) = \sqrt{x + 1}$ expanded about $a = 0$. This is not the only possible choice.

n	$f^{(n)}(x)$	$f^{(n)}(0)$
0	$(x + 1)^{1/2}$	1
1	$\frac{1}{2}(x + 1)^{-1/2}$	1/2
2	$-\frac{1}{4}(x + 1)^{-3/2}$	-1/4
3	$\frac{3}{8}(x + 1)^{-5/2}$	+3/8
4	$-\frac{15}{16}(x + 1)^{-7/2}$	-15/16

It seems reasonable that

$$f^{(n)}(x) = (-1)^{n + 1}\left(\frac{1 \cdot 3 \cdot 5 \cdots (2n - 3)}{2^n}\right)(x + 1)^{-\frac{2n - 1}{2}}$$

for $n \geq 2$. To be surer, we can calculate $f^{(5)}(x)$ as $\frac{105}{32}(x + 1)^{-9/2}$ and verify that this result fits our formula. Thus

$$f(x) = \sqrt{x + 1} = 1 + \frac{x}{2} + \sum_{n = 2}^{\infty} (-1)^{n + 1}\left(\frac{1 \cdot 3 \cdot 5 \cdots (2n - 3)}{2^n}\right) \frac{x^n}{n!} \,.$$

Before we make our approximation, we should determine values of x for which this series converges. We will use the Ratio Test. We want to determine x such that

$$\lim_{n \to \infty} \left|\frac{a_{n + 1}}{a_n}\right| < 1,$$

where a_n is the nth term in the series.

$$\lim_{n \to \infty} \left| \frac{(-1)^{n+2}\left(\dfrac{1 \cdot 3 \cdots (2n-1)}{2^{n+1}}\right) \dfrac{x^{n+1}}{(n+1)!}}{(-1)^{n+1}\left(\dfrac{1 \cdot 3 \cdots (2n-3)}{2^n}\right) \dfrac{x^n}{n!}} \right| = \lim_{n \to \infty} \left| \frac{2n-1}{2} \cdot \frac{x}{n+1} \right| = |x|.$$

So when $|x| < 1$ the above Taylor series converges to $\sqrt{1+x}$.

Now we need to determine n such that

$$R_n = \left| f^{(n)}(c) \frac{x^n}{n!} \right| < .001 \qquad\qquad (c \text{ between } 0 \text{ and } x)$$

for the selected value of x.

If we use $x = -\frac{1}{2}$ so that $\sqrt{1+x} = \frac{\sqrt{2}}{2}$, we really want $R_n < \frac{1}{2}(.001)$. In this case, we have

$$\left| f^{(n)}(c) \frac{x^n}{n!} \right| = \left| (-1)^{n+1} \frac{1 \cdot 3 \cdot 5 \cdots (2n-3)}{2^n}(c+1)^{-\frac{2n-1}{2}} \left(-\frac{1}{2}\right)^n \frac{1}{n!} \right|$$

$$= \frac{1 \cdot 3 \cdot 5 \cdots (2n-3)}{2^{2n}n!} \cdot \frac{1}{(1+c)^{(2n-1)/2}} \qquad\qquad \left(-\frac{1}{2} < c < 0\right)$$

$$\leq \frac{1 \cdot 3 \cdot 5 \cdots (2n-3)}{2^{2n} \cdot n!} \cdot 2^{(2n-1)/2}$$

$$= \frac{1 \cdot 3 \cdot 5 \cdots (2n-3)}{2^{(n+1/2)}n!} < \frac{1 \cdot 3 \cdot 5 \cdots (2n-3)}{2^n \cdot n!} = A.$$

On the other hand, if we use $x = \frac{1}{49}$ so that $\sqrt{1+x} = \frac{5}{7}\sqrt{2}$, we really want $R_n < \frac{5}{7}(.001)$. In this case, we have

$$\left| f^{(n)}(c) \, \frac{x^n}{n!} \right| = \left| (-1)^{n+1} \frac{1 \cdot 3 \cdot 5 \cdots (2n-3)}{2^n}(c+1)^{-\frac{2n-1}{2}} \left(\frac{1}{49}\right)^n \frac{1}{n!} \right|$$

$$= \frac{1 \cdot 3 \cdot 5 \cdots (2n-3)}{2^n \cdot n! \, 7^{2n}} \frac{1}{(1+c)^{(2n-1)/2}} \qquad (0 < c < \frac{1}{49})$$

$$\leq \frac{1 \cdot 3 \cdot 5 \cdots (2n-3)}{2^n \, n! \, 7^{2n}} = B$$

Clearly, B is smaller than A and far fewer terms in the series will be needed if we use $x = \frac{1}{49}$. If $n = 2$, $B < .00005$ so $\sqrt{2}$ to an accuracy of .001 is given by

$$\frac{7}{5}[1 + \frac{1}{2}(\frac{1}{49})] = \frac{99}{70} = 1.4143.$$

A good bit of the work in this example could have been avoided if we had relied on the Binomial series as given in Section 13.11. The Binomial series is especially useful in Probability and Statistics.

Example 4. Determine the Taylor series about the point $a = \pi/4$ for the function $f(x) = x \sin 2x$.
Solution.

n	$f^{(n)}(x)$	$f^{(n)}(\pi/4)$
0	$x \sin 2x$	$\pi/4$
1	$\sin 2x + 2x \cos 2x$	1
2	$-4x \sin 2x + 4 \cos 2x$	$-\pi$
3	$-12 \sin 2x - 8x \cos 2x$	-12
4	$16x \sin 2x - 32 \cos 2x$	4π
5	$80 \sin 2x + 32x \cos 2x$	$+80$
6	$-64x \sin 2x + 192 \cos 2x$	-16π
7	$-448 \sin 2x - 128x \cos 2x$	-448
8	$256x \sin 2x - 1024 \cos 2x$	64π
9	$2304 \sin 2x + 512x \cos 2x$	$+2304$

The general form for $f^{(n)}(\pi/4)$ needs to be examined for the two separate cases of when n is even and when n is odd. For n even, the terms alternate in sign and increase by a factor of 4.

$$a_{2n} = (-1)^n (\tfrac{\pi}{4})(4)^n.$$

For n odd, the terms alternate in sign and once we factor out all the twos we find

$$a_{2n + 1} = (-1)^n (2n + 1)(4)^n.$$

Thus,

$$f(x) = \sum_{n = 0}^{\infty} f^{(n)}(\pi/4) \frac{(x - \pi/4)^n}{n!}$$

$$= \sum_{k = 0}^{\infty} \left(f^{(2k)}(\pi/4) \frac{(x - \pi/4)^{2k}}{(2k)!} + f^{(2k + 1)}(\pi/4) \frac{(x - \pi/4)^{2k + 1}}{(2k + 1)!} \right)$$

$$= \sum_{k = 0}^{\infty} \left((-1)^k (\pi/4) \cdot 4^k \cdot \frac{(x - \pi/4)^{2k}}{(2k)!} + (-1)^k (2k + 1) \cdot 4^k \cdot \frac{(x - \pi/4)^{2k + 1}}{(2k + 1)!} \right).$$

It may be interesting to simplify this expression.

$$= \sum_{k = 0}^{\infty} (-1)^k \frac{(x - \pi/4)^{2k}}{(2k)!} 4^k (\tfrac{\pi}{4} + x - \tfrac{\pi}{4})$$

$$= x \sum_{k = 0}^{\infty} (-1)^k \frac{(2x - \pi/2)^{2k}}{(2k)!}$$

$$= x \cos(2x - \pi/2) = x \sin 2x.$$

It appears that we could have solved this problem by multiplying the series expansions for x and sin 2x which we should already know.

SECTIONS 13.9-10

A Taylor series expansion for a function $f(x)$ is only useful if we know the interval (or region) of convergence for the series (that is, those values of x for which R_n approaches zero). More generally, a power series is only useful if we know where it converges. The radius of convergence may be found by applying either the Ratio Test or the Root Test. For the Ratio Test we want to find those values of x for which

$$\lim_{n \to \infty} \left| \frac{f^{(n+1)}(a)(x-a)^{n+1}/(n+1)!}{f^{(n)}(a)(x-a)^n/n!} \right| = \lim_{n \to \infty} \left| \frac{f^{(n+1)}(a)}{f^{(n)}(a)} \right| \frac{|x-a|}{n+1} < 1$$

The series diverges for those values of x, if any, for which this limit is greater than 1. The radius of convergence may be zero (the series converges only for $x = a$), infinite (the series converges for all real numbers x), or some positive number R (the series converges for x between $a - R$ and $a + R$). In this third case, we need to decide on the convergence or divergence of the series at the endpoints $x = a - R$ and $x = a + R$. The two specific series that arise at these endpoints are frequently closely related. One may be the alternating form of the other. It really is a toss up as to which of the series to treat first. Whichever way your coin lands, there is no point trying to use either the Ratio Test or Root Test to determine the convergence of these series. The reason is simple. The question of endpoint behavior arose because the Root Test or Ratio Test was inconclusive at these points. Finally, if some of the coefficients in the Taylor series expansion are zero we look at the ratio of successive non-zero terms (see Example 2).

Example 1. Determine the region of convergence for the Taylor series

$$\sum_{n=0}^{\infty} (-1)^n \frac{(x-1)^n}{n \cdot 2^n} .$$

Solution.

$$\lim_{n \to \infty} \left| \frac{(-1)^{n+1}(x-1)^{n+1}/(n+1)2^{n+1}}{(-1)^n(x-1)^n/n\,2^n} \right| = \lim_{n \to \infty} \frac{n}{2(n+1)} |x-1| = \frac{1}{2}|x-1| \, .$$

The series converges for $\frac{1}{2}|x-1| < 1$ and diverges for $\frac{1}{2}|x-1| > 1$. Had the question been to find the radius of convergence, we would be done. The radius is 2. We need to examine the endpoints in order to completely specify the region of convergence. When $\frac{1}{2}|x-1| = 1$ we have $x = 3$ and $x = -1$.

For $x = 3$ the series becomes

$$\sum_{n=0}^{\infty} (-1)^n \frac{1}{n}$$

which converges. (Why?) For $x = -1$ the series is the Harmonic series which diverges. The region of convergence is $-1 < x \le 3$.

Example 2. Determine the region of convergence for the Taylor series

$$\sum_{n=0}^{\infty} (-1)^n \frac{x^{3n}}{n^2} \, .$$

Solution.

$$\lim_{n \to \infty} \left| \frac{(-1)^{n+1}x^{3n+3}/(n+1)^2}{(-1)^n x^{3n}/n^2} \right| = \lim_{n \to \infty} \left(\frac{n}{n+1}\right)^2 |x|^3 = |x^3| \, .$$

The series converges for $|x^3| < 1$ or $-1 < x < 1$. If $x = -1$, we have the p-series with $p = 2$. For $x = 1$, we find the alternating form of this series. The region of convergence is $-1 \le x \le 1$.

Call the series in this example $g(x)$. Can you show that

$$x \cdot g'(x) + 3 \log|1 + x^3| = 0?$$

Taylor series may be used to approximate integrals and derivatives.

Example 3. Estimate $\int_0^{1/2} \sin(x^2)dx$ to an accuracy of .0001.

<u>Solution</u>. If the error in approximating $f(x)$ is ε, then the error incurred by using this estimate to approximate

$$\int_a^b f(x)dx$$

is less than or equal to $(b - a)\varepsilon$. Recall the Maclaurin series for $\sin x$. We know that

$$\sum_{k=0}^{n-1} (-1)^k \frac{(x^2)^{2k+1}}{(2k+1)!}$$

is an estimate for $\sin(x^2)$ that is in error by

$$\varepsilon = \frac{(c^2)^{2n+1}}{(2n+1)!}$$

for some $0 < c < x$. Certainly then

$$\varepsilon < \frac{1}{(2n+1)!} \left(\frac{1}{4}\right)^{2n+1}.$$

We want $\varepsilon(b - a) = \dfrac{\varepsilon}{2} < .0001$. But,

$$\frac{1}{2} \cdot \varepsilon < \frac{1}{2} \cdot \frac{1}{(2n+1)!} \left(\frac{1}{4}\right)^{2n+1} < .0001$$

for $n \geq 2$. Thus,

$$\int_0^{1/2} \sin(x^2)dx \approx \int_0^{1/2} \left(\frac{x^2}{1} - \frac{x^6}{6}\right)dx = \frac{x^3}{3} - \frac{x^7}{42}\bigg|_0^{1/2} = \frac{1}{24}\left(1 - \frac{1}{224}\right) = \frac{223}{5376} = .0415.$$

<u>Example 4</u>. Approximate the derivative of $\arctan x$ at $x = .1$ to an accuracy of .0001.

<u>Solution</u>. The Maclaurin series expansion for $\arctan x$ is

$$\sum_{n=0}^{\infty} (-1)^n \frac{x^{2n+1}}{2n+1}, \qquad -1 \leq x \leq 1.$$

We may differentiate this series term-by-term to find

$$\frac{d}{dx}(\arctan x) = \sum_{n=0}^{\infty} (-1)^n x^{2n}, \qquad -1 < x < 1.$$

For $0 < c < .1$ the polynomial approximation

$$1 - x^2 + x^4$$

is in error by

$$|(-1)^3 c^6| < 10^{-6}.$$

Thus, $\arctan (.1) \approx 1 - \frac{1}{100} + \frac{1}{10000} = .9901.$

We may check this result since

$$\frac{d}{dx}(\arctan x) = \frac{1}{1 + x^2}$$

and $\frac{d}{dx}[\arctan (.1)] = \frac{100}{101} = .\overline{9900}$ (repeating decimal).

Vivos **Vectors**

14

Comments on the material in this Chapter are restricted to solutions of the following exercises from the text.

p. 577 (Section 14.1): 7

p. 588 (Section 14.3): 14b

p. 595 (Section 14.4): 9, 10, 18, 20

p. 601 (Section 14.5): 9d, 12, 14b

p. 608 (Section 14.6): 16, 20, 21

p. 614 (Section 14.7): 12, 16, 23, 24b

p. 618 (Section 14.9): 3, 4b, 4d, 6, 7, 13, 14

Following our previous practice the solutions include comments which may help you solve related problems. A vector is indicated by a bar as in $\overline{a}$. Some terminology from Linear Algebra is introduced and will be referenced in the Concluding Comments for Chapters 14 and 15.

SECTION 14.1

Solution for Exercise 7 (p. 577). First, you should verify that the coordinates for points S and R are labeled correctly. For example, line segment PR is parallel to the y-axis, so the x (and z) coordinates of any two points on this line segment are the same. The length of line segment PR is $b_2 - a_2$. If point U is two-thirds of the way from P to R, then the coordinates of U are $(a_1, a_2 + \frac{2}{3}(b_2 - a_2), a_3)$.

We may appeal to similar figures and reason that the coordinates of a point T two-thirds of the way from P to Q are

$$(a_1 + \frac{2}{3}(b_1 - a_1),\ a_2 + \frac{2}{3}(b_2 - a_2),\ a_3 + \frac{2}{3}(b_3 - a_3)).$$

Does it seem reasonable that any point on the line segment PQ may be recorded as

$$(x,\ y,\ z) = (a_1 + t(b_1 - a_1),\ a_2 + t(b_2 - a_2),\ a_3 + t(b_3 - a_3))$$

for $0 \le t \le 1$? This idea is among those developed later in the Chapter.

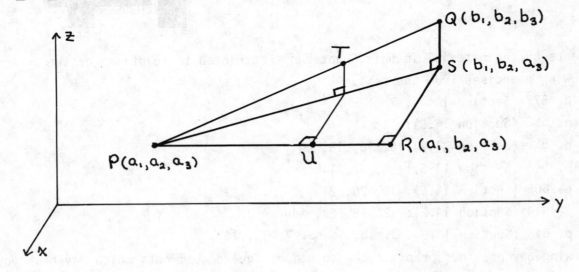

SECTION 14.3

Solution for Exercise 14b (p. 588). We use the fact that two vectors: (m, n, p) and (r, s, t), are equal iff $m = r$, $n = s$, and $p = t$. The statement

$$\overline{d} = A\overline{a} + B\overline{b} + C\overline{c} \tag{1}$$

translates into the system of equations:

$$4 = A - B - 3C$$
$$-1 = A + 3B$$
$$1 = A + 2B + C.$$

The solution for this system is $A = 26/7$, $B = -11/7$, $C = 3/7$.

We may substitute these values into equation (1) and rewrite it as

$$26\overline{a} - 11\overline{b} + 3\overline{c} - 7\overline{d} = \overline{0}.$$

This statement means that the vectors $\bar{a}$, $\bar{b}$, $\bar{c}$, and $\bar{d}$ are <u>linearly</u> <u>dependent</u>. Notice that each vector may be expressed as a linear combination of the other three. Equation (1) is one such example and the equation

$$\bar{a} = \frac{11}{26}\bar{b} - \frac{3}{26}\bar{c} + \frac{7}{26}\bar{d}$$

is another.

A set of vectors is not always linearly dependent. The vectors $\bar{u}_1$, $\bar{u}_2$, $\cdots$, $\bar{u}_n$ are said to be <u>linearly</u> <u>independent</u> iff the only solution of

$$c_1\bar{u}_1 + c_2\bar{u}_2 + \cdots + c_n\bar{u}_n = \bar{0}$$

is $c_1 = c_2 = \cdots = c_n = 0$. Easily, the vectors $\bar{i}$, $\bar{j}$, and $\bar{k}$ are linearly independent since

$$c_1\bar{i} + c_2\bar{j} + c_3\bar{k} = (c_1, c_2, c_3) = (0, 0, 0)$$

necessitates that $c_1 = c_2 = c_3 = 0$. It is not quite so clear that the vectors $\bar{i} + \bar{j}$, $2\bar{i} - \bar{k}$, and $\bar{j} + 3\bar{k}$ are linearly independent. The equation

$$c_1(\bar{i} + \bar{j}) + c_2(2\bar{i} - \bar{k}) + c_3(\bar{j} + 3\bar{k}) = \bar{0}$$

may be written as

$$(c_1 + 2c_2)\bar{i} + (c_1 + c_3)\bar{j} + (-c_2 + 3c_3)\bar{k} = \bar{0}$$

which is true only for $c_1 = c_2 = c_3 = 0$. On the other hand, the vectors $\bar{i} + \bar{j}$, $2\bar{i} - \bar{k}$, and $\bar{j} + \bar{k}$ are linearly dependent since

$$c_1(\bar{i} + \bar{j}) + c_2(2\bar{i} - \bar{k}) + c_3(\bar{j} + \bar{k}) = \bar{0}$$

is true when $c_1 = c_3 = 1$ and $c_2 = -1$.

In this Chapter and the next we handle vectors in 3-space. No collection of <u>four or more</u> such vectors can be independent. One can find no more than three linearly independent vectors in 3-space. In fact, if we have 3 independent vectors in 3-space (such as $\bar{i}$, $\bar{j}$, and $\bar{k}$ or the three noted above), then any other vector in 3-space may be expressed as a linear combination of these three vectors. This is but the tip of the iceberg known as Linear Algebra. The idea of independence provides a vehicle for generalizing the notion of dimension. Further comments in this vein appear in the Concluding Comments for Chapters 14 and 15.

SECTION 14.4

<u>Solution for Exercise 9 (p. 595)</u>. We want to prove that $\overline{a} \cdot \overline{b} = \overline{a} \cdot \overline{c}$ does not necessarily imply that $\overline{b} = \overline{c}$. Most students consider proofs to consist only of general arguments. Such is not the case. This problem requires a counterexample to <u>prove</u> that the statement "$\overline{a} \cdot \overline{b} = \overline{a} \cdot \overline{c}$ implies $\overline{b} = \overline{c}$" is not true. Sometimes the most difficult proofs to construct are those by counterexample for one is looking for a specific exception to a statement. In this case the solution is not so difficult.

We could use $\overline{a} = (2, 1, 3)$, $\overline{b} = (3, 0, -2)$ and $\overline{c} = (1, 1, -1)$. Then, $\overline{a} \cdot \overline{b} = 6 + 0 - 6 = 0$ and $\overline{a} \cdot \overline{c} = 2 + 1 - 3 = 0$, but $\overline{b} \neq \overline{c}$. You should be able to provide other (counter)examples. Notice that if we let $\overline{b} + \overline{c} = \overline{0}$ with $\overline{a} \cdot \overline{b} = \overline{0}$, many counterexamples are readily found.

It may be easier to visualize geometrically why this statement fails. One possible diagram is given on the right. Vectors $\overline{b}$ and $\overline{c}$ lie in the same plane.

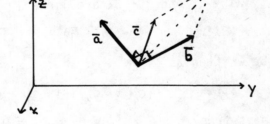

<u>Solution for Exercise 10 (p. 595)</u>. We can verify that

$$4(\overline{a} \cdot \overline{b}) = ||\overline{a} + \overline{b}||^2 - ||\overline{a} - \overline{b}||^2 \tag{1}$$

by appealing to the law of cosines.

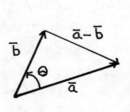

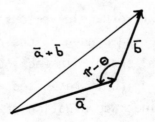

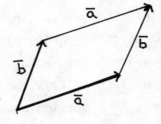

Diagram 1   Diagram 2   Diagram 3

From Diagram 1 we have

$$||\overline{a} - \overline{b}||^2 = ||\overline{a}||^2 + ||\overline{b}||^2 - 2||\overline{a}|| \, ||\overline{b}|| \cos \theta, \qquad (2)$$

and from Diagram 2 we have

$$||\overline{a} + \overline{b}||^2 = ||\overline{a}||^2 + ||\overline{b}||^2 - 2||\overline{a}|| \, ||\overline{b}|| \cos(\pi - \theta). \qquad (3)$$

Now, $\cos(\pi - \theta) = -\cos \theta$, so when we subtract equation (2) from equation (3) we obtain

$$||\overline{a} + \overline{b}||^2 - ||\overline{a} - \overline{b}||^2 = 4||\overline{a}|| \, ||\overline{b}|| \cos \theta. \qquad (4)$$

But, equations (4) and (1) are the same once we remember that

$$\overline{a} \cdot \overline{b} = ||\overline{a}|| \, ||\overline{b}|| \cos \theta.$$

A natural question arises. How does equation (1) help us show that

$$\overline{a} \perp \overline{b} \quad \text{iff} \quad ||\overline{a} + \overline{b}|| = ||\overline{a} - \overline{b}||? \qquad (5)$$

To see the connection, we need to rewrite equation (1) as

$$4\overline{a} \cdot \overline{b} = (||\overline{a} + \overline{b}|| - ||\overline{a} - \overline{b}||)(||\overline{a} + \overline{b}|| + ||\overline{a} - \overline{b}||). \qquad (6)$$

As you should verify, $||\overline{a} + \overline{b}|| + ||\overline{a} - \overline{b}|| = 0$ iff $\overline{a} = \overline{b} = 0$ in which case statement (5) is trivially true. Presuming that $||\overline{a} + \overline{b}|| + ||\overline{a} - \overline{b}|| \neq 0$, it is clear from equation (6) that $4\overline{a} \cdot \overline{b} = 0$ iff $||\overline{a} + \overline{b}|| - ||\overline{a} - \overline{b}|| = 0$, which is equivalent to statement (5).

Our problem has an interesting geometric interpretation. The vectors $\overline{a} + \overline{b}$ and $\overline{a} - \overline{b}$ are the diagonals of a parallelogram. Statement (5) says that a parallelogram is a square iff the diagonals are of equal length.

Solution for Exercise 18 (p. 595). Let us assume that the length of an edge of the cube is 1 unit. We shall position one of the cube's vertices at the origin and have each of the positive coordinate axes coincide with an edge of the cube. We select

$$\overline{i} + \overline{j} + \overline{k}$$

as the diagonal of the cube and $\overline{i} + \overline{k}$

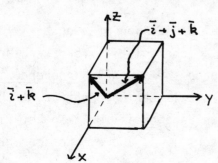

as the diagonal of the face in the xz-plane. The angle between these diagonals
is computed as follows:

$$\cos \theta = \frac{(\overline{i} + \overline{j} + \overline{k}) \bullet (\overline{i} + \overline{k})}{||\overline{i} + \overline{j} + \overline{k}|| \cdot ||\overline{i} + \overline{k}||} = \frac{2}{\sqrt{3} \cdot \sqrt{2}} = \frac{\sqrt{6}}{3},$$

so that

$$\theta \simeq \arccos(0.8165) \simeq 0.62 \text{ radians.}$$

Solution for Exercise 20 (p. 595). Following the hint in the text, we shall solve
$(\overline{a} - s\overline{b}) \bullet (\overline{b}) = 0$ for s (note that s is a scalar). We obtain

$$s = \frac{\overline{a} \bullet \overline{b}}{\overline{b} \bullet \overline{b}} = \frac{\overline{a} \bullet \overline{b}}{||\overline{b}||^2} = \frac{||\overline{a}|| \cos \theta}{||\overline{b}||}.$$

Recall that

$$\text{comp}_{\overline{b}} \, \overline{a} = ||\overline{a}|| \cos \theta$$

is the scalar component of $\overline{a}$ in the direction of $\overline{b}$. Since $\dfrac{\overline{b}}{||\overline{b}||}$ is a unit vector
in the direction of $\overline{b}$, we can see how

$$s\overline{b} = (\text{comp}_{\overline{b}} \, \overline{a}) \frac{\overline{b}}{||\overline{b}||} = ||\overline{a}|| \cos \theta \frac{\overline{b}}{||\overline{b}||}$$

is the projection of $\overline{a}$ on $\overline{b}$.

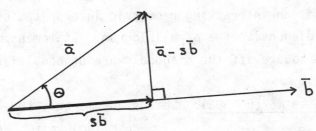

The vector $\overline{a} - s\overline{b}$ is sometimes called the orthogonal projection of $\overline{a}$ on $\overline{b}$.

These ideas are very useful. It was remarked in the solution for Exercise
14b from Section 14.3 that if three vectors in 3-space are linearly independent,
then any other vector in 3-space may be expressed as a linear combination of those
three vectors. Such a set of three vectors is known as a basis for 3-space. The
standard basis consists quite naturally of the vectors $\overline{i}$, $\overline{j}$, and $\overline{k}$. Notice that

these vectors are perpendicular (orthogonal) to one another. Such is not always
true of vectors in a basis. However, using the simple idea of a projection, we
can right things. (Angle-wise)!

In 2-space, commonly referred to as a plane, two non-zero vectors form a
basis iff they are not parallel. Suppose $\overline{a}$ and $\overline{b}$ are two such vectors. As shown
above, we may write $\overline{a}$ as the sum of its projection on $\overline{b}$ and its orthogonal
projection on $\overline{b}$. That is,

$$\overline{a} = s\overline{b} + (\overline{a} - s\overline{b}).$$

Thus, if an arbitrary vector $\overline{v}$ is expressed in terms of $\overline{a}$ and $\overline{b}$, viz.,

$$\overline{v} = \alpha \cdot \overline{a} + \beta\overline{b},$$

we may reexpress $\overline{v}$ as the sum of two perpendicular vectors ($\overline{a} - s\overline{b}$ and $\overline{b}$), viz.,

$$\overline{v} = \alpha(\overline{a} - s\overline{b}) + (\alpha s + \beta)\overline{b}.$$

This type of operation is particularly useful in physical settings where, for
example, we may consider the new basis or reference frame of $\overline{a} - s\overline{b}$ and $\overline{b}$ to be
related to the standard basis of $\overline{i}$ and $\overline{j}$ by a rotation of axes.

SECTION 14.5

Solution for Exercise 9d (p. 601). Lines ℓ_1 and ℓ_2 intersect if and only if the
values of x, y and z are the same for some t and u. Thus, we will sove the system:

$$3 + t = 1$$
$$1 - t = 4 + u$$
$$5 + 2t = 2 + u.$$

From the first two equations we learn that t = -2 and u = -1. We must verify that
these values also satisfy the third equation. They do! (If they hadn't, then
the lines ℓ_1 and ℓ_2 would intersect the vertical line (x = 1 and y = 3) at
different heights, i.e. for different values of z.)

The point of intersection is (1,3,1). The angle of intersection is given by

$$\theta = \arccos\left(\frac{\overline{a}\bullet\overline{b}}{||\overline{a}||\ ||\overline{b}||}\right)$$

with $\overline{a} = (1,-1,2)$ and $\overline{b} = (0,1,1)$. We leave the arithmetic to you.

Solution for Exercise 12 (p. 601). It should be sufficient for you to reread the
solution that was given for Exercise 7 from Section 14.1.

Solution for Exercise 14b (p. 601). From the diagram we want to determine t so
that

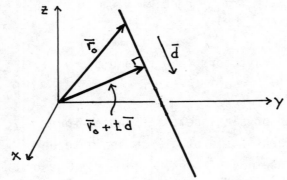

$$(\overline{r}_0 + t\overline{d})\bullet(\overline{d}) = 0.$$

If t_0 is that value of t, then

$$||\overline{r}_0 + t_0\ \overline{d}||$$

is the required distance. Our computations yield the following for

$$\overline{r}_0 = (\sqrt{3},0,0) \text{ and } \overline{d} = \frac{\sqrt{3}}{3}(1,1,1).$$

$$(\overline{r}_0 + t\overline{d})\bullet(\overline{d}) = (\sqrt{3} + t\tfrac{\sqrt{3}}{3},\ t\tfrac{\sqrt{3}}{3},\ t\tfrac{\sqrt{3}}{3})\bullet(\tfrac{\sqrt{3}}{3}, \tfrac{\sqrt{3}}{3}, \tfrac{\sqrt{3}}{3}) = 1 + \tfrac{t}{3} + \tfrac{t}{3} + \tfrac{t}{3} = 0$$

Thus, t = -1 and

$$||\overline{r}_0 + t\overline{d}|| = \sqrt{\left(3 - \tfrac{\sqrt{3}}{3}\right)^2 + \left(-\tfrac{\sqrt{3}}{3}\right)^2 + \left(-\tfrac{\sqrt{3}}{3}\right)^2} = \sqrt{\tfrac{4}{3} + \tfrac{1}{3} + \tfrac{1}{3}} = \sqrt{2}.$$

The problem may be done another way. The coordinates of any point on the line
are given by

$$x = \sqrt{3} + t\ \frac{\sqrt{3}}{3} \qquad\qquad y = t\ \frac{\sqrt{3}}{3} \qquad\qquad z = t\ \frac{\sqrt{3}}{3}.$$

The distance of this point from the origin is $\sqrt{x^2 + y^2 + z^2}$ or, in terms of t,

$$\sqrt{3 + 2t + t^2} = \sqrt{(t + 1)^2 + 2}.$$

The right hand side of the last equation is clearly minimal (value of $\sqrt{2}$) when
t = -1. (Hmm!)

SECTION 14.6

Solution for Exercise 16 (p. 608). Three vectors $\bar{v}_1$, $\bar{v}_2$, $\bar{v}_3$ are coplanar iff there exist scalars a, b, c not all zero such that

$$a\bar{v}_1 + b\bar{v}_2 + c\bar{v}_3 = \bar{0}. \tag{1}$$

Here $\bar{v}_1 = (0,1,-1)$, $\bar{v}_2 = (3,-1,2)$ and $\bar{v}_3 = (3,-2,3)$. Equation (1) gives rise to the system:

$$3b + 3c = 0$$
$$a - b - 2c = 0$$
$$-a + 2b + 3c = 0$$

which you should verify as having the solution a = 1, b = -1, c = 1. Thus, $\bar{v}_1 - \bar{v}_2 + \bar{v}_3 = \bar{0}$ and the vectors are coplanar. Notice that in the language we introduced in the solution for Exercise 14b of Section 14.3, we may say that three vectors in 3-space are coplanar iff they are linearly dependent.

Solution for Exercise 20 (p. 608). The vector (A,B,C) is normal to the plane Ax + By + Cz + D = 0. Hence, the direction numbers for any line perpendicular to the plane are A, B, C. The line through (x_0, y_0, z_0) may be written as

$$\frac{x - x_0}{A} = \frac{y - y_0}{B} = \frac{z - z_0}{C}$$

provided that A, B and C are non-zero.

Solution for Exercise 21 (p. 608). Recall that the three vectors $\bar{r}_0 - \bar{R}_0$, $\bar{d}$ and $\bar{D}$ are coplanar iff there exist scalars a, b, c not all zero such that

$$a(\bar{r}_0 - \bar{R}_0) + b\bar{d} + c\bar{D} = \bar{0}. \tag{2}$$

First, we shall assume that the vectors are coplanar. If a = 0, then equation (2) tells us that $b\bar{d} + c\bar{D} = \bar{0}$. This means that the lines are parallel, contrary to what was given. Consequently, $a \neq 0$ and we may rewrite equation (2) as

$$\bar{r}_0 + \frac{b}{a}\bar{d} = \bar{R}_0 - \frac{c}{a}\bar{D}.$$

Clearly, the lines intersect when t = b/a and u = -c/a.

As for the converse, if the lines intersect then

$$\overline{r}_0 + t\overline{d} = \overline{R}_0 + u\overline{D}$$

or
$$(\overline{r}_0 - \overline{R}_0) + t\overline{d} - u\overline{D} = \overline{0}$$

for some t and u. But this is just equation (2) with a = 1, b = t, and c = -u. Since a, b, c are not all zero, the vectors are coplanar.

SECTION 14.7

<u>Solution for Exercise 12 (p. 614).</u> Definition 14.7.1 is too messy to memorize. The determinant form of the definition (14.7.2) is easier to use once one is aware of a convenient trick for evaluating a 3 by 3 determinant. We reproduce the first two columns to the right of the determinant.

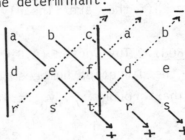

The value of the determinant is found by adding the products of the elements along the descending diagonals:

$$aet + bfr + cds,$$

and then subtracting the products of the elements along the ascending diagonals:

$$-rec - sfa - tdb.$$

For example,

$$\begin{vmatrix} 1 & 2 & 3 \\ 2 & 0 & 1 \\ 4 & -1 & 2 \end{vmatrix} = \begin{vmatrix} 1 & 2 & 3 \\ 2 & 0 & 1 \\ 4 & -1 & 2 \end{vmatrix} \begin{matrix} 1 & 2 \\ 2 & 0 \\ 4 & -1 \end{matrix} = + (0) + (8) + (-6) - (0) - (-1) - (8) = -5$$

You should be cautioned that this "arrow technique" does not work for determinants larger than a 3 by 3.

Now to the problem at hand.

$$(\overline{i} - 3\overline{j} + \overline{k}) \times (4\overline{i} + \overline{k}) = \begin{vmatrix} \overline{i} & \overline{j} & \overline{k} \\ 1 & -3 & 1 \\ 4 & 0 & 1 \end{vmatrix} =$$

$$+ (-3\overline{i}) + (4\overline{j}) + (0\overline{k}) - (-12\overline{k}) - (0\overline{i}) - (\overline{j}) = -3\overline{i} + 3\overline{j} + 12\overline{k}.$$

So, $(2\overline{i} + \overline{j}) \bullet (-3\overline{i} + 3\overline{j} + 12\overline{k}) = -6 + 3 = -3.$

<u>Solution for Exercise 16 (p. 614)</u>. First we show that $\overline{b} \times \overline{a} = -\overline{a} \times \overline{b}$. Let
$\overline{a} = (a_1, a_2, a_3)$ and $\overline{b} = (b_1, b_2, b_3)$. Then,

$$\overline{b} \times \overline{a} = \begin{vmatrix} \overline{i} & \overline{j} & \overline{k} \\ b_1 & b_2 & b_3 \\ a_1 & a_2 & a_3 \end{vmatrix}$$

$$= (b_2 a_3 - a_2 b_3)\overline{i} + (b_3 a_1 - b_1 a_3)\overline{j} + (b_1 a_2 - a_1 b_2)\overline{k}$$

$$= -(a_2 b_3 - a_3 b_2)\overline{i} - (a_3 b_1 - a_1 b_3)\overline{j} - (a_1 b_2 - a_2 b_1)\overline{k}$$

$$= -\overline{a} \times \overline{b}$$

Next, if we let $\overline{a} = \overline{b}$, we have

$$\overline{a} \times \overline{a} = -\overline{a} \times \overline{a}$$

so that $2\overline{a} \times \overline{a} = \overline{0}$ or $\overline{a} \times \overline{a} = \overline{0}$ as desired.

<u>Solution for Exercise 23 (p. 614)</u>. We shall use the facts that

$$\overline{a} \times \overline{b} = -\overline{b} \times \overline{a} \qquad \text{and} \qquad \overline{a} \times \overline{a} = \overline{0}.$$

Then, $(\overline{a} + \overline{b}) \times (\overline{a} - \overline{b}) = (\overline{a} + \overline{b}) \times \overline{a} - (\overline{a} + \overline{b}) \times \overline{b}$

$$= \overline{a} \times \overline{a} + \overline{b} \times \overline{a} - \overline{a} \times \overline{b} - \overline{b} \times \overline{b}$$

$$= \overline{0} - \overline{a} \times \overline{b} - \overline{a} \times \overline{b} - \overline{0} = -2(\overline{a} \times \overline{b}).$$

Solution for Exercise 24b (p. 614). Suppose that $P(x, y, z)$ is any point in the desired plane. The vectors $\overrightarrow{P_1P}$, $\overrightarrow{P_2P}$, $\overrightarrow{P_3P}$ are coplanar so that

$$\overrightarrow{P_1P} \cdot [\ \overrightarrow{P_2P} \times \overrightarrow{P_3P}\ ] = 0.$$

For the current problem, we have

$$(x - 1, y - 1, z - 1) \cdot [(x - 2, y + 2, z + 1) \times (x, y - 2, z - 1)] = 0.$$

After considerable algebra we find

$$x + y - z = 1$$

as the equation of the plane.

Let us look at another way to solve the problem. Vectors $\overrightarrow{P_1P_2} = (1, -3, -2)$ and $\overrightarrow{P_1P_3} = (-1, 1, 0)$ lie in the plane. Let $\overline{N} = (a, b, c)$ be a normal to the plane. Then,

$$\overline{N} \cdot \overrightarrow{P_1P_2} = \overline{N} \cdot \overrightarrow{P_1P_3} = 0.$$

In particular, we have the system of equations:

$$a - 3b - 2c = 0$$

$$-a + b \qquad = 0$$

for which the solution is $a = -c$, $b = -c$ and $c = c$. Consequently, $\overline{N} = c(-1, -1, 1)$. We shall select $c = 1$ so the equation of the plane is $\overline{N} \cdot \overrightarrow{P_2P} = 0$ or

$$-(x - 2) - (y + 2) + (z + 1) = 0.$$

As expected this second derivation of the plane leads to the same result.

Here is yet another solution. This is a new idea which generalizes the vector parametrization of a line, viz.,

$$\overline{r}(t) = \overline{r}_0 + t\overline{d}.$$

Suppose that $\overline{D}_1$ and $\overline{D}_2$ are non-parallel vectors. Consider the collection of all vectors which may be expressed as linear combinations of $\overline{D}_1$ and $\overline{D}_2$, that is,

$$s\overline{D}_1 + t\overline{D}_2 \qquad \text{with s, t as any real scalars.}$$

The vectors $\overline{D}_1$, $\overline{D}_2$, $s\overline{D}_1 + t\overline{D}_2$ are coplanar for all choices of s and t. Physically this is clear; here is the algebra.

Let $\overline{D}_1 = (a, b, c)$ and $\overline{D}_2 = (m, n, p)$. Then,

$$\overline{D}_1 \cdot [\overline{D}_2 \times (s\overline{D}_1 + t\overline{D}_2)] = \overline{D}_1 \cdot [\overline{D}_2 \times s\overline{D}_1 + \overline{0}] = s\overline{D}_1 \cdot (\overline{D}_2 \times \overline{D}_1)$$

$$= s(a, b, c) \cdot \begin{vmatrix} \overline{i} & \overline{j} & \overline{k} \\ m & n & p \\ a & b & c \end{vmatrix}$$

$$= s(a, b, c) \cdot (nc - bp, ap - mc, mb - na)$$

$$= s(nca - abp + apb - mcb + mbc - nac) = 0$$

Conversely, any vector in the plane which contains the non-parallel vectors $\overline{D}_1$ and $\overline{D}_2$ may be written as a linear combination of them.

Consequently, we offer the following <u>two parameter</u> representation of a plane:

$$\overline{P}(s,t) = \overline{r}_0 + s\overline{D}_1 + t\overline{D}_2.$$

You should compare the following diagram to Figures 14.5.2 and 14.5.3 in the text.

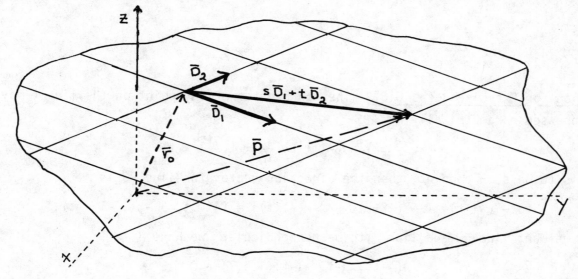

For the current problem we may let $\overline{D}_1 = \overrightarrow{P_1P_2}$, $\overline{D}_2 = \overrightarrow{P_1P_3}$ and $\overline{r}_0 = \overrightarrow{OP_1}$. Thus,

$$\overline{D}_1 = (1,-3,-2), \qquad \overline{D}_2 = (-1,1,0), \qquad \overline{r}_0 = (1,1,1)$$

and

$$P(s,t) = (1,1,1) + s(1,-3,-2) + t(-1,1,0).$$

A scalar representation of the plane is given by

$$x = 1 + s - t$$
$$y = 1 - 3s + t$$
$$z = 1 - 2s$$

We "eliminate" s and t from this system as follows. From the last two equations we find that

$$\frac{1}{2}(1 - z) = s \qquad \text{and} \qquad y - 1 + 3(\frac{1 - z}{2}) = t.$$

When we substitute these quantities in the equation for x, we find

$$x = 1 + \frac{1}{2}(1 - z) - y + 1 - \frac{3}{2}(1 - z)$$

which reduces to

$$x + y - z = 1.$$

Holy expletive deleted!

SECTION 14.9

Solution for Exercise 3 (p. 618). The coordinates of any point on the line ℓ may be expressed parametrically as

$$(x, y, z) = (-1 + t, -2 + t, -1 + t).$$

If Q is to be such a point, then the direction vector for line PQ is

$$(3 - (-1+t), 1 - (-2+t), -2 - (-1+t)) = (4 - t, 3 - t, -1 - t).$$

Since we want line PQ and line ℓ to be perpendicular, we have

$$(4-t, 3-t, -1-t) \cdot (1,1,1) = 0$$

or t = 2. Thus, Q = (1,0,1).

Another approach to this problem is suggested by the solution of Exercise 14 given later. Hint: can you use the distance formula to find Q?

<u>Solution for Exercise 4b (p. 618)</u>. A normal to a plane is a direction vector for a line perpendicular to that plane. Sometimes the direction numbers of such a line are referred to as the attitude numbers for the plane. It is useful to keep this intimate connection between lines and planes in mind. A mason's or plasterer's tool is a good model to help you get a handle on the subject.

The direction numbers for the given line are 2, 3, -4 so the attitude numbers for the desired plane are 2, 3, -4. The equation of this plane is

$$2(x - 2) + 3(y - 1) - 4(z + 3) = 0.$$

<u>Solution for Exercise 4d (p. 618)</u>. Given that the two lines lie in the plane, we may substitute values for t and u to find three points in the plane. For $t = 0$, $t = 1$ and $u = 1$, we obtain the points $(-1,2,1)$, $(1,5,2)$ and $(0,1,3)$ respectively. The general form for a plane is $Ax + By + Cz + D = 0$. Substitution of the three points in this equation yields,

$$-A + 2B + C + D = 0$$
$$A + 5B + 2C + D = 0$$
$$B + 3C + D = 0.$$

Subtracting the third equation from the first two, we have

$$-A + B - 2C = 0$$
$$A + 4B - C = 0$$

Combining these equations, we find

$$5B - 3C = 0 \qquad \text{or} \qquad B = \frac{3}{5} C.$$

We may now substitute back to discover that

$$A = C - 4B = -\frac{7}{5} C$$

and

$$D = -B - 3C = -\frac{18}{5} C.$$

Now select any non-zero value for C. A reasonable choice is $C = -5$. The equation of the plane

$$C(-\frac{7}{5} x + \frac{3}{5} y + z - \frac{18}{5}) = 0$$

becomes

$$7x - 3y - 5z + 18 = 0.$$

Of course, we could have used one of the techniques exampled in the solution for Exercise 24b of Section 14.7. Since the lines lie in the plane, any normal $\overline{N} = (a,b,c)$ to the plane is perpendicular to the direction vector of each line. Thus,

$$(a,b,c) \cdot (2,3,1) = 0 \qquad \text{and} \qquad (a,b,c) \cdot (1,-1,2) = 0.$$

The solution of this system is $a = a$, $b = \dfrac{-3}{7} a$, $c = \dfrac{-5}{7} a$. The plane is given by

$$(x + 1) - \frac{3}{7}(y - 2) - \frac{5}{7}(z - 1) = 0,$$

which agrees with the previous result.

Solution for Exercise 6 (p. 618). The natural approach is to solve the equations of the planes simultaneously.

$$2x + y - 3z + 6 = 0$$
$$x + 4y - 5z - 7 = 0$$

We may **subtract** twice the second equation from the first to find

$$-7y + 7z + 20 = 0 \qquad \text{or} \qquad y = z + \frac{20}{7}.$$

Substitution of this result into the equation of the second plane gives

$$x = -4y + 5z + 7 = z - \frac{31}{7}.$$

The line of intersection is given by $x = -\dfrac{31}{7} + t$, $y = \dfrac{20}{7} + t$, $z = t$.

Here is another way to approach the problem. Since the line lies in both planes, its direction vector, $\overline{N} = (a,b,c)$, is perpendicular to normals for the planes. Thus,

$$(a,b,c) \cdot (2,1,-3) = 0 \qquad \text{and} \qquad (a,b,c) \cdot (1,4,-5) = 0.$$

A solution of this system is $N = (1,1,1)$. Now we need to find a point on the line. Trial and error is sufficient. We substitute $x = 1$ in the equations of the planes.

$$y - 3z + 8 = 0 \qquad \text{and} \qquad 4y - 5z - 6 = 0.$$

The solution of this system is y = 58/7, z = 38/7. The equations of the line are x = 1 + u, y = $\frac{58}{7}$ + u, z = $\frac{38}{7}$ + u. This agrees with the earlier result once we notice that t = u + $\frac{38}{7}$.

<u>Solution for Exercise 7 (p. 618)</u>. First we write the line as x = -2 + 3t, y = 1 + 2t, z = -6 + t. When we substitute these expressions in the equation of the plane

$$2(-2 + 3t) + (1 + 2t) - 3(-6 + t) + 6 = 0,$$

we find that the line intersects the plane when t = $- \frac{21}{5}$. The point of inter-section is $(- \frac{73}{5}, - \frac{37}{5}, - \frac{51}{5})$.

<u>Solution for Exercise 13 (p. 618)</u>. The line segment which marks off the distance from the point P(4,6,-4) to the line ℓ joining (2,2,1) and (4,3,-1):

$$x = 2 + 2t, \qquad\qquad y = 2 + t, \qquad\qquad z = 1 - 2t,$$

lies in the plane which passes through P and is perpendicular to line ℓ. Consequently, a direction vector for ℓ; viz., (2,1,-2), may serve as a normal to the plane. The equation of the plane is

$$2(x - 4) + (y - 6) - 2(z + 4) = 0.$$

Line ℓ intersects this plane when

$$2(2 + 2t - 4) + (2 + t - 6) - 2(1 - 2t + 4) = 0$$

or t = 2. The coordinates of this point are (6,4,-3). Its distance from P is 3 units.

   Another solution to this problem does not require that we find the plane containing the line segment. The distance from P to <u>any</u> point on line ℓ is given by

$$D(t) = \sqrt{(2 + 2t - 4)^2 + (2 + t - 6)^2 + (1 - 2t + 4)^2}$$
$$= \sqrt{9t^2 - 36t + 45} = \sqrt{9(t - 2)^2 + 9} .$$

Clearly, D(t) is minimal when t = 2 and D(2) = 3 units.

Solution for Exercise 14 (p. 618).   The use of vector methods in elementary geometry yields some elegant and simple proofs. Once we recognize that

$$\bar{a} + \bar{v} = \bar{b} + \bar{c}$$

it follows rather quickly that

$$\bar{r} = \frac{1}{2}\,\bar{a} + \frac{1}{2}\,\bar{v}$$

$$= \frac{1}{2}\,\bar{b} + \frac{1}{2}\,\bar{c} = \bar{t}$$

and

$$\bar{u} = \frac{1}{2}\,\bar{v} - \frac{1}{2}\,\bar{c}$$

$$= \frac{1}{2}\,\bar{b} - \frac{1}{2}\,\bar{a} = \bar{s}.$$

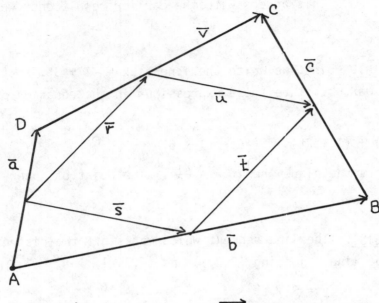

$$\overrightarrow{AD} = \bar{a} \qquad\qquad \overrightarrow{BC} = \bar{c}$$

$$\overrightarrow{AB} = \bar{b} \qquad\qquad \overrightarrow{DC} = \bar{v}$$

Vector Calculus

15

Most of the material in this Chapter is mechanical, albeit important, in nature. There are ample examples in the text. Consequently, most of the exercises selected for solution are non-mechanical or permit additional comment. We conclude with some comments which pull together the work done in this and the previous Chapter.

SECTION 15.1

Solution for Exercise 15 (p. 625). The vector functions under consideration are defined in terms of a single parameter. Consequently, many properties of vector functions may be established by considering the functions by components and then appealing to earlier results for a function of a single variable.

Let $\overline{f}(t) = f_1(t)\overline{i} + f_2(t)\overline{j} + f_3(t)\overline{k}$. Since two vectors are equal if and only if their respective components are equal, the condition $\overline{f}'(t) = \overline{0}$ for t in I tells us that $f_1'(t) = f_2'(t) = f_3'(t) = 0$ for t in I. We conclude from Theorem 4.2.3 that there exist constants c_1, c_2, c_3 such that $f_1(t) = c_1$, $f_2(t) = c_2$, $f_3(t) = c_3$ for all t in I. In other words, $\overline{f}(t) = c_1\overline{i} + c_2\overline{j} + c_3\overline{k} = \overline{c}$ for all t in I.

317

<u>Solution for Exercise 17 (p. 625).</u> All that is needed here is the linearity of the integral (Formula 5.6.3):

$$\int_a^b [\alpha \, f(x) + \beta \, g(x)]dx = \alpha \int_a^b f(x)dx + \beta \int_a^b g(x)dx.$$

Again the argument centers about reposing the problem so that earlier results may be used. We leave the routine writing of details to you.

<u>Solution for Exercise 18 (p. 625).</u> We use the definitions of limit and norm to restate the problem:

$$\lim_{t \to t_0} \overline{f}(t) = L \qquad\qquad \text{iff} \qquad\qquad \lim_{t \to t_0} ||\overline{f}(t) - \overline{L}|| = 0,$$

as follows.

$$\lim_{t \to t_0} f_1(t) = \ell_1, \qquad \lim_{t \to t_0} f_2(t) = \ell_2, \qquad \lim_{t \to t_0} f_3(t) = \ell_3 \qquad\qquad (1)$$

if and only if

$$\lim_{t \to t_0} \sqrt{[f_1(t) - \ell_1]^2 + [f_2(t) - \ell_2]^2 + [f_3(t) - \ell_3]^2} = 0. \qquad\qquad (2)$$

Because the square root function is continuous, statement (2) is the same as

$$\lim_{t \to t_0} \{[f_1(t) - \ell_1]^2 + [f_2(t) - \ell_2]^2 + [f_3(t) - \ell_3]^2\} = 0. \qquad\qquad (3)$$

Statement (3) is an easy consequence of (1). Conversely, suppose that statement (3) is true. We appeal to the fact that a sum of squares of real numbers is zero iff each of those numbers is zero. That is, $a^2 + b^2 + c^2 = 0$ iff $a = b = c = 0$. Consequently, each of the limits

$$\lim_{t \to t_0} [f_i(t) - \ell_i]^2 \qquad\qquad (i = 1,2,3)$$

exists and is zero. In other words, we have statement (1).

This problem brings a question to mind. Is it true that

$$\lim_{t \to t_0} ||\overline{f}(t)|| = ||\lim_{t \to t_0} \overline{f}(t)||?$$

The answer is no. Let $f_2(t) = f_3(t) = 0$ and look back, if necessary, to Chapter 2 for a good choice for $f_1(t)$.

SECTION 15.2

Solution for Exercise 23 (p. 630). From the definition of the norm it follows that $||\overline{f}(t)||$ is constant if and only if $\overline{f}(t) \cdot \overline{f}(t)$ is constant. However, $\overline{f}(t) \cdot \overline{f}(t)$ is constant if and only if $\frac{d}{dt}[\overline{f}(t) \cdot \overline{f}(t)] = 0$. But, $\frac{d}{dt}[\overline{f}(t) \cdot \overline{f}(t)] = 2\overline{f}(t) \cdot \overline{f}'(t)$ and we are done.

This problem admits to a reasonable geometric interpretation. If the norm of the position vector $\overline{f}(t)$ is constant, say r, then the curve or path traced out by the tip of $\overline{f}(t)$ lies on the surface of the sphere of radius r centered at the origin. To say that $\overline{f}(t) \cdot \overline{f}'(t) = 0$ means the instantaneous change, $\overline{f}'(t)$, in the path is perpendicular to the radius, $\overline{f}(t)$. That is, the direction of the instantaneous motion lies in the plane tangent to the sphere.

Solution for Exercise 25 (p. 630). The key step in evaluating $(\overline{f} \times \overline{g})'(t)$ by the difference quotient:

$$\lim_{h \to 0} \frac{1}{h}\{\overline{f}(t + h) \times \overline{g}(t + h) - \overline{f}(t) \times \overline{g}(t)\}, \tag{1}$$

is to subtract and add the quantity $\overline{f}(t) \times \overline{g}(t + h)$ within the brackets. We regroup the terms in statement (1) as follows.

$$\lim_{h \to 0} \left\{ \frac{\overline{f}(t + h) - \overline{f}(t)}{h} \times \overline{g}(t + h) + \overline{f}(t) \times \frac{\overline{g}(t + h) - \overline{g}(t)}{h} \right\}$$

Our informal derivation is complete once we notice that as $h \to 0$

$$\frac{\overline{f}(t + h) - \overline{f}(t)}{h} \to \overline{f}'(t), \quad \overline{g}(t + h) \to \overline{g}(t), \quad \frac{\overline{g}(t + h) - \overline{g}(t)}{h} \to \overline{g}(t).$$

This informal derivation may be dressed up as a formal argument by donning mathematical ties and tails (ϵ's and δ's). The necessary changes are style, not approach.

SECTION 15.3

<u>Solution for Exercise 8 (p. 635)</u>. We are given that $\bar{r}(t) = c\bar{r}'(t)$ for some scalar c. It certainly seems reasonable that when the direction of change, $\bar{r}'(t)$, is parallel to the position vector (which emanates from the origin), then the tangent line will pass through the origin. The supportive calculations are just as reasonable. The vector parametrization of the tangent line at $t = t_0$ is given by

$$\bar{L}(s) = \bar{r}(t_0) + s[\bar{r}'(t_0)].$$

But, $\bar{r}(t) = c\bar{r}'(t)$ so that

$$\bar{L}(s) = (c + s)\,\bar{r}'(t_0)$$

and line L passes through the origin when $s = -c$.

<u>Solution for Exercise 9 (p. 635)</u>. Curves $\bar{r}_1(t)$ and $\bar{r}_2(u)$ intersect at those points for which

$$u = e^t, \quad 2\sin(t + \pi/2) = 2 \quad \text{and} \quad t^2 - 2 = u^2 - 3$$

for some u and t. From the second of these equations we see that we must have $t = 2n\pi$ for some integer n. The simplest choice of n is $n = 0$. When we substitute $t = 0$ in both the first and third equations, we find $u = 1$. The point of intersection is $\bar{r}_1(0) = \bar{r}_2(1) = (1,2,-2)$. The angle of intersection of the curves is defined as the angle of intersection of their respective tangent lines at $(1,2,-2)$. Evaluation of the quantity

$$\cos\theta = \frac{\bar{r}_1'(0)\cdot\bar{r}_2'(1)}{||\bar{r}_1'(0)||\ ||\bar{r}_2'(1)||}$$

gives us $\cos\theta = 1/\sqrt{5}$ or θ as approximately 1.11 radians.

Do these curves intersect in another point? We have already discovered from the second equation that $t = 2n\pi$, where n is an integer. Substitution of the first equation in the third gives us

$$t^2 + 1 = e^{2t}.$$

This type of equation is called transcendental since it involves the transcendental function e^{2t}. Such equations frequently occur in mathematical models of the real world. Unfortunately, transcendental equations are usually very difficult to solve. However, there is but one solution (t = 0) this time. Notice that $e^{2t} < 1 < t^2 + 1$ for t < 0 is a simple inequality. When t > 0 we have $e^{2t} > t^2 + 1$. (Why? Compare the derivatives and remember the equality for t = 0).

<u>Solution for Exercise 10 (p. 635)</u>. The circular helix is illustrated in Figure 15.4.2 of the text. The thread on a bolt is a good model. The general helix is a three-dimensional spiral. It occurs all around us. The DNA molecule, the basic building block of all life forms, is shaped like a double helix, i.e., two inter-locking helices. The scrolls atop Corinthian columns from Greece's Age of Reason provide another, intriguing, model. Finally, the path of the Earth's revolutions as it follows the sun's journey through space is helical.

As for the problem, we may verify part (a) by noticing that

$$[x(t)]^2 + [y(t)]^2 = (a \cos \omega t)^2 + (a \sin \omega t)^2 = a^2.$$

The direction vector for a line parallel to the z-axis is $\bar{k} = (0,0,1)$. Consequently,

$$\cos[\theta(t)] = \frac{\bar{r}'(t) \cdot \bar{k}}{||\bar{r}'(t)|| \; ||\bar{k}||} = \frac{b\omega}{\sqrt{a^2 + b^2\omega^2}}.$$

SECTION 15.4

<u>Solution for Exercise 2 (p. 643)</u>. This problem is straightforward. Two quick calculations show that $||\bar{r}'(t)|| = |a|$ and $\bar{r}''(0) = -a\bar{i}$. This routine problem and the results take on added interest once the $\bar{r}(t)$ is described. Can you eliminate the parameter t to find a Cartesian representation of the curve? You should

uncover a circle of radius a with center at the point (b,c).

Solution for Exercise 5 (p. 643). We have

$$x = 2 \cos 2t = 2(\cos^2 t - \sin^2 t) = 2(2 \cos^2 t - 1) = 2(2(\tfrac{y}{3})^2 - 1).$$

Rearranging terms we find the desired $4y^2 - 9x = 18$. However, the entire parabola is not traced out. In fact, $-2 \le x \le 2$ and $-3 \le y \le 3$. The arc is traced precisely once for each interval in t of length π. Hence the motion is periodic with period 2π (why?).

   The velocity and acceleration vectors are given by

$$\overline{r}'(t) = (-4 \sin 2t)\overline{i} + (-3 \sin t)\overline{j}$$

and

$$\overline{r}''(t) = (-8 \cos 2t)\overline{i} + (-3 \cos t)\overline{j},$$

respectively. We have $\overline{r}'(t) = 0$ when $t = n\pi$. Because of the curve's periodicity noted above, it suffices to compute $\overline{r}''(0) = -8\overline{i} - 3\overline{j}$ and $\overline{r}''(\pi) = -8\overline{i} + 3\overline{j}$. The path and these acceleration vectors are sketched below.

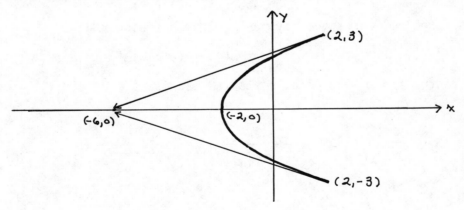

Solution for Exercise 9 (p. 643). An object of constant velocity has an accelera-tion of zero. The torque about the origin must be zero as

$$\overline{\tau}(t) = \overline{r}(t) \times \overline{F}(t) = \overline{r}(t) \times m\overline{a}(t) = \overline{r}(t) \times \overline{0} = \overline{0}.$$

However, torque is the derivative relative to time of the angular momentum. Thus, $\overline{L}'(t) = \overline{\tau}(t) = \overline{0}$ and $\overline{L}(t)$ must be a constant.

Clearly, the path described by the object above is a straight line, unless the velocity is zero. In that event the object's gear lever is stuck in Park. However, it is not necessary for the path to be a straight line for us to obtain no torque and a constant angular momentum. The curve in Exercise **6** provides an example. As you should verify, $\overline{a}(t) = \overline{r}''(t) = \omega^2 \overline{r}(t)$ so that

$$\overline{\tau}(t) = \overline{r}(t) \times \overline{F}(t) = \overline{r}(t) \times \omega^2 m \overline{r}(t) = \overline{0}.$$

(Remember that $\overline{u} \times \overline{u} = \overline{0}$ for any vector $\overline{u}$.) The path described in Exercise 5 might only be familiar to you if you have studied the hyperbolic functions. We may write

$$\overline{r}(t) = \frac{1}{2} a \cosh(\omega t) \overline{i} + \frac{1}{2} b \sinh(\omega t) \overline{j}.$$

In any event, we may eliminate the parameter t to obtain

$$\left(\frac{x}{a/2}\right)^2 - \left(\frac{y}{b/2}\right)^2 = 1.$$

Only the right hand branch of this hyperbola is traced out (and but once).

SECTION 15.5

Solution for Exercise 1 (p. 649). Rather quickly we have that

$$\ell(C) = \int_0^{2\pi} \sqrt{(-a \sin t)^2 + (a \cos t)^2 + (b)^2} \; dt$$

$$= \int_0^{2\pi} \sqrt{a^2 + b^2} \; dt = 2\pi \sqrt{a^2 + b^2}.$$

The points joined by this arc are $(a,0,0)$ and $(a,0,2\pi b)$ for $t = 0$ and 2π respectively. The length of the line segment joining these points is

$$2\pi b \leq 2\pi \sqrt{a^2 + b^2}.$$

It would seem the arc has taken the long way around. However, we picked this problem because it is so easy and allowed us to revisit the circular helix. Exercise 8 contains a new twist; it is also a helix--but traced out on the surface of a circular cone (see Figure 16.2.4).

Solution for Exercise 9 (p. 649). As formulated, this problem will require that we integrate secant cubed. Gulp! We will change the problem. This is a technique some of our "more innovative" students have taught us. However, our motives are pure and above reproach. Let us change the x component of $\bar{r}(t)$ to yield

$$\bar{r}(t) = t^2\bar{i} + (t^2 - 2)\bar{j} + (1 - t^2)\bar{k} .$$

Then,

$$\ell(C) = \int_0^2 \sqrt{(2t)^2 + (2t)^2 + (-2t)^2}\ dt = \int_0^2 2\sqrt{3}t\ dt = 4\sqrt{3}\ \text{units}.$$

As in the previous solution, we observe that the endpoints of the arc are (0,-2,1) and (4,2,-3) for t = 0 and 2 respectively. The distance between these points is $4\sqrt{3}$ units. That's a familiar number. Apparently, $\bar{r}(t)$ must be a line even though the vector parametrization does not appear linear. What does this mean? The answer lies in $\bar{r}'(t)$.

$$\bar{r}'(t) = 2t\bar{i} + 2t\bar{j} - 2t\bar{k} = t(2,2,-2).$$

Notice that for t > 0 the direction of $\bar{r}'(t)$ is constant, only its magnitude changes. The same comment applies for t < 0 but the direction is reversed. Thus $\bar{r}(t)$ represents a ray rather than a line. An object whose motion is given by $\bar{r}(t)$ comes in along the line x = y + 2 = 1 - z through (4,2,-3), slows for an instantaneous pit stop at (0,-2,1), and then retraces its path with ever-increasing velocity.

Solution for Exercise 10 (p. 649). The key observation is that the arc length of a curve, say from t to t + h, is defined as the least upper bound of the lengths of all polygonal approximations to this arc. This least upper bound is denoted here by

$$\ell = s(t + h) - s(t).$$

Consequently, ℓ is greater than or equal to the length of any polygonal approximation. The length of the simplest approximation (one line segment) is

$$||\bar{r}(t + h) - \bar{r}(t)||$$

so

$$||\overline{r}(t + h) - \overline{r}(t)|| \le \ell.$$

Additionally, ℓ is not greater than any upper bound for this collection of polygonal approximations. As shown completely in the early stages of the proof for Theorem 15.5.1,

$$\ell \le \int_t^{t + h} ||\overline{r}'(t)|| \, dt.$$

SECTION 15.6

Solution for Exercises 10 and 11 (p. 657). Curvature is a measure of the change of the tangent to a plane curve; it is not concavity. Both curvature and concavity come from second derivatives, but these derivatives are taken relative to two different variables. Concavity measures a change without regard to the slope of the tangent. Curvature measures variations in the direction of the tangent vector. This measurement is only meaningful if the magnitude of the tangent vector is standardized. Consequently, curvature is measured relative to arc length so that the magnitude of the tangent vector is always one.

In Exercise 10, we may easily parametrize the ray by

$$\overline{R}(t) = (x_0, y_0, z_0) + t(x_1 - x_0, y_1 - y_0, z_1 - z_0) = \overline{u} + t\overline{v} \qquad (t \ge 0).$$

However, in an arc length parametrization of any curve we want any unit increase in the parameter to generate a unit increase in the arc length. That is, we want $s(t + 1) - s(t) = 1$ for all t. Thus, we must write

$$\overline{R}(t) = \overline{u} + t \frac{\overline{v}}{||\overline{v}||}, \qquad (t \ge 0) \qquad (1)$$

Notice that $||\overline{R}(t + 1) - \overline{R}(t)|| = 1$ for all t. In other words,

$$||\overline{R}'(t)|| = ||\overline{v}||/||\overline{v}|| = 1.$$

If in addition we permit $t < 0$, equation (1) determines a line. The curvature of this line is zero since $\overline{R}'(t)$ is constant and $||\overline{R}''(t)|| = 0$. This resolves Exercise 11.

CONCLUDING COMMENTS ON VECTORS

In these last two chapters we have examined the fundamental properties and the basic calculus of vectors in 2-space and 3-space that are of particular value for elementary physics. We have seen that a vector is a quantity characterized by magnitude and direction. At times we have restricted our attention to position vectors; namely, vectors whose tails are fixed at the origin and whose tips indicate a position. Sometimes the distinction between the use of free standing vectors and position vectors may not have been clear. This is not critical. All the properties and processes carry through in either case. The difference really comes at the interpretive level, i.e., what we are using the vector to represent.

For example, we considered the calculus of the vector representation for a curve in 3-space. Here the discussion was implicitly concerned with position vectors (see Figure 1). Such a curve might represent the motion of some object through space subject to certain gravitational forces (or   pollen in air currents; an oil spill subject to eddies, tides, and winds; a golf ball on a green subject to the breaks and foibles of Bermuda grass; etc.). This restriction to position vectors is not necessary.

Possibly we are looking at a stream or collection of curves or paths of which any one is representative (see Figure 2). On the other hand, it is possible that variations in position might change the curve (see Figure 3). In this instance, we use more than one parameter in the vector representation. Some of the parameters select a path while the remaining parameters indicate a position on this selected path.

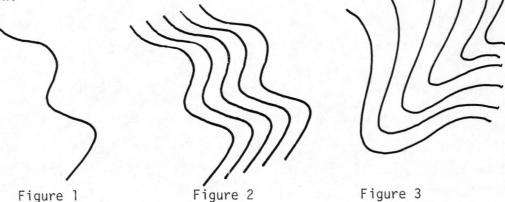

Figure 1   Figure 2   Figure 3

For example, the two-parameter position vector

$$\overline{R}(s,t) = s(\cos t \; \overline{i} + \sin t \; \overline{j})$$

may be used to indicate linear radiation from a point at an angle $0 \leq t \leq 2\pi$ and a distance $s \geq 0$. Alternately, $\overline{R}(s,t)$ could be seen as a collection of concentric circular ripples in a pool with $s \geq 0$ indicating the radius of a ripple or wave

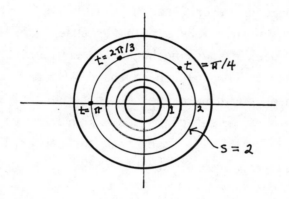

and t specifying points on a particular ripple.

We have seen that the vector $2\overline{i} - 3\overline{j} + 7\overline{k}$ may be written as $(2,-3,7)$. Further, the vector function $\overline{f}(t) = 2t\overline{i} - t^2\overline{j}$ could be written as $f(t) = (2t,-t^2)$. Why don't we just talk about ordered pairs and ordered triples instead of vectors? Well, this is really a question of semantics. For, vectors in their most general form are defined as ordered pairs, ordered triples, or ordered n-tuples: $(a_1, a_2, \cdots, a_n)$, of objects. The arrows we have used are but one useful model or representation for an abstraction called a vector.

A complete discussion of this abstract entity is best left for the course in Linear Algebra which follows the Calculus sequence. We take but a brief look. You will recall that at most three vectors in 3-space may be linearly independent. Further, at most two vectors in 2-space (or the plane) may be linearly independent. This correspondence of the threes and twos is not a coincidence. Linear independence is a way of getting at the notion of dimension.

Let us look at it another way. In 3-space we have three (independent) directions or variables. If we write down <u>one</u> equation, say, $2x - y + 3z = 6$, we obtain a plane, which is a <u>two</u>-dimensional object. The equation acts as a restraint, expressing one variable in terms of the other two. If we write down <u>two</u> equations, say $x + 1 = y - 2 = z + 4$, we obtain a line, which is a <u>one</u>-dimensional object. If we record <u>three</u> equations, say, $x = 1$, $y = 2$, $z = 5$, we obtain a point. Notice that in each case the dimensionality of the object and the number of equations sums to three.

Dimensionality and restraints need not be physical. They may be used to indicate the scope of or degrees of freedom in a problem. The max-min word problems we studied in Chapter 4 were single variable or one-dimensional problems even though we may have been discussing higher dimensional objects like area or volume. That is, restraints were placed on the dimension or elements in the problem so that everything could be expressed in terms of a single variable which was free to roam over a certain set of values. There are higher dimensional problems. They are harder. We will study them in Chapter 17. For example, in order to minimize the surface area of a rectangular solid given only its volume we need to consider a function of two variables:

$$S(x,y) = 2(xy + \frac{V}{x} + \frac{V}{y}).$$

Finally, the objects we use for components in a vector do not need to be numbers. The vectors need not have an "arrow model." For example, any function which satisfies the equation $f''(x) = f(x)$ may be written as a linear combination of the linearly independent functions e^x and e^{-x}. That is, $f(x) = ae^x + be^{-x}$, where a and b are any real numbers. The functions e^x and e^{-x} are linearly independent since $c_1 e^x + c_2 e^{-x}$ is zero for all numbers x only if $c_1 = c_2 = 0$. The functions e^x and e^{-x} act as a basis for the collection of all functions for which $f''(x) = f(x)$. Many more of the ideas about vectors you have seen in a physical or "arrow" context may be generalized or abstracted to serve in this otherwise seemingly unrelated area of differential equations.

Functions of Several Variables

16

The domain of a multivariate function is identified in much the same way that it was in the one variable case. Namely, unless it is already specified, the domain of the function is considered to be the set of all values for which the function rule makes sense. The most obvious ways in which exceptions arise are values of the variables which would result in the illegal operations of division by zero or the taking of even (e.g., square) roots of a negative number. The range of a function consists of all values assumed by the function on its domain.

Example 1. Determine the domain and range and sketch the level curves for

$$f(x,y) = y/x^2.$$

Solution. The domain consists of the entire plane except for the y-axis. That is, Dom(f) = {(x,y): x, y ϵ R; x $\neq$ 0}. The range of f consists of all real numbers, since y/x^2 = C may be solved for x and y given any value for C. The level curves for f consist of the x-axis and the parabolas $y = Cx^2$ for any real number C with the point (0,0) omitted in each case (why?).

In interpreting level curves you should remember that all points which generate the same function value (the height of the surface above the point (x,y)) lie on the same level curve. In this example, all the points on the level curve labeled C = 4 lie four units above the xy-plane and all points on the level curve for f(x,y) = C = -20 lie 20 units below the xy-plane.

329

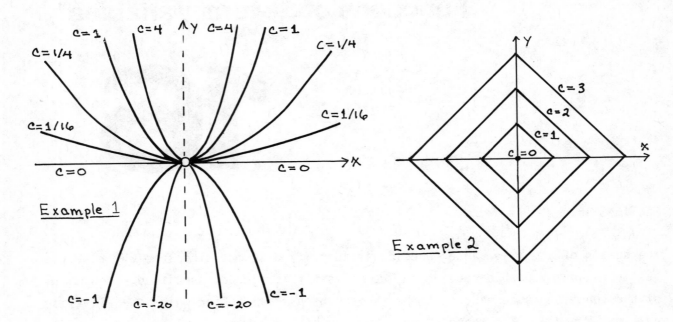

$c=1$ $c=4$ $\uparrow y$ $c=4$ $c=1$
$c=1/4$ $c=1/4$

$c=1/16$ $c=1/16$

$c=0$ $c=0$ $\rightarrow x$

Example 1

$c=-1$ $c=-20$ $c=-20$ $c=-1$

$\uparrow y$
$c=3$
$c=2$
$c=1$
$c=0$ $\rightarrow x$

Example 2

Example 2.   Determine the domain and range and sketch the level curves for
$f(x,y) = |x| + |y|$.
Solution. Clearly, Dom(f) = all real numbers. The range of f consists of all non-
negative real numbers (since $|x| \geq 0$, $|y| \geq 0$ and $|x| + |y| = C \geq 0$ has the
solution x = C, y = 0 among others).

   We now want to graph $|x| + |y| = C$ for some non-negative values of C. Notice
that $|x| + |y| = C$ requires that x and y are between -C and C and that (x,y) lies
on one of the lines x + y = C, x - y = C, -x + y = C, -x - y = C. We obtain the
above sketch from which it is clear that $f(x,y) = |x| + |y|$ is an upside-down
infinitely tall Egyptian pyramid. It is still under construction by the Army Corps
of Engineers!

Example 3.   The paraboloid of revolution $x^2 + y^2 = 4z$ and the plane z - 2y + x = 4
intersect in a space curve C. Find the equation for the projection of C onto the
xy-plane.
Solution. If we substitute specific values for x and y in an equation for a
surface and solve for z, this value of z is the height of the surface above the
xy-plane at the point (x,y,0). Naturally, a point lies on the curve C iff these
heights are equal for each of the intersecting surfaces. The projection of C in

the xy-plane may be found, then, by solving the equations simultaneously to eliminate z. We have

$$\frac{1}{4}(x^2 + y^2) = z = 2y - x + 4$$

which may be rewritten as

$$(x + 2)^2 + (y - 4)^2 = 36,$$

a circle of radius 6 centered at $(-2,4,0)$.

The matter of projections will come up in Chapter 18 when we discuss the volume of solids bounded by surfaces like those categorized in Section 16.2. Not surprisingly, we may find the projection of a space curve C in the yz-plane or the xz-plane as well. (How?)

SECTIONS 16.4-6

There are a variety of notations for partial derivatives. Each of the following represents the partial derivative of $z = f(x,y)$ with respect to x:

$$D_x f, \quad f'_x, \quad f_x, \quad \frac{\partial f}{\partial x}, \quad \text{and} \quad \frac{\partial z}{\partial x}.$$

We are partial to the fourth and fifth options. Each of the following represents the second partial of $z = f(x,y)$ with respect to x (that is, the partial derivative with respect to x): <u>of</u> the partial derivative of f with respect to x.

$$D_{xx}^2 f, \quad f''_x, \quad f_{xx}, \quad \frac{\partial^2 f}{\partial x^2}, \quad \text{and} \quad \frac{\partial^2 z}{\partial x^2}.$$

Again, the last two notations are our favorites.

A function of two variables has two first order partial derivatives and four second order partial derivatives. Each of the following represents the partial derivative with respect to x <u>of</u> the partial derivative of $z = f(x,y)$ with respect to y:

$$D_{xy}^2 f, \quad f''_{xy}, \quad f_{xy}, \quad \frac{\partial^2 f}{\partial x \partial y}, \quad \text{and} \quad \frac{\partial^2 z}{\partial x \partial y}.$$

Notice the order of differentiation and the order of the symbols in the notation.

Predictably, we shall use the last two forms. If you believe that a function of n variables has n first order partial derivatives, n^2 second order partials ($n^2 - n$ of which are "mixed") and n^3 third order partials, you are ready for the first example.

Example 1. Calculate all the first and second order partials for

$$f(x,y) = x^3 + x^2 y^5 + y^7.$$

Solution. $\dfrac{\partial f}{\partial x} = 3x^2 + 2xy^5$ $\dfrac{\partial f}{\partial y} = 5x^2 y^4 + 7y^6$

$\dfrac{\partial^2 f}{\partial y \partial x} = 10xy^4$ $\dfrac{\partial^2 f}{\partial x \partial y} = 10xy^4$

$\dfrac{\partial^2 f}{\partial x^2} = 6x + 2y^5$ $\dfrac{\partial^2 f}{\partial y^2} = 20x^2 y^3 + 42y^5$

                                                                                        Q.E.D.

Notice how easily this is done. The partial derivative of any multivariate expression with respect to a specified variable is simply found by treating all the other variables as if they were constants and then applying the ordinary rules of differentiation. Our next example supports this technique.

Example 2. Using the definition, calculate $\dfrac{\partial f}{\partial x}$ for $f(x,y) = x^2 + xy^3 - 4y$.
Solution.

$\dfrac{\partial f}{\partial x} = \lim_{h \to 0} \dfrac{f(x + h, y) - f(x,y)}{h}$

$= \lim_{h \to 0} \dfrac{[(x + h)^2 + (x + h)y^3 - 4y] - [x^2 + xy^3 - 4y]}{h}$

$= \lim_{h \to 0} \dfrac{2xh + h^2 + hy^3}{h} = \lim_{h \to 0} (2x + h + y^3) = 2x + y^3.$

                                                                                        Q.E.D.

Notice how y^3 and $-4y$ behave essentially like constants.

The mixed partials ($\dfrac{\partial^2 f}{\partial y \partial x}$ and $\dfrac{\partial^2 f}{\partial x \partial y}$) in Example 1 are equal. In fact, it is a rarity for you to encounter a function whose mixed partials are not equal, even though the continuity of $\frac{\partial f}{\partial x}$, $\frac{\partial f}{\partial y}$, and $\frac{\partial^2 f}{\partial y \partial x}$ (or $\frac{\partial^2 f}{\partial x \partial y}$) is needed to guarantee this equality. Consequently, the next example should be a real treat.

Example 3. Calculate the mixed partials (at the origin) for

$$f(x,y) = \begin{cases} xy(\dfrac{x^2 - y^2}{x^2 + y^2}), & (x,y) \neq (0,0) \\[4mm] 0 & (x,y) = (0,0) \end{cases}$$

Solution.

$$\frac{\partial}{\partial x} f(0,y) = \lim_{h \to 0} \frac{f(0 + h, y) - f(0,y)}{h} = \lim_{h \to 0} y(\frac{h^2 - y^2}{h^2 + y^2}) = -y$$

and

$$\frac{\partial}{\partial y} f(x,0) = \lim_{h \to 0} \frac{f(x,0 + h) - f(x,0)}{h} = \lim_{h \to 0} x(\frac{x^2 - h^2}{x^2 + h^2}) = x$$

Thus,

$$\frac{\partial^2 f(0,0)}{\partial y \partial x} = -1 \neq +1 = \frac{\partial^2 f(0,0)}{\partial x \partial y}$$

W.O.W.

The first partials of a function of two variables admit to a very nice geometric interpretation. See Figures 16.4.1 and 16.4.2 and the discussion in the text. The notion of a tangent plane for a function of two variables is analogous to the notion of a tangent line for a function of one variable. The tangent plane need not exist at a given point. If it does exist, then the tangent lines with slopes of $\frac{\partial f}{\partial x}$ and $\frac{\partial f}{\partial y}$ determine the plane.

What causes us some dismay is the fact that even though two different lines which intersect will determine a unique plane, the existence of $\frac{\partial f}{\partial x}$ and $\frac{\partial f}{\partial y}$ is not enough to guarantee the existence of a tangent plane. More formally, even though $\frac{\partial f}{\partial x}$ and $\frac{\partial f}{\partial y}$ may exist at a point (x_0, y_0) for some function f, f need not be differentiable at (x_0, y_0). As you might well suspect, f is differentiable at (x_0, y_0) iff

$$\lim_{(x,y) \to (x_0, y_0)} \frac{f(x,y) - f(x_0, y_0)}{h} \text{ exists,}$$

where h is the distance from (x_0,y_0) to (x,y). The difficulty arises from the fact that there are an infinite number of paths by which (x,y) may approach (x_0,y_0). In the single variable case there are only two paths of approach to a limit.

The notion of continuity of a two variable function involves the same complication. A careful reading of Section 16.6 in the text should help you appreciate this point. In particular, you should carefully consider the examples which demonstrate that:

       (i) a function may be continuous in each variable separately

           and yet fail to be continuous

and

      (ii) the existence of first partials does not guarantee continuity.

Gradients; Extreme Values; Differentials

17

SECTIONS 17.1-2

We write the gradient for a function of two variables, $f(x,y)$, as

$$\nabla f = \frac{\partial f}{\partial x}\, \overline{i} + \frac{\partial f}{\partial y}\, \overline{j}, \tag{1}$$

and for a function of three variables, $f(x,y,z)$, we write

$$\nabla f = \frac{\partial f}{\partial x}\, \overline{i} + \frac{\partial f}{\partial y}\, \overline{j} + \frac{\partial f}{\partial z}\, \overline{k}. \tag{2}$$

In each case ∇f is easy to compute. On the other hand, the meaning of the gradient may not be quite so easy to comprehend. The gradient is a vector which points in the direction of greatest increase in the value of the function. This is a simple statement. We present several examples below to insure that you understand what this simple statement means.

We consider the two variable case first. Suppose you are a mountain climber. Your compass has been converted into a gradient machine. Whenever held level, the "needle" will point in the direction you should step so that your step is as "up" as possible. Now, if you are scaling a sheer cliff, the matter of "up" is rather clear since "down" is "intuitively obvious." For the less venturesome climber, the gradient machine is needed to determine which horizontal direction will result in the "most up." For example, it is not always obvious to the golfer whether or not his putt is uphill, or more critically, the extent to which it is a sidehill lie. "The direction of greatest increase" is telling us how to change the domain variables x and y in order to effect the greatest increase in the function value $z = f(x,y)$.

335

It may help to draw the gradient in the xy-plane along with some level curves. Consider our mountainside to be the plane whose equation is z = -2x.

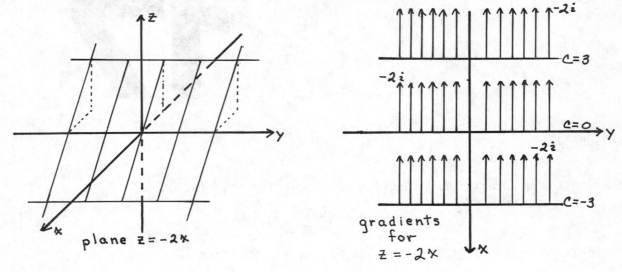

We would expect the "direction of greatest increase" to be parallel to the negative x-axis, as this is the direction of greatest tilt in the plane. From equation (1) we find that

$$\nabla f = -2\bar{i} + 0\bar{j}.$$

If, instead, we were trying to climb out of a hole in the shape of the paraboloid of revolution z = f(x,y) = $x^2 + y^2$, the diagram implies we should head directly away from the origin to effect the quickest escape.

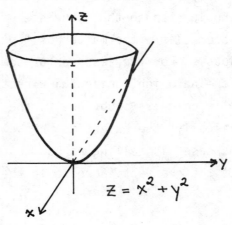

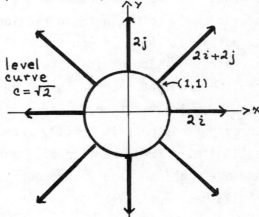

This is verified by the fact that

$$\nabla f = 2x\overline{i} + 2y\overline{j} \ .$$

A more complicated example, which is still easy to visualize, is the elliptic paraboloid $z = 4x^2 + y^2$.

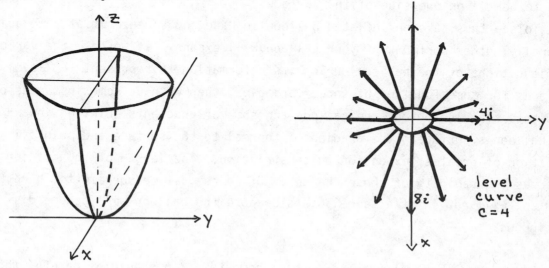

We compute the gradient as

$$\nabla f = 8x\overline{i} + 2y\overline{j}.$$

Notice in particular for x and y greater than zero that as long as $x > \frac{1}{4} y$ the gradient will be more in line with the x-axis than the y-axis. This supports our feel for the surface since the slope of ascent in the direction of the x-axis is "steeper" than that in the direction of the y-axis.

If it seems that each gradient is perpendicular (normal) to the level curve, your eyes are not deceiving you. This perpendicularity is proven in Section 17.4. It provides a foundation for our discussion of tangent planes. But, for the moment, notice how this observation reinforces our notion that the gradient points in the "direction of greatest increase" in the function. If we traverse the surface so that the projection in the xy-plane of our path is a level curve, there is no increase or decrease in our altitude. Common sense suggests that a right-turn or a left-turn would produce the path of greatest ascent (or descent).

What about the 3-variable case? Well, level surfaces are somewhat difficult
to draw. But, we can suggest a physical model. A porcupine is a decent level
surface whose quills make admirable gradients. A collection of porcupines each of
which is swallowed whole by a larger porcupine is a prickly but palatable model
for a function of three variables complete with gradients. The gradients don't
have to be all on one side of the surface.

To see this take an inflated balloon in hand and squeeze it at some point.
Naturally, other portions of your balloon will expand. If we take the various
configurations of the balloon during its deformation to represent the level
surfaces for some function of three variables, then we have some instances of
gradients sticking out of the balloon and some of gradients pointing inward from
the balloon's surface. The gradient is normal to the surface and points in the
direction of greatest expansion or constriction. The length of the gradient
measures the rapidity of these changes. Of course, if we squeeze too hard, the
gradient will not exist at some points because the balloon will become discontinuous.

Directional derivatives are easy to compute. You should be careful that the
vector you use to indicate the direction has unit length.

Example 1. Find the directional derivative of $f(x,y) = 4x^2 + y^2$ at (1,2) in the
direction of (-3,5).
Solution. The vector from (1,2) to (-3,5) is $-4\overline{i} + 3\overline{j}$ and has length 5. The unit
vector in the specified direction is $\overline{u} = \frac{1}{5}(-4\overline{i} + 3\overline{j})$. The directional derivative
is given by

$$f'_{\overline{u}}(x,y) = \nabla f(x,y)\cdot\overline{u} = (8x\overline{i} + 2y\overline{j})\cdot\overline{u},$$

so that

$$f'_{\overline{u}}(1,2) = (8\overline{i} + 4\overline{j})\cdot(-\frac{4}{5}\overline{i} + \frac{3}{5}\overline{j}) = -4.$$

Example 2. Find the directional derivative of $f(x,y) = 4x^2 + y^2$ at (1,2) in the
direction of (0,4).
Solution. It is important that we take the difference of these coordinates in the

proper order lest we get $-\overline{u}$ rather than $\overline{u}$. Here we have

$$\overline{u} = \frac{1}{\sqrt{5}}(-\overline{i} + 2\overline{j})$$

so that

$$f'_{\overline{u}}(1,2) = (8\overline{i} + 4\overline{j})\bullet(-\frac{1}{\sqrt{5}}\overline{i} + \frac{2}{\sqrt{5}}\overline{j}) = 0.$$

What does this mean? Remember that "the directional derivative represents the component of the gradient vector in that direction." Consequently, at (1,2) it seems the gradient is perpendicular to $\overline{u}$. From our earlier comments it may be reasonable to guess that if we are at (1,2) and moving in the direction of $\overline{u}$, we are moving along a level curve. Let us check this. The level curve through (1,2) is $4x^2 + y^2 = 8$. Since $\frac{dy}{dx} = -\frac{4x}{y}$ for $y \neq 0$, the tangent to this curve at (1,2) is

$$y - 2 = -2(x - 1) \qquad \text{or} \qquad y = -2x + 4.$$

Notice that this tangent line passes through (0,4). Ahem. We are indeed at this instant moving along a level curve.

SECTION 17.3

Almost all of the parametrized treatments of multivariate functions in this text are restricted to the case of a single parameter. Much of the work carries over rather readily to two or more parameters but the details can be cumbersome and uninteresting. For example, you will normally see the chain rule as embodied in Formulas 17.3.3 and 17.3.4. However, lest you overlook it we will example the more general statement you are asked to establish in Exercise 15 (p. 716).

<u>Example 1.</u> Evaluate $\frac{\partial u}{\partial s}$ and $\frac{\partial u}{\partial t}$ when $s = 1$, $t = 2$ given that $u = x^2 - y^3 + 4z$, $x = 3st$, $y = s^2 + t$, $z = -7s + t^3$.
<u>Solution.</u> The general forms for $\frac{\partial u}{\partial s}$ and $\frac{\partial u}{\partial t}$ are

$$\frac{\partial u}{\partial s} = \frac{\partial u}{\partial x}\frac{\partial x}{\partial s} + \frac{\partial u}{\partial y}\frac{\partial y}{\partial s} + \frac{\partial u}{\partial z}\frac{\partial z}{\partial s} \qquad\qquad (1)$$

and

$$\frac{\partial u}{\partial t} = \frac{\partial u}{\partial x}\frac{\partial x}{\partial t} + \frac{\partial u}{\partial y}\frac{\partial y}{\partial t} + \frac{\partial u}{\partial z}\frac{\partial z}{\partial t} \ . \tag{2}$$

Notice for example how equation (1) when written as

$$\frac{\partial}{\partial s}[f(\overline{r}(s,t))] = \nabla f(\overline{r}(s,t)) \cdot \frac{\partial}{\partial s}[\overline{r}(s,t)]$$

parallels the chain rule statement in Theorem 17.3.2. Substituting the specifics
in equations (1) and (2), we have

$$\frac{\partial u}{\partial s} = (2x)(3t) + (-3y^2)(2s) + (4)(-7)$$

and

$$\frac{\partial u}{\partial t} = (2x)(3s) + (-3y^2)(1) + (4)(3t^2).$$

Finally, when s = 1 and t = 2, we obtain x = 6, y = 3 and z = 1 from the original
equations. Thus, $\partial u/\partial s = -10$ and $\partial u/\partial t = 57$.

We close with a comment on notation. If F is a function of the single
variable t, then we may write dF/dt instead of $\partial F/\partial t$. Such is usually done. Also,
$\partial x/\partial x = 1$, as you would expect. Perhaps you can now solve Exercise 24 (p. 717).
We shall use this result in the next Section.

SECTION 17.4

From some of the diagrams and comments made in Sections 17.1-2 it should come as
no surprise that the gradient at a point for a function of two (three) variables
is perpendicular to the level curve (surface) which passes through that point.

In the two-variable case some students seem to confuse Formulas 17.4.2 and
17.4.3 for the normal and tangent lines, respectively. It seems natural to suspect
that the slope dy/dx is given by $(\partial f/\partial y)/(\partial f/\partial x)$. This is poor detective work,
for you were asked to show in Exercise 24 of Section 17.3 that

$$\frac{dy}{dx} = - \frac{\partial f/\partial x}{\partial f/\partial y} \qquad (\partial f/\partial y \neq 0). \tag{1}$$

Indeed, this is the slope for the tangent line as given in 17.4.3. Perhaps a diagram will help.

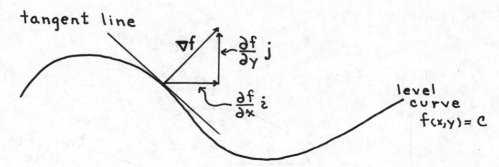

Clearly, the slope of the <u>normal</u> line as given by ∇f is (∂f/∂y)/(∂f/∂x). With the slope of the tangent line as the "negative reciprocal" equation (1) is upheld again.

Similarly, some students have a tendency to mishandle Formulas 17.4.5 and 17.4.6. The problem is that a surface is sometimes given implicitly as

$$f(x,y,z) = 0$$

and sometimes explicitly as

$$z = g(x,y).$$

Of course, the two cases agree since we could write

$$f(x,y,z) = g(x,y) - z = 0.$$

A diagram may help here too.

Consider a surface $z = g(x,y)$ with a tangent plane at some point. If we slice this surface with a vertical plane $P(x = x_0)$ perpendicular to the x-axis, we describe a space curve on the surface. The intersection of P and the tangent plane yields the tangent line to the curve. As noted back in Section 16.4 the "slope" of this tangent line is ∂g/∂y. Equivalently, a direction vector for this line is $(0,1,\partial g/\partial y)$. Similarly, a direction vector for the tangent line cut out by a plane $y = y_0$ is $(1,0,\partial g/\partial x)$. Both of these lines lie in the tangent plane. So if $\bar{u} = (a,b,c)$ is a normal to the plane, we must have

$$\bar{u} \cdot (0,1,\partial g/\partial y) = 0 = \bar{u} \cdot (1,0,\partial g/\partial x).$$

Clearly,

$$\overline{u} = (\partial g/\partial x, \ \partial g/\partial y, \ -1)$$

meets these conditions since

$$\overline{u} = \nabla f = (\partial f/\partial x, \ \partial f/\partial y, \ \partial f/\partial z)$$

for $f(x,y,z) = g(x,y) - z = 0$.

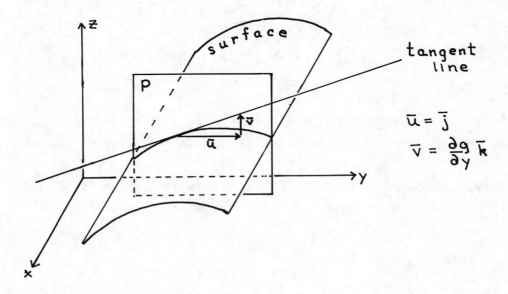

In the event that the surface is specified implicitly by $f(x,y,z) = c$, can you show that direction vectors for these two tangent lines are

$$\left(0, \ 1, \ - \frac{\partial f/\partial y}{\partial f/\partial z}\right) \qquad \text{and} \qquad \left(1, \ 0, \ - \frac{\partial f/\partial x}{\partial f/\partial z}\right)$$

respectively? You could appeal to our earlier discussion of the two variable case, since y or x, respectively, is held constant along the space curves. Or, you could use the directional derivative. For example, the component of the gradient that lies in the plane $x = x_0$ is $(0, \ \partial f/\partial y, \ \partial f/\partial z)$. This component is perpendicular to the tangent line so a direction vector for the tangent is $\left(0, \ 1, \ - \frac{\partial f/\partial y}{\partial f/\partial z}\right)$.

Example 1. The surfaces $z = 4x^2 + y^2$ and $x^2 + y^2 + 3 = \frac{1}{8} z^2$ intersect in a curve which passes through the point $(1,2,8)$. What are the equations for the line of

intersection of the respective tangent planes to the two surfaces at this point?
<u>Solution</u>. We rewrite the elliptic paraboloid $z = 4x^2 + y^2$ as

$$f(x,y,z) = 4x^2 + y^2 - z = 0.$$

The gradient

$$\nabla f = (8x, 2y, -1)$$

is normal to the tangent plane. The normal at $(1,2,8)$ is $(8,4,-1)$ and the equation of the tangent plane is

$$8(x - 1) + 4(y - 2) - (z - 8) = 0. \tag{1}$$

Let us rewrite the hyperboloid of two sheets $x^2 + y^2 + 3 = \frac{1}{8} z^2$ as

$$f(x,y,z) = x^2 + y^2 - \frac{1}{8} z^2 + 3 = 0.$$

Then,

$$\nabla f = (2x, 2y, -\frac{z}{4})$$

so that $(2,4,-2)$ is normal to the tangent plane at $(1,2,8)$. The equation for this tangent plane is

$$(x - 1) + 2(y - 2) - (z - 8) = 0. \tag{2}$$

We solve planes (1) and (2) simultaneously to find the line of intersection:

$$x = 1 + 2t \qquad y = 2 - 7t \qquad z = 8 - 12t.$$

It was not necessary to find the equations for the tangent planes. Their line of intersection is a line which is orthogonal (perpendicular) to the normal to each tangent plane and passes through the point $(1,2,8)$. This problem is now easy to solve using the cross product. Since the vector

$$\begin{vmatrix} \bar{i} & \bar{j} & \bar{k} \\ 8 & 4 & 1 \\ 2 & 4 & -2 \end{vmatrix} = -4\bar{i} + 14\bar{j} + 24\bar{k}$$

is **orthogonal** to both normals,

$$x = 1 - 4u \qquad y = 2 + 14u \qquad z = 8 + 24u$$

is the required line. You should verify that this agrees with the previous
solution.

SECTION 17.5

Even for simple problems the algebra that results from testing stationary points
by using the definitions for extrema is almost prohibitive. Fortunately, other
techniques that parallel the tests for the single variable case are developed in
the next Section. Solutions for Exercises 2 and 13 (p. 731) are offered below.

<u>Solution for Exercise 2 (p. 731).</u> We identify the stationary points for

$$f(x,y) = x^2 + 2y^2 - 4y,$$

an elliptic paraboloid, by setting the first partials equal to zero.

$$\frac{\partial f}{\partial x} = 2x = 0 \qquad \text{and} \qquad \frac{\partial f}{\partial y} = 4y - 4 = 0.$$

The only stationary point is (0,1) and f(0,1) = -2. If $f(x,y) \geq -2$ for all (x,y)
near (0,1), we have a local minimum at (0,1). If $f(x,y) \leq -2$ for all (x,y) near
(0,1), we have a local maximum at (0,1). If neither inequality holds for all
(x,y) near (0,1), we have a saddle point. Thus, we will compare f(h,1 + k) and
f(0,1), where x = 0 + h, y = 1 + k and h, k are "small" so that (x,y) is near
(0,1).

$$f(h,1 + k) = h^2 + 2(1 + k)^2 - 4(1 + k) = h^2 + 2k^2 - 2$$

Rather easily

$$h^2 + 2k^2 - 2 \geq -2$$

so there is a local minimum of -2 at (0,1).

<u>Solution for Exercise 13 (p. 731).</u> As above, we begin with

$$\frac{\partial f}{\partial x} = 2x + y + 2 = 0 \qquad \text{and} \qquad \frac{\partial f}{\partial y} = x + 2 = 0.$$

The only stationary point is (-2,2). We want to compare f(-2 + h, 2 + k) and f(-2,2) = 1 for small h and k.

$$f(-2 + h, 2 + k) = (-2 + h)^2 + (-2 + h)(2 + k) + 2(-2 + h) + 2(2 + k) + 1$$

$$= h^2 + hk + 1 = h^2 + hk + f(-2,2)$$

Now is $h^2 + hk$ never negative? If so we have a minimum at (-2,2). If $h^2 + hk$ is not positive for small h and k we have a maximum. However, $h^2 + hk$ is both positive and negative for arbitrarily small h and k, since

$$h^2 + hk > 0 \qquad \text{when } k = h.$$

and $\qquad\qquad\qquad h^2 + hk < 0 \qquad \text{when } k = -2h.$

There is a saddle point at (-2,2).

SECTION 17.6

The Second-Partials Test makes life much easier. We restate it for easy reference as you read the examples.

Let (x_0, y_0) be a stationary point of $f(x,y)$, and let $A = \dfrac{\partial^2 f}{\partial x^2}(x_0, y_0)$,

$$B = \frac{\partial^2 f}{\partial y \partial x}(x_0, y_0), \quad C = \frac{\partial^2 f}{\partial y^2}(x_0, y_0), \quad D = B^2 - AC.$$

If $D > 0$, we have a saddle point at (x_0, y_0).

If $D < 0$ and $A > 0$, we have a minimum at (x_0, y_0).

If $D < 0$ and $A < 0$, we have a maximum at (x_0, y_0).

If $D = 0$, then anything could occur at (x_0, y_0).

In connection with the inconclusive case of $D = 0$, we note that the last example in the text merits more than a single reading. Extensive tests involving higher order partials can be made to resolve the case of $D = 0$. The story is not simple. We leave it untold. You may also wonder why the min-max test when $D < 0$ is made with A. We could also use C. For, if A and C do not agree in sign, then $D \geq 0$ and we do not have a reason for looking at the sign of A (or C). Following a couple of examples, solutions for Exercises 19, 22, and 23 (p. 736) are given.

<u>Example 1</u>. Examine the function $f(x,y) = x^3 - 4xy^2$ at the origin.

<u>Solution</u>. First some calculations. At the origin, $\frac{\partial f}{\partial x} = \frac{\partial f}{\partial y} = 0$. So we consider

$$\frac{\partial^2 f}{\partial x^2} = 6x, \qquad \frac{\partial^2 f}{\partial y \partial x} = -8y, \qquad \frac{\partial^2 f}{\partial y^2} = -8x.$$

Thus,

$$D = 64y^2 + 48x^2.$$

When $x = y = 0$, we have $D = 0$. All is not lost. Notice that

$$f(x,y) = x(x - 2y)(x + 2y)$$

so that the function is zero along the lines $x = 0$, $x = 2y$, and $x = -2y$. These
lines divide the plane into six wedges or pie-shaped regions. In each of these
regions the function is always positive or always negative. By evaluating some
function values we discover the function to be positive in the darkened regions of
our diagram and negative in the white regions.

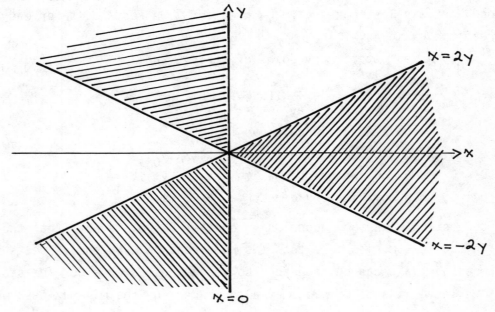

Clearly, there is neither a maximum nor a minimum at the stationary point (0,0).
Accordingly, we call it a saddle point, even though this saddle would only fit a
horse of another color.

Example 2. Determine the extremes for the function $f(x,y) = x^2 - 2xy + y - x$ on the closed region bounded by the curves $y = x^2$, $y = -x$ and $x = 4$.

Solution. This problem is unlike any of those in the text. However, it is a

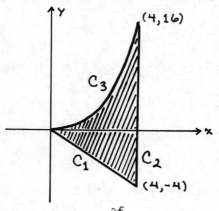

direct analog of the single variable case. We treat the interior of the region and its boundary separately. Recall how we considered the endpoints separately in the single variable case. Here we just happen to have lots of "endpoints." We proceed in the usual fashion for the interior. We calculate

$$\frac{\partial f}{\partial x} = 2x - 2y - 1 \qquad \text{and} \qquad \frac{\partial f}{\partial y} = -2x + 1$$

so that when $\partial f/\partial x = \partial f/\partial y = 0$ we have $x = 1/2$, $y = 0$. This stationary point $(\frac{1}{2},0)$ lies in the interior of our region, so we calculate

$$\frac{\partial^2 f}{\partial x^2} = 2 \qquad\qquad \frac{\partial^2 f}{\partial y \partial x} = -2 \qquad\qquad \frac{\partial^2 f}{\partial y^2} = 0.$$

At the point $(\frac{1}{2},0)$, the value of D is 4. We have a saddle point. The function f is continuous on this closed region so it has absolute extrema. Apparently, they must occur on the boundary.

Now, what about the boundary? We can proceed in a number of ways. We could examine $f(x,y)$ subject to the side conditions that x and y satisfy the equations for the boundaries. Side conditions are discussed in the next Section. Or, for each of the pieces of the boundary we could substitute into $f(x,y)$. For example, on the parabolic arc $y = x^2$ we would examine $f(x,x^2)$ for $0 \le x \le 2$. This approach only works if the formulas for the boundary are relatively simple. A third option is to obtain a parametrization of the boundary, say $\overline{r}(t) = x(t)\overline{i} + y(t)\overline{j}$ for t in some interval. Then, we appeal to the chain rule

$$f'(t) = (\frac{\partial f}{\partial x}\,\overline{i} + \frac{\partial f}{\partial y}\,\overline{j})\bullet\overline{r}'(t)$$

and analyze f as a function of the single variable t.

The actual computation of f'(t) is easier here if first we write f in terms of t. There are three pieces to the boundary. On $C_1(y = -x)$ we could write $x(t) = t$, $y(t) = -t$ for $0 \leq t \leq 4$. This gives us

$$f_1(t) = f(x(t), y(t)) = 3t^2 - 2t.$$

On $C_2(x = 4)$ we could let $x(t) = 4$, $y(t) = t$ for $-4 \leq t \leq 16$. This gives us

$$f_2(t) = f(x(t), y(t)) = -7t + 12.$$

Finally, on $C_3(y = x^2)$ we can let $x(t) = t$, $y(t) = t^2$ for $0 \leq t \leq 4$. We find

$$f_3(t) = -2t^3 + 2t^2 - t.$$

We need to examine each of these three functions. And, we should not forget the "endpoints" at $(0,0)$, $(4,-4)$ and $(4,16)$. The following calculations will help.

$f_1'(t) = 6t - 2$	$f_1''(t) = 6$	$0 < t < 4$
$f_2'(t) = -7$	$f_2''(t) = 0$	$-4 < t < 16$
$f_3'(t) = -6t^2 + 4t - 1$	$f_3''(t) = -12t + 4$	$0 < t < 4$

Notice that only f_1 has a local extreme. It occurs for $t = 1/3$, or the point $(1/3, -1/3)$, and is a minimum: $f(1/3, -1/3) = -1/3$.

There remain the endpoints. Since $f_3'(4^-) < 0$, there is an endpoint minimum for f_3 at $(4,16)$. Further, $f_2'(16^-) < 0$, so $(4,16)$ is also a minimum for f_2. Thus, f has a minimum at $(4,16)$: $f(4,16) = -100$. At $(4,-4)$ we find $f_1'(4^-) > 0$ and $f_2'(-4^+) < 0$ so there is a maximum: $f(4,-4) = 40$. Finally, $f_1'(0^+) < 0$ and $f_3'(0^+) < 0$, so there is also a local maximum at $(0,0)$: $f(0,0) = 0$.

Solution for Exercise 19 (p. 736). Clearly, the lengths of the sides of the solid are x,y,z. We want to maximize xyz subject to the restraint that $x + y + z = 1$. We consider the function

$$f(x,y) = xy(1 - x - y) \qquad x,y > 0.$$

First,

$$\frac{\partial f}{\partial x} = y - 2xy - y^2 = y(1 - 2x - y)$$

and

$$\frac{\partial f}{\partial y} = x - x^2 - 2xy = x(1 - x - 2y).$$

We should remember that x and y are positive, so when we set $\partial f/\partial x = 0 = \partial f/\partial y$ the only stationary point is $(\frac{1}{3}, \frac{1}{3})$. Next,

$$\frac{\partial^2 f}{\partial x^2} = -2y, \qquad \frac{\partial^2 f}{\partial y \partial x} = 1 - 2x - 2y, \qquad \frac{\partial^2 f}{\partial y^2} = -2x$$

so that

$$D = (1 - 2x - 2y)^2 - 4xy.$$

At $(\frac{1}{3}, \frac{1}{3})$ the value of D is $-1/3$ and $\partial^2 f/\partial x^2 < 0$ so f is maximal. The maximum volume is $f(1/3, 1/3) = 1/27$ cubic unit.

Solution for Exercise 22 (p. 736). It might seem that there are three variables here: x, y, and θ. However, the perimeter is a constant P, so we should be able to eliminate one of the variables. First, we identify some of the other lengths in the figure. Remember that the triangle is isosceles. Thus,

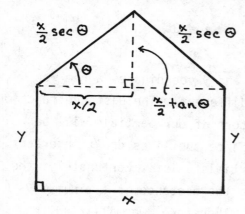

$$P = x + 2y + 2(\frac{x}{2} \sec \theta)$$

which we may solve for y.

$$y = \frac{1}{2}(P - x - x \sec \theta).$$

Now we are in a position to specify an area function

$$f(x,y) = xy + \frac{1}{2}(x)(\frac{x}{2} \tan \theta) = \frac{x}{2}(P - x - x \sec \theta) + \frac{x^2}{4} \tan \theta.$$

The domain for f is $0 < \theta < \pi/2$ and $0 < x < \frac{P}{1 + \sec \theta}$. Predictably, we find

$$\frac{\partial f}{\partial x} = \frac{P}{2} - x - x \sec \theta + \frac{x}{2} \tan \theta$$

and

$$\frac{\partial f}{\partial \theta} = -\frac{x^2}{2} \sec \theta \tan \theta + \frac{x^2}{4} \sec^2\theta.$$

Within the domain of f, the only time

$$\frac{\partial f}{\partial \theta} = \frac{x^2}{4} \sec \theta(\sec \theta - 2 \tan \theta)$$

is zero occurs for $\theta = \pi/6$. From $\partial f/\partial x = 0$ and $\theta = \pi/6$, we obtain $x = (2 - \sqrt{3})P$. Some easy calculations lead to

$$D = \frac{x^2}{4} \sec^2\theta(\tan^2\theta - 2 - 2 \sec \theta + \tan \theta)$$

and

$$\frac{\partial^2 f}{\partial x^2} = -1 - \sec \theta + \frac{1}{2} \tan \theta.$$

So, f has a minimum for $\theta = \pi/6$ and $x = P(2 - \sqrt{3})$. The dimensions of the perimeter

are base $x = (2 - \sqrt{3})P$, sides $y = (\frac{3 - \sqrt{3}}{6})P$ and roof $\frac{x}{2} \sec \theta = (\frac{2\sqrt{3} - 3}{3})P$.

Solution for Exercise 23 (p. 736). We begin by recording the parametric form for the lines. We have

$$x = t \qquad\qquad y = 2t \qquad\qquad z = 3t$$

and
$$x = s \qquad\qquad y = 2 + s \qquad\qquad z = s.$$

Notice that the lines do not intersect. If the x and y coordinates agree we obtain t = s = 2 for which the z coordinates are different. The distance function involves the square root function so the denominators of our partials will be cluttered with massive square root expressions. Since the lines don't intersect these square roots are always positive and the partials exist. Fortunately, the square root function is an increasing function, so the distance is a minumum whenever the square of the distance is a minimum. Therefore, we will find the minimum of

$$f(s,t) = (t - s)^2 + (2t - 2 - s)^2 + (3t - s)^2 \qquad\qquad \text{(all real } s,t\text{),}$$

which represents the square of the distance between two arbitrary points on the respective lines. It may help to rewrite f as:

$$f(s,t) = 14t^2 + 3s^2 - 12st - 8t + 4s + 4.$$

Then,

$$\frac{\partial f}{\partial s} = 6s - 12t + 4 \qquad \text{and} \qquad \frac{\partial f}{\partial t} = 28t - 12s - 8.$$

For $\partial f/\partial s = \partial f/\partial t = 0$, we obtain $t = 0$, $s = 2/3$ as the only stationary point. As you should verify, we have $D = -24$, $\partial^2 f/\partial s^2 = 6$ and a minimum. This minimum distance is $\sqrt{f(2/3,0)} = 2\sqrt{2}$.

SECTION 17.7

The method of Lagrange multipliers may be extended to functions involving more than three variables and problems with more than one side condition. We are faced then with the mammoth task of solving the system that results from

$$\nabla f = \lambda_1 \nabla g_1 + \lambda_2 \nabla g_2 + \cdots \lambda_n \nabla g_n.$$

We shall not encounter such problems. Instead, we confine ourselves to providing solutions for Exercises 11, 15, 18, and 19 (p. 743).

Solution for Exercise 11 (p. 743). We wish to maximize

$$f(x,y,z) = 2x + 3y + 5z$$

subject to the restraint

$$g(x,y,z) = x^2 + y^2 + z^2 - 19 = 0.$$

The condition $\nabla f = \lambda \nabla g$ for stationary points gives us the system

$$2 = 2\lambda x \qquad\qquad 3 = 2\lambda y \qquad\qquad 5 = 2\lambda z.$$

Clearly, $\lambda \neq 0$ so we may solve for x,y,z in terms of λ and use $g(x,y,z) = 0$ to find

$$(\frac{2}{2\lambda})^2 + (\frac{3}{2\lambda})^2 + (\frac{5}{2\lambda})^2 = 19 \qquad\qquad \text{or} \qquad\qquad \lambda = \pm\ 1/\sqrt{2}.$$

We have $(\sqrt{2}, \frac{3}{2}\sqrt{2}, \frac{5}{2}\sqrt{2})$ and $(-\sqrt{2}, -\frac{3}{2}\sqrt{2}, -\frac{5}{2}\sqrt{2})$ as stationary points. The value of f at these points is $19\sqrt{2}$ and $-19\sqrt{2}$, respectively. Of these two values, $19\sqrt{2}$ is the maximum.

Why is it the maximum for f subject to the given side restraint? We give a geometric argument. The equation $f(x,y,z) = c$ for a constant c is a plane. The effect of the restraint $g(x,y,z) = 0$ is to require that the plane and the sphere intersect. The maximum (and minimum) value occurs then when the plane is tangent to the sphere.

<u>Solution for Exercise 15 (p. 743)</u>. Let (x,y) be any point on the parabola. We want to minimize

$$f(x,y) = \sqrt{(x - 0)^2 + (y - 1)^2}$$

subject to the restraint

$$g(x,y) = x^2 - 4y = 0. \tag{1}$$

From the condition $\nabla f = \lambda \nabla g$ we obtain

$$\frac{x}{\sqrt{x^2 + (y - 1)^2}} = 2\lambda x \qquad \text{and} \qquad \frac{(y - 1)}{\sqrt{x^2 + (y - 1)^2}} = -4\lambda.$$

Substituting the second equation in the first, we find

$$\frac{2x}{\sqrt{x^2 + (y - 1)^2}} = \frac{-(y - 1)x}{\sqrt{x^2 + (y - 1)^2}} . \tag{2}$$

The only solution to equations (1) and (2) is $x = y = 0$. The required minimum (why?) distance is 1.

Our procedure yields a general result that may reinforce your feeling for $\nabla f = \lambda \nabla g$. Namely, to identify the distance from any point to any curve we need only determine the normals to the curve through the given point. In more geometric terms, the distance formula $f(x,y)$ really embodies a collection of concentric circles. The distance problem requires that we find the smallest such circle which touches the given curve. But, then the circle and the curve share a tangent line. Ah, but then they must also share normals and $\nabla f = \lambda \nabla g$ (as direction vectors for these normals).

<u>Solution for Exercise 18 (p. 743)</u>. Let (x,y,z) be any point on the plane. We want to minimize

$$f(x,y,z) = \sqrt{x^2 + y^2 + z^2}$$

subject to the restraint

$$g(x,y,z) = Ax + By + Cz + D = 0. \tag{3}$$

Our computations will be greatly simplified if instead we treat the equivalent problem of minimizing

$$F(x,y,z) = x^2 + y^2 + z^2.$$

From $\nabla F = \lambda \nabla g$ we derive

$$2x = \lambda A \qquad\qquad 2y = \lambda B \qquad\qquad 2z = \lambda C \tag{4}$$

Substitution for x,y,z in equation (3) gives us

$$\lambda(A^2 + B^2 + C^2) + 2D = 0.$$

Solving this equation for λ, we substitute back in equations (4) to find

$$x = -AD/s \qquad\qquad y = -BD/s \qquad\qquad z = -CD/s,$$

where $s = A^2 + B^2 + C^2$. This is the only stationary point and from the nature of the problem we know the minimum exists. Thus, the required distance is

$$f\left(-\frac{AD}{s}, -\frac{BD}{s}, -\frac{CD}{s}\right) = \sqrt{\frac{D^2(A^2 + B^2 + C^2)}{s^2}} = \sqrt{\frac{D^2}{s}} = \frac{|D|}{\sqrt{A^2 + B^2 + C^2}}$$

where we have remembered that $\sqrt{D^2} = |D|$.

Solution for Exercise 19 (p. 743). We draw your attention to the fact that the point you are asked to find here is the same as the point you were asked to identify in Exercise 15 of Section 17.5; that is, the sum of the squares of the distances from P to the vertices is also minimal. The algebra involved is otherwise unenlightening.

SECTION 17.9

We remarked in Section 4.9 while studying the differential for single variable

354

functions that it is legitimate to write $dy = f'(x)dx$. This latitude with the notation allows us to look upon the differential as the derivative of $y = f(x)$ with respect to an unspecified variable. We may do the same for the differential in the multivariate case. We may write

$$df = \frac{\partial f}{\partial x} dx + \frac{\partial f}{\partial y} dy + \frac{\partial f}{\partial z} dz$$

for the function $f(x,y,z)$. Then if x,y,z are functions of t we have the chain rule

$$\frac{df}{dt} = \frac{\partial f}{\partial x} \frac{dx}{dt} + \frac{\partial f}{\partial y} \frac{dy}{dt} + \frac{\partial f}{\partial z} \frac{dz}{dt} .$$

On the other hand, if x,y,z are functions of s and t we would have statements like

$$\frac{\partial f}{\partial s} = \frac{\partial f}{\partial x} \frac{\partial x}{\partial s} + \frac{\partial f}{\partial y} \frac{\partial y}{\partial s} + \frac{\partial f}{\partial z} \frac{\partial z}{\partial s} .$$

A differential equation is an equation involving a function and its derivatives. For example, any time you evaluate the integral

$$y = F(x) = \int_a^x f(t)dt$$

you are solving the differential equation

$$\frac{dy}{dx} = F'(x) = f(x).$$

This differential equation is called <u>separable</u> since we may separate the x's and y's as

$$dy = f(x)dx$$

and integrate both sides of the equation to find y. Section 6.8 discussed the linear differential equation

$$\frac{dy}{dx} + P(x)y = Q(x).$$

In this Section we are concerned with reconstructing, if possible, a function $f(x,y)$ from its gradient

$$P(x,y)\overline{i} + Q(x,y)\overline{j}.$$

This is precisely the activity involved in solving what are known as <u>exact</u> differential equations. We may write the differential equation

$$P(x,y) + Q(x,y)\frac{dy}{dx} = 0$$

as

$$P(x,y)dx + Q(x,y)dy = 0. \tag{1}$$

This differential equation is exact iff there is some function $f(x,y) = c$ such that

$$\frac{\partial f}{\partial x} = P(x,y) \qquad \text{and} \qquad \frac{\partial f}{\partial y} = Q(x,y)$$

so that (1) takes the form

$$\frac{\partial f}{\partial x} dx + \frac{\partial f}{\partial y} dy = 0.$$

As developed in the text, the test used to determine exactness is

$$\frac{\partial P}{\partial y} = \frac{\partial^2 f}{\partial y \partial x} = \frac{\partial^2 f}{\partial x \partial y} = \frac{\partial Q}{\partial x} .$$

We close with an example.

Example 1. Solve the differential equation

$$(2y - 4xy)\frac{dy}{dx} + (3x^2 - 2y^2 + 3) = 0$$

subject to the condition $y(0) = 2$.

Solution. Since

$$\frac{\partial}{\partial y}(3x^2 - 2y^2 + 3) = -4y = \frac{\partial}{\partial x}(2y - 4xy),$$

the differential equation

$$(3x^2 - 2y^2 + 3)dx + (2y - 4xy)dy = 0$$

is exact. The solution is $f(x,y) = c$ where

$$\frac{\partial f}{\partial x} = 3x^2 - 2y^2 + 3 \qquad \text{and} \qquad \frac{\partial f}{\partial y} = 2x - 4xy. \tag{2}$$

Integrating the first of these two equations with respect to x, we find

$$f(x,y) = x^3 - 2xy^2 + 3x + g(y), \tag{3}$$

where g(y) is some function in y. Next we differentiate equation (3) with respect to y and compare the result with the second equation in (2). Thus,

$$-4xy + g'(y) = 2y - 4$$

so that

$$g'(y) = 2y \qquad \text{or} \qquad g(y) = y^2 + k$$

and

$$f(x,y) = c = x^3 - 2xy^2 + 3x + y^2.$$

The constant k is absorbed in the arbitrary constant c. Since $y = 2$ when $x = 0$, we discover that $c = 4$. The required solution is

$$x^3 - 2xy^2 + 3x + y^2 = 4.$$

Double and Triple Integrals

18

This Chapter focuses on multiple integrals. The hardest part of this subject is in visualizing three dimensional objects for which one seeks an integral representation for the volume. Some comments aimed at alleviating this problem area are made later. Everything else we shall see in connection with this subject is straightforward and mechanical. Unless you truly master these mechanical points as they come up, you may be lost along the way.

Some complicated notation is needed to define multiple integrals. You can help insure that the remaining sections are easy to read by mastering the multiple-sigma notation. Additional examples are given below.

Example 1. $\displaystyle\sum_{i=2}^{5} \sum_{j=2}^{3} j^2 = \sum_{i=2}^{5} (2^2 + 3^2) = 4(2^2 + 3^2)$.

Example 2. $\displaystyle\sum_{j=2}^{3} \sum_{i=2}^{5} j^2 = \sum_{j=2}^{3} j^2 \sum_{i=2}^{5} 1 = \sum_{j=2}^{3} j^2(4) = 4 \cdot 2^2 + 4 \cdot 3^2$.

Example 3. $\displaystyle\sum_{j=1}^{3} \sum_{i=1}^{2} (i + j) = \sum_{j=1}^{3} [(1 + j) + (2 + j)] = \sum_{j=1}^{3} (3 + 2j)$

$$= 5 + 7 + 9 = 21.$$

Example 4. $\displaystyle\sum_{i=1}^{4} \sum_{j=1}^{3} (i + j) = \sum_{i=1}^{4} \left(\sum_{j=1}^{3} i + \sum_{j=1}^{3} j \right) = \sum_{i=1}^{4} (3i + 6)$

$$= 3 \sum_{i=1}^{4} i + 6 \sum_{i=1}^{4} 1 = 3(10) + 6(4) = 54.$$

357

The next two examples go beyond what you need to know about multiple-sigma notation.

Example 5. $\displaystyle\sum_{j=1}^{4}\sum_{i=1}^{j} i = \sum_{i=1}^{1} i + \sum_{i=1}^{2} i + \sum_{i=1}^{3} i + \sum_{i=1}^{4} i$

$\qquad\qquad\qquad = 1 + (1 + 2) + (1 + 2 + 3) + (1 + 2 + 3 + 4)$

$\qquad\qquad\qquad = 1 + 3 + 6 + 10 = 20$

or, $\displaystyle\sum_{j=1}^{4}\sum_{i=1}^{j} i = \sum_{j=1}^{4} \frac{j(j+1)}{2} = \frac{1}{2} \sum_{j=1}^{4} (j^2 + j) = \frac{1}{2} \sum_{j=1}^{4} j^2 + \frac{1}{2} \sum_{j=1}^{4} j$

$\qquad\qquad\qquad = \frac{1}{2}(1 + 4 + 9 + 16) + \frac{1}{2}(1 + 2 + 3 + 4)$

$\qquad\qquad\qquad = 15 + 5 = 20.$

Example 6. $\displaystyle\sum_{j=1}^{4}\sum_{i=1}^{j} a_{ij} = \sum_{i=1}^{4}\sum_{j=i}^{4} a_{ij}.$ Can you provide the details?

SECTION 18.2

Solutions for Exercises 5, 6 and 8 are given below. You should be aware that the extrema for a function do not necessarily occur at the corners of the rectangles in the partition, even though it is rigged that way in the numerical problems to keep the calculations simple.

Solution for Exercises 5 and 6 (p. 769). We label the rectangles of the partition as follows.

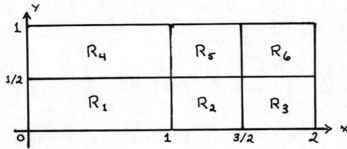

The data required for the computation of $L_f(P)$ and $U_f(P)$ for $f(x,y) = x + 2y$ are given in the table below. Notice that for this function on this region the minimum and maximum values for $f(x,y)$ on a rectangle in the partition are found at the lower left-hand and upper right-hand corners, respectively. Consequently,

i	1	2	3	4	5	6
min value of f on R_i	0	1	3/2	1	2	5/2
max value of f on R_i	2	5/2	3	3	7/2	4
Area of R_i	1/2	1/4	1/4	1/2	1/4	1/4

so that

$$L_f(P) = 0(\tfrac{1}{2}) + 1(\tfrac{1}{4}) + \tfrac{3}{2}(\tfrac{1}{4}) + 1(\tfrac{1}{2}) + 2(\tfrac{1}{4}) + \tfrac{5}{2}(\tfrac{1}{4}) = \tfrac{9}{4}$$

and

$$U_f(P) = 2(\tfrac{1}{2}) + \tfrac{5}{2}(\tfrac{1}{4}) + 3(\tfrac{1}{4}) + 3(\tfrac{1}{2}) + \tfrac{7}{2}(\tfrac{1}{4}) + 4(\tfrac{1}{4}) = \tfrac{23}{4} .$$

Solution for Exercise 8 (p. 769). On each rectangle

$$R_{ij} = \{(x,y) : x \in [x_{i-1}, x_i], \ y \in [y_{j-1}, y_j]\}.$$

The function $f(x,y) = x + 2y$ has a maximum $M_{ij} = x_i + 2y_j$. Thus,

$$U_f(P) = \sum_{i=1}^{m} \sum_{j=1}^{n} M_{ij} \cdot (\text{area of } R_{ij})$$

$$= \sum_{i=1}^{m} \sum_{j=1}^{n} (x_i + 2y_j)(x_i - x_{i-1})(y_j - y_{j-1}).$$

SECTIONS 18.3-4

The evaluation of iterated integrals simply reciprocates the technique of partial differentiation. The "inner" integral is the first to be evaluated. For example, to calculate

$$xy + \frac{y^2}{2}$$
$$= x^2y + \frac{xy^2}{2} - xy + \frac{y^2}{2}$$

$$I = \int_2^3 \int_1^x (x + y)\,dy\,dx, \tag{1}$$

we first compute

$$\int_1^x (x + y)\,dy = \left(xy + \frac{y^2}{2}\right)\bigg|_1^x = \frac{3x^2}{2} - x - \frac{1}{2},$$

where we have treated x as a constant and integrated with respect to y. Then,

$$I = \int_2^3 \left(\frac{3x^2}{2} - x - \frac{1}{2}\right)dx = \left(\frac{x^3}{2} - \frac{x^2}{2} - \frac{x}{2}\right)\bigg|_2^3 = \frac{13}{2}.$$

$$\frac{3x^3}{3} - \frac{x^2}{2} - \frac{x}{2}$$

Consider the planar region Ω in Diagram 1.

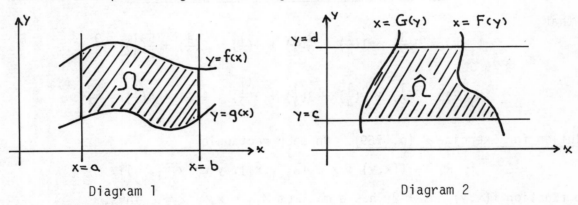

Diagram 1   Diagram 2

The area of region Ω may be represented as

$$\int_a^b \int_{g(x)}^{f(x)} 1 \cdot dy\,dx, \tag{2}$$

which easily simplifies to

$$\int_a^b (y)\bigg|_{g(x)}^{f(x)} dx = \int_a^b [f(x) - g(x)]\,dx. \tag{3}$$

Expression (2) may also be interpreted as the volume of the solid of constant height 1 with base Ω. Further, expression (2) may be interpreted as the mass of a planar plate in the shape of region Ω with uniform density of one unit. More generally, for positive $F(x,y)$

$$\int_a^b \int_{g(x)}^{f(x)} F(x,y)dy \, dx \qquad\qquad (4)$$

may be thought of as the volume of a solid with variable height F(x,y) and base Ω
or as the mass of a planar plate in the shape of region Ω with variable density
given by F(x,y).

Equation (3) should suggest to you that the determination of the limits of
integration for repeated (iterated) integrals requires nothing more than the
ability to determine upper and lower boundaries and intervals as if you were
evaluating the area of region Ω.

We may also integrate first in the x direction and then in the y direction.
Thus,

$$\int_c^d \int_{G(y)}^{F(y)} 1 \cdot dx \, dy = \int_c^d [F(y) - G(y)]dy$$

represents the area of region $\hat{\Omega}$ in Diagram 2.

Perhaps the problem that gives students the most difficulty is the
determination of the limits of integration. It bears repeating that you already
know how to do this since it essentially entails finding boundary curves and
intervals as you did with area problems in Chapter 5. Expression (4) may be
viewed as the evaluation of a sum involving F(x,y) over a region determined by
the limits or bounds a, b, g(x), f(x). The integrand tells you <u>what</u> you are
summing and the limits of integration tell you <u>where</u>. In one sense the what and
where are independent. They are related, but in practice it may appear that they
are not because first we think about what and then we think about where.

For example, suppose we are trying to find the volume of a solid which is
bounded above by z = A(x,y) and bounded below by z = B(x,y). Our integrand is
the height of the solid given by F(x,y) = A(x,y) - B(x,y). We know we will have
one or more integrals that look like

$$\iint F(x,y)dy \, dx \quad \text{or} \quad \cdot \iint F(x,y)dx \, dy.$$

The <u>projection</u> of this solid in the xy-plane determines the region of integration. This region of integration dictates our limits of integration. In effect, we may ignore F(x,y) at this point, since all we have left to do is the same thing we would do if we were finding the area for the region of integration. And, instead of writing

$$\int_a^b [\text{upper curve} - \text{lower curve}]dx,$$

we fill in the blanks by writing, say,

$$\int_a^b \int_{\text{lower curve}}^{\text{upper curve}} F(x,y)dy\ dx.$$

For the most part, the rest of this Chapter is devoted to bringing home this point in different contexts.

We do more than just "set up" and evaluate integrals. Sometimes we want to change the order of integration for a given integral. It is <u>not</u> simply a matter of interchanging the dx and dy. We must also consider what changes, sometimes dramatic, are needed in the limits of integration.

Look back at the integral in (1). The region of integration is that area bounded above by y = x and below by y = 1 on the interval [2,3].

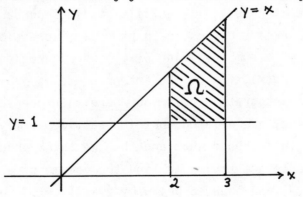

Since the integrand x + y is positive on this region, Ω, we may interpret I as the volume of the solid bounded above by the plane z = x + y and below by the plane z = 0 over the region Ω. We may also consider this as the volume of the solid between z = x + y + y^2 and z = y^2 over Ω since x + y = (x + y + y^2) - (y^2). In

either case Ω is the same;   the limits of integration are unaffected. Suppose
now we wanted to reverse the order of integration. This is really an area
problem. Since we would write

$$\int_1^2 [(3) - (2)]dy + \int_2^3 [(3) - (y)]dy$$

to represent the area of Ω, we write

$$\int_1^2 \int_2^3 (x + y) \, dx \, dy + \int_2^3 \int_y^3 (x + y) \, dx \, dy \qquad (5)$$

to revise the order of integration in (1).

   Finally, the fact that the repeated (or iterated) integrals in (1) and (5) and
the double integral $\iint\limits_{\Omega} (x + y)dA$ exist and are all equal under certain conditions

is provided by an upper level result sometimes called Fubini's Theorem, the proof
of which is quite abstract. Solutions for Exercises 9, 11, 12, 19, 25 and 26
(p. 781) appear below.

Solution for Exercise 9 (p. 781). It is true here that $\sqrt{xy} = \sqrt{x} \cdot \sqrt{y}$ since
neither x nor y is negative.

$$\iint\limits_{\Omega} \sqrt{xy} \; dxdy = \int_0^1 \int_x^{\sqrt{x}} \sqrt{xy} \; dy \, dx$$

$$= \int_0^1 \left(\frac{2}{3} x^{1/2} y^{3/2} \right) \Big|_x^{\sqrt{x}} dx$$

$$= \frac{2}{3} \int_0^1 (x^{5/4} - x^2) dx$$

$$= \frac{2}{3} \left(\frac{4}{9} x^{9/4} - \frac{x^3}{3} \right) \Big|_0^1 = 2/27.$$

Alternately, we could have started with

$$\int_0^1 \int_{y^2}^y \sqrt{xy} \; dx \, dy.$$

Solution for Exercise 11 (p. 781).

$$\iint_{\Omega} (4 - y^2)dx\, dy = \int_{-2}^{2} \int_{y^2/2}^{(8 - y^2)/2} (4 - y^2)dx\, dy$$

$$= \int_{-2}^{2} x(4 - y^2) \Big|_{y^2/2}^{(8 - y^2)/2} dy$$

$$= \int_{-2}^{2} (4 - y^2)[\frac{8 - y^2}{2} - \frac{y^2}{2}]dy$$

$$= \int_{-2}^{2} (4 - y^2)^2 dy$$

$$= (16y - \frac{8}{3} y^3 + \frac{y^5}{5}) \Big|_{-2}^{2} = 512/15.$$

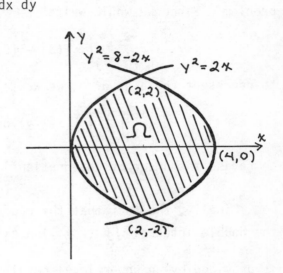

Or,

$$\iint_{\Omega} (4 - y^2)dx\, dy = \int_{0}^{2} \int_{-\sqrt{2x}}^{\sqrt{2x}} (4 - y^2)dy\, dx + \int_{2}^{4} \int_{-\sqrt{8 - 2x}}^{\sqrt{8 - 2x}} (4 - y^2)dy\, dx$$

which you should verify as also having the value 512/15. Notice why integrating first in the y direction requires <u>two</u> repeated integrals.

Solution for Exercise 12 (p. 781).

$$\iint_{\Omega} e^{x^2} dx\, dy = \int_{0}^{2} \int_{0}^{x/2} e^{x^2} dy\, dx$$

$$= \int_{0}^{2} (ye^{x^2}) \Big|_{0}^{x/2} dx = \int_{0}^{2} \frac{xe^{x^2}}{2} dx$$

$$= (\frac{1}{4} e^{x^2}) \Big|_{0}^{2} = \frac{e^4 - 1}{4} .$$

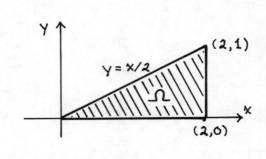

Or,

$$\iint_{\Omega} e^{x^2} dx\, dy = \int_{0}^{1} \int_{2y}^{2} e^{x^2} dx\, dy,$$

which we are <u>unable</u> to evaluate. The integral

$$\int_a^b e^{x^2} dx$$

must be approximated numerically (e.g., by Simpson's Rule) as it is not possible
to determine a function F such that

$$\int e^{x^2} dx = F(x) + C.$$

Section 18.5 of the Supplement will consider

$$\int_{-\infty}^{\infty} e^{-x^2} dx,$$

which is of central importance for Probability and Statistics.

<u>Solution for Exercise 19 (p. 781)</u>. When not explicitly stated, the region of
integration Ω may be found as the projection of the solid on the xy-plane. By
noting the intersection of the plane $\frac{x}{2} + \frac{y}{3} + \frac{z}{4} = 1$ with the other bounding
surfaces for the solid (in this case the coordinate planes), we may determine the
region of integration Ω as:

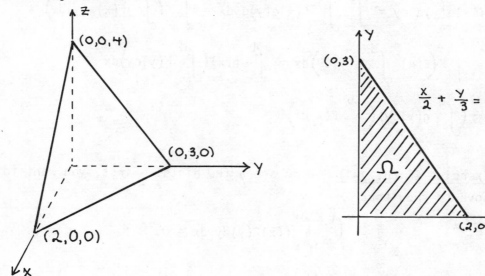

Consequently,

$$\text{Volume} = \iint_{\Omega} 4(1 - \frac{x}{2} - \frac{y}{3})dx\ dy = \int_{0}^{2} \int_{0}^{3(1 - \frac{x}{2})} 4(1 - \frac{x}{2} - \frac{y}{3})dy\ dx$$

$$= 4 \int_{0}^{2} [(1 - \frac{x}{2})y - \frac{y^2}{6}]\Big|_{0}^{3(1 - \frac{x}{2})} dx = 4 \int_{0}^{2} [\frac{3}{2}(1 - \frac{x}{2})^2]dx$$

$$= -4(1 - \frac{x}{2})^3 \Big|_{0}^{2} = 4$$

Or, we could write

$$\text{Volume} = \int_{0}^{3} \int_{0}^{2(1 - \frac{y}{3})} 4(1 - \frac{x}{2} - \frac{y}{3})dx\ dy.$$

<u>Solution for Exercise 25 (p. 781)</u>. The key step in the solution comes from recognizing that if the variables t and z are independent, then the functions F(t) and G(z) are independent. Thus,

$$\int F(t) \cdot G(z)dt = G(z) \int F(t)dt,$$

as G(z) is essentially a constant relative to the integration in t. Since x and y are independent, we have that

$$\iint_{R} f(x)g(y)dxdy = \int_{a}^{b} \int_{c}^{d} f(x)g(y)dydx = \int_{a}^{b} \left(\int_{c}^{d} f(x)g(y)dy \right)dx$$

$$= \int_{a}^{b} \left(f(x) \int_{c}^{d} g(y)dy \right)dx = \int_{a}^{b} f(x)\left(\int_{c}^{d} g(y)dy \right)dx$$

$$= \left(\int_{c}^{d} g(y)dy \right)\left(\int_{a}^{b} f(x)dx \right).$$

<u>Solution for Exercise 26 (p. 781)</u>. Here are a few hints. First, you want to use Exercise 25 above to show that

$$I = \int_{0}^{1} \int_{0}^{1} f(x)f(y)dy\ dx = 0.$$

Then, you want to show that

$$I = \int_0^1 \int_y^1 f(x)f(y)dx\ dy + \int_0^1 \int_0^y f(x)f(y)\ dx\ dy$$

and that the two integrals on the right are equal. For this last step you might consider what happens when you change the order of integration in the second integral.

SECTION 18.5

Sometimes students are initially confused and persist in omitting what looks like an "extra r" in the polar coordinate double integral formula

$$\iint_\Omega F(r,\theta)r\ dr\ d\theta. \qquad (1)$$

↖ ⌣ not "extra"

This "extra r" arises because the area of polar rectangles really is found by considering sectors of circles.

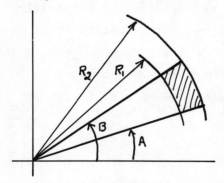

Area of "representative polar rectangle" is

$$\frac{B - A}{2\pi}(\pi R_2^2 - \pi R_1^2) = \frac{B - A}{2}(R_2^2 - R_1^2).$$

<u>not</u>

$$(B - A)(R_2 - R_1)$$

If the region Ω is given by

$$\{(r,\theta) : 0 \le r \le f(\theta),\ \theta_1 \le \theta \le \theta_2\}$$

and $z = F(r,\theta)$ is positive on Ω, then expression (1) represents the volume of the solid bounded above by $z = F(r,\theta)$ and below by $z = 0$ on Ω. We may rewrite (1) as

$$\int_{\theta_1}^{\theta_2} \int_0^{f(\theta)} F(r,\theta)r\ dr\ d\theta. \qquad (2)$$

The limits of integration over Ω are determined by the same procedure by which we determined inner and outer boundaries and intervals of θ when calculating polar areas. Just as in the case of rectangular coordinates, expression (2) represents the area of region Ω when $F(r,\theta) = 1$. In fact, noting that

$$\int_{\theta_1}^{\theta_2} \int_0^{f(\theta)} 1 \cdot r \; dr \; d\theta = \int_{\theta_1}^{\theta_2} \left(\frac{r^2}{2}\right)\Big|_0^{f(\theta)} d\theta = \frac{1}{2} \int_{\theta_1}^{\theta_2} [f(\theta)]^2 d\theta$$

may help you remember that "extra r". (Extra only in the sense that it is unexpected. It certainly is not unnecessary.) Finally, expression (2) is subject to the same interpretations involving volume, mass, and density that were made in the rectangular case.

We found that for double integrals in rectangular coordinates we could usually employ either order of integration (x first then y, or conversely). Sometimes one of the representations was easier to evaluate. The situation with polar coordinates is significantly different. It is normally the case for double integrals in polar coordinates that we can <u>not</u> reverse the order of integration from $\iint f(r,\theta)r \; dr \; d\theta$ to $\iint f(r,\theta)r \; d\theta \; dr$ because of real difficulties in determining the new limits of integration. Essentially stuck with the single order $\iint f(r,\theta)r \; dr \; d\theta$, we necessarily convert a hard-to-handle polar integral to rectangular coordinates. On the other hand, it is sometimes useful to convert from rectangular to polar coordinates.

Solutions for Exercises 1, 7, 12 and 14 (p. 790) are given after the following aside which relates to Probability and Statistics and may be of interest.

You should refresh your acceptance of the fact that

$$\int_a^b \int_c^d g(y)f(x) \; dx \; dy = \left(\int_a^b f(x)dx\right)\left(\int_c^d g(y)dy\right),$$

if x and y are independent. This point was resolved in the solution for Exercise 25 (p. 782). We shall also need to use some "improper integrals" (Section 12.7).

The <u>normal distribution</u> is indispensable in Probability and Statistics. Associated theorems tell us if we take large enough random samples from any collection (e.g., grades on an exam), then the results will be normally distributed. Such a result is the source of the absurd question: "Will you grade on a curve?". Actually, the student should ask: "Will you raise the grades?", for if he and his comrades are normal, their exam results will graph approximately as a bell-shaped or normal curve. This curve is characterized by $y = e^{-x^2/2}$. As is always the case, we want the area under the curve to be equal to 1. The reason for this is that if

$$\int_a^b f(x)dx = 1,$$

then

$$\int_a^x f(x)dx = p$$

tells us that the probability is p that a value (distributed according to the function $y = f(x)$) lies between a and x. Therefore, our interest is focused on finding a value for c such that

$$c \int_{-\infty}^{\infty} e^{-\frac{x^2}{2}} dx = 1.$$

The integral in question cannot be handled in the usual way. That is to say, we are going to embark on something exciting. Let

$$I = \int_{-\infty}^{\infty} e^{-\frac{x^2}{2}} dx = \int_{-\infty}^{\infty} e^{-\frac{y^2}{2}} dy,$$

where <u>x and y are independent</u>. Therefore,

$$I^2 = (\int_{-\infty}^{\infty} e^{-\frac{x^2}{2}} dx)(\int_{-\infty}^{\infty} e^{-\frac{y^2}{2}} dy) = \int_{-\infty}^{\infty} \int_{-\infty}^{\infty} e^{-\frac{1}{2}(x^2 + y^2)} dx\, dy.$$

"Conventional" techniques of integration still don't apply, so we will convert to polar coordinates in the conventional way!

$$I^2 = \int_{-\infty}^{\infty} \int_{-\infty}^{\infty} e^{-\frac{1}{2}(x^2 + y^2)} dx\, dy = \int_0^{2\pi} \int_0^{\infty} e^{-\frac{1}{2}r^2} \cdot r\, dr\, d\theta$$

$$= \int_0^{2\pi} \left\{ \lim_{\rho \to \infty} \int_0^\rho e^{-r^2/2} r \, dr \right\} d\theta = \int_0^{2\pi} \left\{ \lim_{\rho \to \infty} (1 - e^{-\rho^2/2}) \right\} d\theta$$

$$= \int_0^{2\pi} 1 \, d\theta = 2\pi.$$

Thus,

$$\frac{1}{\sqrt{2\pi}} \int_{-\infty}^\infty e^{-\frac{x^2}{2}} dx = 1.$$

Solution for Exercise 1 (p. 790).

$\text{Area} = \iint_\Omega r \, dr \, d\theta$

$= \int_{-\pi/3}^{\pi/3} \int_{3\sec\theta/4}^{3/2} r \, dr \, d\theta$

$= \int_{-\pi/3}^{\pi/3} \frac{1}{2} \left(\frac{9}{4} - \frac{9}{16} \sec^2\theta \right) d\theta$

$= \left(\frac{9}{8}\theta - \frac{9}{32} \tan\theta \right) \Big|_{-\pi/3}^{\pi/3} = \frac{3\pi}{4} - \frac{9\sqrt{3}}{16} \, .$

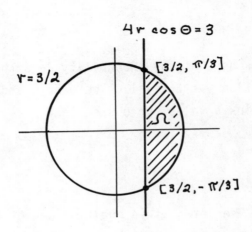

Solution for Exercise 7 (p. 790). Relying on the symmetry of the region Ω, we have

$\text{Area} = \iint_\Omega r \, dr \, d\theta$

$= 2 \int_0^{\pi/2} \int_{1/(1+\cos\theta)}^1 r \, dr \, d\theta$

$= 2 \int_0^{\pi/2} \frac{1}{2} \left[1 - \frac{1}{(1+\cos\theta)^2} \right] d\theta$

$= \frac{\pi}{2} - \int_0^{\pi/2} \frac{1}{(2\cos^2\frac{\theta}{2})^2} d\theta$

$= \frac{\pi}{2} - \frac{1}{4} \int_0^{\pi/2} (1 + \tan^2\frac{\theta}{2}) \sec^2\frac{\theta}{2} \, d\theta = \frac{\pi}{2} - \frac{1}{4} \left(2\tan\frac{\theta}{2} + \frac{2}{3}\tan^3\frac{\theta}{2} \right) \Big|_0^{\pi/2} = \frac{\pi}{2} - \frac{2}{3} \, .$

Solution for Exercise 12 (p. 790).

$$\text{Volume} = \iint\limits_{\Omega} (1 - x^2 - y^2)dx\,dy$$

$$= \iint\limits_{\Omega} (1 - r^2)r\,dr\,d\theta$$

$$= \int_{-\pi/2}^{\pi/2} \int_{0}^{\cos\theta} (r - r^3)dr\,d\theta$$

$$= \int_{-\pi/2}^{\pi/2} \left(\frac{\cos^2\theta}{2} - \frac{\cos^4\theta}{4}\right)d\theta$$

$$= \int_{-\pi/2}^{\pi/2} \left\{\frac{1}{4}(1 + \cos 2\theta) - \frac{1}{16}\left[1 + 2\cos 2\theta + \frac{1}{2}(1 + \cos 4\theta)\right]\right\}d\theta$$

$$= \int_{-\pi/2}^{\pi/2} \left[\frac{5}{32} + \frac{1}{8}\cos 2\theta - \frac{1}{32}\cos 4\theta\right]d\theta = 5\pi/32.$$

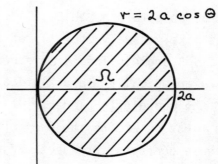

$(x - \frac{1}{2})^2 + y^2 = \frac{1}{4}$

$r = \cos\theta$

In rectangular coordinates we have

$$\text{Volume} = \int_{0}^{1} \int_{-S}^{S} (1 - x^2 - y^2)dy\,dx \qquad \text{with} \qquad S = \sqrt{\frac{1}{4} - (x - \frac{1}{2})^2},$$

which you should evaluate. The substitution $x - \frac{1}{2} = \frac{1}{2}\sin t$ will help.

Solution for Exercise 14 (p. 790). Since the cone is the upper surface, we select $z = \sqrt{x^2 + y^2} = r$. The polar form for the circle $x^2 + y^2 = 2ax$ is $r = 2a\cos\theta$ on $[-\pi/2, \pi/2]$.

$$\text{Volume} = \iint\limits_{\Omega} \sqrt{x^2 + y^2}\, r\,dr\,d\theta$$

$$= \int_{-\pi/2}^{\pi/2} \int_{0}^{2a\cos\theta} r^2\,dr\,d\theta$$

$$= \frac{8a^3}{3} \int_{-\pi/2}^{\pi/2} \cos^3\theta\,d\theta$$

$$= \frac{8a^3}{3}\left(\sin\theta - \frac{\sin^3\theta}{3}\right)\Bigg|_{-\pi/2}^{\pi/2} = \frac{32a^3}{9}.$$

$r = 2a\cos\theta$

$2a$

SECTIONS 18.6-7

There are six possible orders of integration for triple integrals. One particular
order of integration may be more difficult to evaluate than another and/or may
require more than one integral to represent the volume of the given solid. Perhaps
the hardest part of "setting up" triple integrals to represent volume is the
determination of the region of integration. This is especially acute when one is
simply given a number of intersecting surfaces. We then need to determine the
projection (or shadow) of the solid in one of the coordinate planes. In
particular, we need to find the projection of the curves in which the surfaces
intersect. This task is far easier than you might suspect. The topic of
projections was covered in Section 16.2. We shall repeat it here.

Suppose that the surfaces $z = f(x,y)$ and $z = g(x,y)$ intersect. Above a given
point in the xy-plane we have an intersection of the surfaces iff

$$f(x,y) = z = g(x,y).$$

Solving the equations simultaneously by eliminating the variable z will give an
equation in x and y. This equation is the projection of the curve in the xy-plane.
Again, the projection of the curve of intersection of $F(x,y,z) = 0$ and
$G(x,y,z) = 0$ into the yz-plane may be found by eliminating the variable x in a
simultaneous solution of these two equations.

The following example illustrates these points. For each of the six possible
orders of integration we will represent the volume of the solid bounded by the
surfaces $z = x^2 + y^2$ and $z = y + 2$. You <u>should</u> recognize these equations as being
those for a paraboloid of revolution and a plane, respectively. Even though you
may not be able to draw this solid, you should have some spatial feeling for what
it looks like; namely, the tip of the paraboloid of revolution has been sliced
off at a 45° angle. This simple recognition and the algebra discussed above is
enough to graphically and analytically represent the projections of this solid in
each of the coordinate planes.

We first determine the projections into each coordinate plane of the curve
where these surfaces intersect. First, we eliminate z by simultaneous solution.
Combining

$$z = x^2 + y^2 \qquad \text{and} \qquad z = y + 2,$$

we have

$$x^2 + y^2 = y + 2$$

so that

$$x^2 + (y - \tfrac{1}{2})^2 = (\tfrac{3}{2})^2$$

is the projection in the xy-plane. Next we eliminate y,

$$z = x^2 + y^2 = x^2 + (z - 2)^2$$

$$z = x^2 + z^2 - 4z + 4$$

so that

$$x^2 + (z - \tfrac{5}{2})^2 = (\tfrac{3}{2})^2$$

is the projection in the xz-plane. Finally, a little thought is all that is
needed to show that

$$z = y + 2 \qquad \qquad \text{for } -1 \le y \le 2$$

is the projection in the yz-plane. Now we are in a position to represent the
projections of the solid into each of the coordinate planes. A sketch of the solid
is thrown in for good measure. You should carefully note the orientation of the
coordinate axes in each diagram. The projections are drawn this way so as to
suggest a view of the solid from the first octant. Triple integral representation
of the volume is now just a matter of patience.

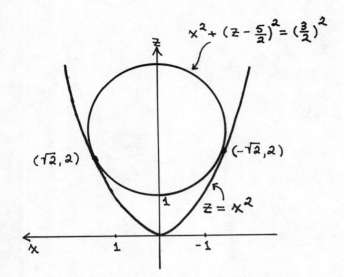

$x^2 + (z - \frac{5}{2})^2 = (\frac{3}{2})^2$

$(\sqrt{2}, 2)$

$(-\sqrt{2}, 2)$

$z = x^2$

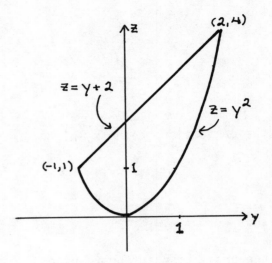

$z = y + 2$

$(2, 4)$

$z = y^2$

$(-1, 1)$

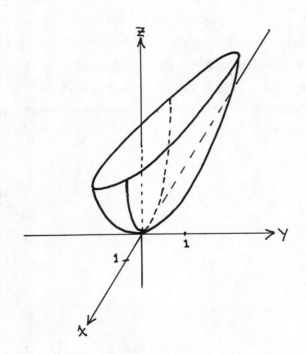

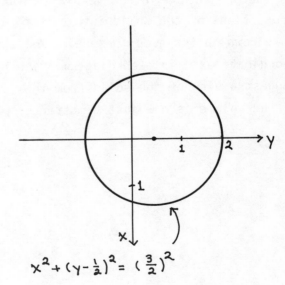

$x^2 + (y - \frac{1}{2})^2 = (\frac{3}{2})^2$

Region of integration in xy-plane so z axis is "up".

$$V = \int_{-\frac{3}{2}}^{\frac{3}{2}} \int_{\frac{1}{2} - \sqrt{\frac{9}{4} - x^2}}^{\frac{1}{2} + \sqrt{\frac{9}{4} - x^2}} \int_{x^2 + y^2}^{y + 2} dz \, dy \, dx$$

$$V = \int_{-1}^{2} \int_{-\sqrt{\frac{9}{4} - (y - \frac{1}{2})^2}}^{\sqrt{\frac{9}{4} - (y - \frac{1}{2})^2}} \int_{x^2 + y^2}^{y + 2} dz \, dx \, dy$$

Region of integration in yz-plane so x-axis is "up".

$$V = \int_{-1}^{2} \int_{y^2}^{y + 2} \int_{-\sqrt{z - y^2}}^{\sqrt{z - y^2}} dx \, dz \, dy$$

$$V = \int_{0}^{1} \int_{-\sqrt{z}}^{\sqrt{z}} \int_{-\sqrt{z - y^2}}^{\sqrt{z - y^2}} dx \, dy \, dz + \int_{1}^{4} \int_{z - 2}^{\sqrt{z}} \int_{-\sqrt{z - y^2}}^{\sqrt{z - y^2}} dx \, dy \, dz$$

Region of integration in xz-plane so y-axis is "up".

$$V = \int_{-\frac{3}{2}}^{\frac{3}{2}} \int_{\frac{5}{2} - \sqrt{\frac{9}{4} - x^2}}^{\frac{5}{2} + \sqrt{\frac{9}{4} - x^2}} \int_{z - 2}^{\sqrt{z - x^2}} dy \, dz \, dx + \int_{-\sqrt{2}}^{\sqrt{2}} \int_{x^2}^{\frac{5}{2} - \sqrt{\frac{9}{4} - x^2}} \int_{-\sqrt{z - x^2}}^{\sqrt{z - x^2}} dy \, dz \, dx$$

$$V = \int_{1}^{4} \int_{-\sqrt{\frac{9}{4} - (z - \frac{5}{2})^2}}^{\sqrt{\frac{9}{4} - (z - \frac{5}{2})^2}} \int_{z - 2}^{\sqrt{z - x^2}} dy \, dx \, dz + \int_{1}^{2} \int_{\sqrt{\frac{9}{4} - (z - \frac{5}{2})^2}}^{\sqrt{z}} \int_{-\sqrt{z - x^2}}^{\sqrt{z - x^2}} dy \, dx \, dz$$

$$+ \int_{1}^{2} \int_{-\sqrt{z}}^{-\sqrt{\frac{9}{4} - (z - \frac{5}{2})^2}} \int_{-\sqrt{z - x^2}}^{\sqrt{z - x^2}} dy \, dx \, dz + \int_{0}^{1} \int_{-\sqrt{z}}^{\sqrt{z}} \int_{-\sqrt{z - x^2}}^{\sqrt{z - x^2}} dy \, dx \, dz$$

Whew!

Next we find the volume of the solid enclosed by $y^2 + z = 4$, $y + z = 2$, $x = 0$ and $x = 2$. This is Exercise 15 on p. 801 of the text. You should do the algebra to verify that the projections exhibited below are correct. Representations of the volume for each of six orders of integrations are given on the next page. You should evaluate at least two of them.

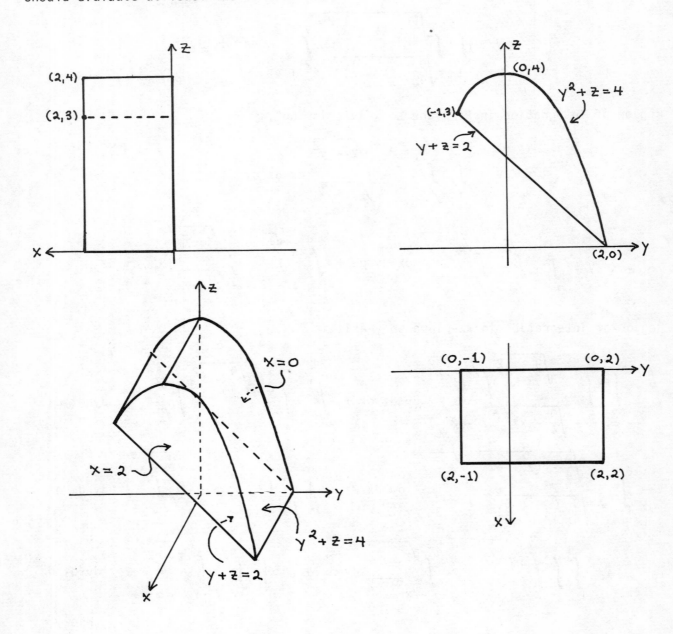

$$\text{Vol} = \int_{-1}^{2} \int_{0}^{2} \int_{2-y}^{4-y^2} dz\ dx\ dy = \int_{0}^{2} \int_{-1}^{2} \int_{2-y}^{4-y^2} dz\ dy\ dx$$

$$\text{Vol} = \int_{0}^{2} \int_{0}^{3} \int_{2-z}^{\sqrt{4-z}} dy\ dz\ dx + \int_{0}^{2} \int_{3}^{4} \int_{-\sqrt{4-z}}^{\sqrt{4-z}} dy\ dz\ dx$$

$$\text{Vol} = \int_{0}^{3} \int_{0}^{2} \int_{2-z}^{\sqrt{4-z}} dy\ dx\ dz + \int_{3}^{4} \int_{0}^{2} \int_{-\sqrt{4-z}}^{\sqrt{4-z}} dy\ dx\ dz$$

$$\text{Vol} = \int_{-1}^{2} \int_{2-y}^{4-y^2} \int_{0}^{2} dx\ dz\ dy$$

$$\text{Vol} = \int_{0}^{3} \int_{2-z}^{\sqrt{4-z}} \int_{0}^{2} dx\ dy\ dz + \int_{3}^{4} \int_{-\sqrt{4-z}}^{\sqrt{4-z}} \int_{0}^{2} dx\ dy\ dz$$

To finish let's do a problem in the other direction. We shall begin with the triple integral representation for some volume, determine the bounding surfaces, and then represent the same volume using a different order of integration.

Consider

$$\int_{-4}^{4} \int_{-\sqrt{4-y}}^{\sqrt{4-y}} \int_{-\sqrt{y+4}}^{\sqrt{y+4}} dz\ dx\ dy.$$

From the limits on the z we have the top and bottom of the parabolic cylinder $z^2 = y + 4$ providing the upper and lower surfaces. The limits on y and x give us the region of integration as shown on the right.

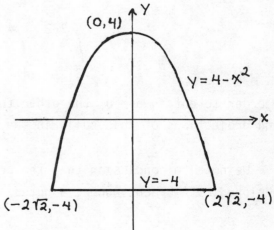

Along the parabolic arc, z is not zero (in z^2 = y + 4) so the sides here are
provided by the parabolic cylinder y = 4 - x^2 (which is parallel to the z-axis).
As for the line y = -4, it is possible that the plane y = -4 provides a flat
side here for the solid. However, along this line z is zero (in z^2 = y + 4), so
this boundary is provided by the parabolic cylinder z^2 = y + 4 turning back under
itself. A suggestive model for this solid is a baseball that has been batted
about a bit. The space curve of intersection for these two parabolic cylinders
looks like the stitching on said ball:

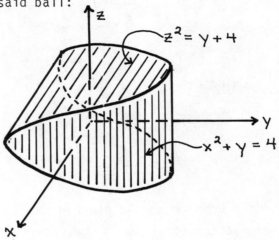

    To interchange the outer two orders of integration (dx dy to dy dx) doesn't
require any tinkering with the limits on z. All we need to do is to refer to the
region of integration sketched above. We have

$$\int_{-2\sqrt{2}}^{2\sqrt{2}} \int_{-4}^{4 - x^2} \int_{-\sqrt{y + 4}}^{\sqrt{y + 4}} dz \ dy \ dx.$$

In order to obtain one of the other four possible orders of integration we have to
find projections of this solid in the other coordinate planes.

    We note in conclusion that the projective diagrams used above were essentially
developed by Albrecht Dürer, a noted artist who was also a significant amateur

mathematician. The interested reader could pursue this appealing topic further
by consulting The World of Mathematics, Vol. 1 (Erwin Panofsky, "Durer as a
Mathematician").

SECTIONS 18.8-10

These Sections are very straightforward. Pappus' Theorem is especially elegant
and deserves particular attention. Since we can add nothing significant to the
text in these Sections and since it would be impossible to out-do Pappus, we offer
an interesting (and clearly optional) problem for you to sink your teeth into.

Consider an apple. It is a sphere. Its color is not important. Now drill
a hole which symmetrically passes through the center of the apple. What can you
say about the volume of the solid remaining? What about the depth of the hole?
Careful, it is less than the diameter of the apple. Finally, what can you say
about the ratio of the surface area of the new solid to its volume? Mathematically,
we are considering the sphere $x^2 + y^2 + z^2 = 1$ and the circular cylinder
$x^2 + y^2 = R^2$ for $0 < R < 1$.

ANSWERS: Where R is the radius of the hole drilled in a sphere of radius 1, the
remaining volume is $\frac{4\pi}{3}(1 - R^2)^{3/2}$, and the depth of the hole is $2\sqrt{1 - R^2}$. The
surface area is $4\pi(R + 1)\sqrt{1 - R^2}$ and is maximal when $R = 1/2$. Finally,

$$\lim_{R \uparrow 1} \frac{\text{surface area}}{\text{volume}} = \infty.$$

Line Integrals and Surface Integrals

The key result linking the differential and integral calculus for single variable functions is The Fundamental Theorem of Calculus. This Chapter focuses upon three analogous results for the vector calculus. Green's Theorem, the Divergence Theorem, and Stokes' Theorem tie together the basic ideas considered in Chapters 17 and 18. Prerequisite to the study of these theorems is a clear understanding of the line integral.

The line integral is given by Formula 19.1.4 as

$$W = \int_C \overline{F}(\overline{r}) \cdot d\overline{r} = \int_a^b [\overline{F}(\overline{r}(t)) \cdot \overline{r}'(t)]dt$$

where the integral is taken over the path $C(t)$ as t varies from a to b. The curve C can be any piecewise smooth curve. The vector form of the line integral is the most natural way to appreciate its physical significance, but the line integral itself is generally easier to evaluate in scalar form.

We begin with the vector function

$$\overline{F}(\overline{r}) = \overline{F}(x,y) = P(x,y)\overline{i} + Q(x,y)\overline{j}$$

and

$$\overline{r} = x\overline{i} + y\overline{j}.$$

Then,

$$\overline{F}(\overline{r}) \cdot d\overline{r} = P(x,y)dx + Q(x,y)dy$$

so that

$$\int_C \overline{F}(\overline{r}) \cdot d\overline{r} = \int_C P(x,y)dx + Q(x,y)dy. \tag{1}$$

The easiest way to evaluate the right hand side of equation (1) is to parametrize x and y as functions of a single variable t. In this case,

$$dx = x'(t)dt \qquad \text{and} \qquad dy = y'(t)dt.$$

Equation (1) may be rewritten as

$$W = \int_C \overline{F}(\overline{r}) \cdot d\overline{r} = \int_a^b [P\big(x(t),\ y(t)\big)x'(t) + Q\big(x(t),\ y(t)\big)y'(t)]dt,$$

which will succumb to the usual single variable techniques of integration.

Line integrals are not restricted to two dimensions. In three dimensions we would write

$$W = \int_a^b [P\big(x(t),y(t),z(t)\big)x'(t) + Q\big(x(t),y(t),z(t)\big)y'(t) + R\big(x(t),y(t),z(t)\big)z'(t)]dt$$

A few examples will clarify the advantage of using this scalar representation.

__Example 1__. Evaluate $\int_C \overline{F}(\overline{r}) \cdot d\overline{r}$ where $\overline{F}(x,y) = y\overline{i} + x^2\overline{j}$ and C is the straight line segment joining the points (0,0) and (1,1).

__Solution__. We know that

$$\overline{r} = x\overline{i} + y\overline{j}$$

and

$$d\overline{r} = dx\overline{i} + dy\overline{j}.$$

Since

$$P(x,y) = y$$

and

$$Q(x,y) = x^2,$$

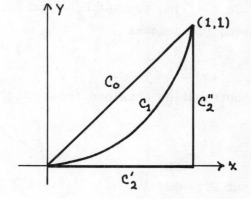

the integral I_0 in question is given by:

$$I_0 = \int_{C_0} P(x,y)dx + Q(x,y)dy = \int_{C_0} ydx + x^2dy. \qquad (2)$$

The curve C_0 given by $y = x$ from (0,0) to (1,1) may be parametrized as:

$$y = t \qquad x = t \qquad\qquad (0 \leq t \leq 1).$$

Therefore,

$$I_0 = \int_0^1 t\,dt + t^2 dt = \int_0^1 (t + t^2)dt = 5/6.$$

Notice that it is not necessary in this example to introduce an additional variable (t) as a parameter. We can parametrize the curve C_0 in terms of x to obtain

$$I_0 = \int_0^1 x\,dx + x^2 dx = 5/6.$$

What if this problem had asked instead that we evaluate the integral along the path C_1 given by $y = x^2$ from (0,0) to (1,1)? As you should verify, we obtain

$$I_1 = \int_{C_1} y\,dx + x^2 dy = \int_0^1 [x^2 dx + x^2(2x\,dx)] = 5/6.$$

Here we have chosen x as our parameter so $dy = 2x\,dx$ along C_1. It _seems_ that changing the path has no affect on the outcome. Let us try one more path.

Consider the polygonal path C_2 from (0,0) to (1,0) and then from (1,0) to (1,1). This curve is not smooth, but it is piecewise smooth. We must integrate separately over each smooth piece of the curve.

On the line segment C_2' from (0,0) to (1,0) we have $y = 0$ so that $dy = 0$. Equation (2) becomes

$$I_2' = \int_{C_2'} 0 = 0.$$

The contribution obtained from integrating along C_2'' is

$$I_2'' = \int_{C_2''} y\,dx + x^2 dy = \int_0^1 dy = 1. \qquad \text{(Why?)}$$

The sum of these two integrals is 1 so our original suspicion is unfounded. Apparently it was only chance that caused the first two paths we chose to yield the same value from (0,0) to (1,1).

Example 2. Evaluate the integral $\int_C \bar{r}\cdot d\bar{r}$ where C is the polygonal path joining the points (1,0,0), (0,0,0), (0,1,0), and (0,1,π/2) in that order.

<u>Solution.</u> There are three pieces, P_1, P_2
and P_3, to the path C. We need to
calculate the integral along each smooth
piece separately and then add the results.

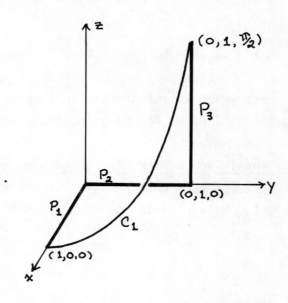

Along P_1 from (1,0,0) to (0,0,0) we
have z = 0 and y = 0. Therefore,
dz = dy = 0, $\overline{r}$ = x$\overline{i}$ and d$\overline{r}$ = dx$\overline{i}$. So,

$$\int_{P_1} \overline{r} \cdot d\overline{r} = \int_{(1,0,0)}^{(0,0,0)} \overline{r} \cdot d\overline{r} = \int_1^0 x dx = -\frac{1}{2} .$$

Along P_2 from (0,0,0) to (0,1,0) we
find that $\overline{r}$ = y$\overline{j}$, d$\overline{r}$ = dy$\overline{j}$, and

$$\int_{P_2} \overline{r} \cdot d\overline{r} = \int_{(0,0,0)}^{(0,1,0)} \overline{r} \cdot d\overline{r} = \int_0^1 y dy = \frac{1}{2} .$$

On the last piece, P_3, from (0,1,0) to (0,1,π/2) we know that y = 1 and x = 0.
Therefore dy = dx = 0, and

$$\int_{P_3} \overline{r} \cdot d\overline{r} = \int_{(0,1,0)}^{(0,1,\pi/2)} \overline{r} \cdot d\overline{r} = \int_0^{\pi/2} z dz = \pi^2/8.$$

The value of the line integral along C from (1,0,0) to (0,1,π/2) is the sum of
these three values: $\pi^2/8$.

Let us try another path. Consider the segment of the helix (labelled C_1 in
the diagram) given by

$$x = \cos t \qquad y = \sin t \qquad z = t,$$

which joins (1,0,0) and (0,1,π/2). In this case the parameter t varies from 0 to
π/2. Since

$$\overline{r} = x\overline{i} + y\overline{j} + z\overline{k} = \cos t \ \overline{i} + \sin t \ \overline{j} + t\overline{k}$$

so that

$$d\overline{r} = (-\sin t)\overline{i} + \cos t \ \overline{j} + \overline{k},$$

we have

$$\int_{C_1} \overline{r} \cdot d\overline{r} = \int_{(1,0,0)}^{(0,1,\pi/2)} \overline{r} \cdot d\overline{r} = \int_0^{\pi/2} (-\sin t \cos t + \sin t \cos t + t) dt = \pi^2/8.$$

Again we seem to have fallen by chance upon the same value. How can we determine conclusively whether or not the value of the integral depends upon the path over which we integrate? This question is partially answered by The Fundamental Theorem for Line Integrals (Theorem 19.2.3). We restate it below.

Fundamental Theorem for Line Integrals

If f is a continuously differentiable function on an open set, containing the piecewise smooth curve C from a to b, then

$$\int_C \nabla f(\overline{r}) \cdot d\overline{r} = \int_a^b [\nabla f(\overline{r}) \cdot \overline{r}'(t)] dt = f(b) - f(a).$$

The theorem tells us that if we can find a function f whose gradient is the vector we are integrating, then we can be sure that the integral is independent of path. The problem of reconstructing a function from its gradient was discussed in Section 17.9. Let us look back at Examples 1 and 2.

We know from Theorem 17.9.1 that a given vector function

$$\overline{F}(x,y) = P(x,y)\overline{i} + Q(x,y)\overline{j}$$

is a gradient if and only if

$$\frac{\partial P}{\partial y}(x,y) = \frac{\partial Q}{\partial x}(x,y).$$

In example 1, $P(x,y) = y$ and $Q(x,y) = x^2$ so that

$$\frac{\partial P}{\partial y}(x,y) = 1 \neq 2x = \frac{\partial Q}{\partial x}(x,y).$$

The conditions of The Fundamental Theorem are <u>not</u> satisfied. Does this assure us that the integral is not independent of path? No!

It is possible to prove the converse of this theorem in two dimensions. Namely,

If a vector function $\overline{F}(\overline{r}) = \overline{F}(x,y)$ is continuous on an open set

S and if $\int_C \overline{F}(\overline{r}) \cdot d\overline{r}$ is independent of path, then a single-valued

scalar function $f(\overline{r})$ exists such that $\overline{F}(\overline{r}) = \nabla f(\overline{r})$ in S.

This converse of The Fundamental Theorem assures us (How?) that the integral in Example 1 is not independent of path. Of course, the counterexample we found is just as reassuring.

How do we deal with Example 2? First we have to derive comparable conditions in three dimensions to those given in Theorem 17.9.1 in order to determine whether or not $\overline{F}(\overline{r})$ is the gradient for some function $f(\overline{r})$.

Assume $\overline{F}(\overline{r}) = \nabla f(\overline{r})$. Then,

$$\nabla f = \frac{\partial f}{\partial x}\overline{i} + \frac{\partial f}{\partial y}\overline{j} + \frac{\partial f}{\partial z}\overline{k} = P(x,y)\overline{i} + Q(x,y)\overline{j} + R(x,y)\overline{k} = \overline{F}$$

and

$$\frac{\partial^2 f}{\partial x \partial y} = \frac{\partial^2 f}{\partial y \partial x}, \qquad \frac{\partial^2 f}{\partial x \partial z} = \frac{\partial^2 f}{\partial z \partial x}, \qquad \text{and} \qquad \frac{\partial^2 f}{\partial y \partial z} = \frac{\partial^2 f}{\partial z \partial y} .$$

We can proceed as in the proof of Theorem 17.9.1 to obtain the three relations,

$$\frac{\partial P}{\partial y} = \frac{\partial Q}{\partial x}, \qquad \frac{\partial Q}{\partial z} = \frac{\partial R}{\partial y}, \qquad \text{and} \qquad \frac{\partial R}{\partial x} = \frac{\partial P}{\partial z} . \tag{3}$$

These conditions assure us of the existence of a function $f(x,y,z)$ whose exact differential is given by $\overline{F}(\overline{r}) \cdot d\overline{r}$. However, conditions (3) do not guarantee that f is single-valued. The question of which supplementary conditions need to be adjoined to the given conditions to assure us that f is single-valued is reserved for an upper-level course. We simply remark that these conditions are sufficient to assure us that $\int_C \overline{F} \cdot d\overline{r}$ is independent of path in the problems we will encounter.

Now we can return to Example 2. Since

$$\overline{r} = x\overline{i} + y\overline{j} + z\overline{k} = P(x,y)\overline{i} + Q(x,y)\overline{j} + R(x,y)\overline{k},$$

we have

$$\frac{\partial P}{\partial y} = 0 = \frac{\partial Q}{\partial x}, \frac{\partial Q}{\partial z} = 0 = \frac{\partial R}{\partial y}, \text{ and } \frac{\partial R}{\partial x} = 0 = \frac{\partial P}{\partial z} .$$

The integral in Example 2 is indeed independent of path. You should verify this by integrating along one more path from $(0,0,0)$ to $(0,1,\pi/2)$; for example, a straight line joining these two points.

One consequence of independence of path deserves special attention. Namely, if C is <u>any</u> simple closed curve (i.e., a closed curve which does not cross itself) then

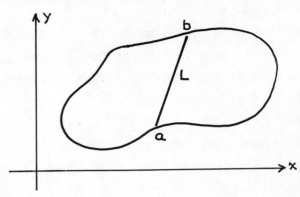

$$\int_C \nabla f(\overline{r}) \cdot d\overline{r} = 0.$$

You are asked to show this in Exercise 16 on p. 828. The diagram to the right may help you provide the details. You will need Formula 19.1.6 of the text.

This intriguing fact will crop up again the the next Section.

SECTION 19.3

Green's Theorem intimately links the idea of a line integral along a simple closed curve with our knowledge of double integrals. But, particular note should be taken of the restriction in Theorem 19.3.1 to integrating only in the <u>counterclockwise direction</u>. If you integrate in the clockwise direction on a closed curve, you will obtain the negative of the value found in the counterclockwise integration. Frequently, the notation $\oint$ is used to indicate the direction of integration. Hence,

$$\oint_C P(x,y)dx + Q(x,y)dy = - \oint_C P(x,y)dx + Q(x,y)dy.$$

Using this notation the conclusion of Green's Theorem may be recorded as

$$\oint_C \overline{F}(\overline{r}) \cdot d\overline{r} = \iint_\Omega [\frac{\partial Q}{\partial x}(x,y) - \frac{\partial P}{\partial y}(x,y)]dxdy.$$

A few examples will help to impress you with the value of this theorem.
<u>Example 1</u>. Let C be the triangle with vertices at $(1,0)$, $(1,1)$, and $(0,0)$. Evaluate

$$\oint_C y^2 dx + x^2 dy.$$

Solution. Using the techniques of
Section 19.1, we would have to
evaluate three line integrals to solve
this problem. You should verify that
the values of these integrals are 0, 1,

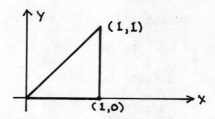

and -2/3. However, if we notice that the hypotheses of Green's Theorem are
satisfied, the problem is easier.

$$\oint y^2 dx + x^2 dy = \int_0^1 \int_y^1 (2x - 2y)dxdy = \int_0^1 (1 - 2y + y^2)dy = 1/3.$$

On the other hand, sometimes we can use Green's Theorem to simplify the
evaluation of a double integral.

Example 2. Find the area enclosed by the
ellipse $x^2 + 4y^2 = 4$.
Solution. If we take

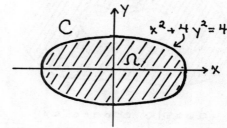

$$P(x,y) = -y \text{ and } Q(x,y) = x$$

we obtain

$$\oint - ydx + xdy = \iint_\Omega (1 - (-1))dxdy = 2 \iint_\Omega dxdy.$$

The polar form of this ellipse is

$$x = 2 \cos \theta \qquad y = \sin \theta \qquad (0 \le \theta \le 2\pi).$$

Using θ as our parameter we find

$$\oint_C - ydx + xdy = \int_0^{2\pi} [(-\sin \theta)(-2 \sin \theta) + (2 \cos \theta)(\cos \theta)]d\theta = \int_0^{2\pi} 2d\theta = 4\pi.$$

Therefore,

$$4\pi = 2 \iint_\Omega dxdy \qquad \text{or} \qquad 2\pi = \iint_\Omega dxdy.$$

This is certainly easier than evaluating the double integral directly.
You might wonder what prompted the choices of $P(x,y) = -y$ and $Q(x,y) = x$.

If you think about it for a moment, these are the simplest choices we could make to force $\frac{\partial Q}{\partial x} - \frac{\partial P}{\partial y}$ to be a nonzero constant. Can you explain why the problem indicates these as the obvious choices?

Green's Theorem can be extended to regions in the plane which are more complicated than simply connected regions (the interior of a simple closed curve). A multiply connected region is one whose boundary consists of several non-intersecting simple closed curves. Both of the regions in Diagram 1 are examples

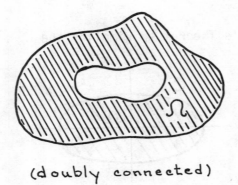
(doubly connected)

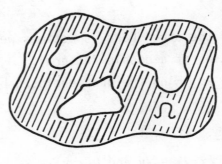
(multiply connected)

DIAGRAM 1

of multiply connected regions. Surprisingly, the extension of Green's Theorem to regions of this type is very simple.

Consider the doubly connected region in Diagram 2. Cut the region with a line segment C joining some point P_0 on C_0 to some point P_1 on C_1 as shown in Diagram 2. The region Ω can now be visualized as a simply connected region with boundary C_0, C, and C_1 where the region's two ends are sewn together along C.

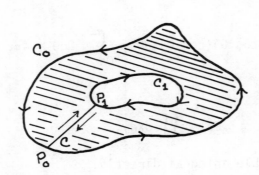

DIAGRAM 2

Visualizing Ω as a simply connected region, we can apply Green's Theorem to the closed curve which makes up its boundary. Notice how the path of integration causes us to traverse C_1 in the clockwise direction and C_0 in the counter-clockwise direction. We obtain

$$\iint_{\Omega} (\frac{\partial Q}{\partial x}(x,y) - \frac{\partial P}{\partial y}(x,y))dxdy = \oint_{C_0} P(x,y)dx + Q(x,y)dy + \int_C P(x,y)dx + Q(x,y)dy$$

$$+ \oint_{C_1} P(x,y)dx + Q(x,y)dy - \int_C P(x,y)dx + Q(x,y)dy,$$

which becomes

$$\iint_{\Omega}(\frac{\partial Q}{\partial x} - \frac{\partial P}{\partial y})dxdy = \oint_{C_0} P(x,y)dx + Q(x,y)dy - \oint_{C_1} P(x,y)dx + Q(x,y)dy. \tag{1}$$

This same technique can be applied to any multiply connected region Ω which has n simple closed curves (or holes), C_1, C_2, $\cdots$, C_n in the interior of C_0. In general we find that

$$\iint_{\Omega} (\frac{\partial Q}{\partial x} - \frac{\partial P}{\partial y})dxdy = \oint_{C_0} P(x,y)dx + Q(x,y)dy - \sum_{i=1}^{n} (\oint_{C_i} P(x,y)dx + Q(x,y)dy)$$

Finally, recall that we indicated that an integral that is independent of path vanishes on every simple closed curve drawn within the region of path independence. Green's Theorem makes this fact easily apparent. Since

$$\int_C P(x,y)dx + Q(x,y)dy$$

is independent of path if and only if

$$\frac{\partial P}{\partial y}(x,y) = \frac{\partial Q}{\partial x}(x,y),$$

the double integral to which it is equal must be 0. What if we apply this to equation (1)? Since the integral is independent of path, we obtain

$$0 = \oint_{C_0} P(x,y)dx + Q(x,y)dy - \oint_{C_1} P(x,y)dx + Q(x,y)dy$$

or

$$\oint_{C_0} P(x,y)dx + Q(x,y)dy = \oint_{C_1} P(x,y)dx + Q(x,y)dy. \qquad (2)$$

This is certainly a startling fact. Let us see how it might be useful.

Example 3.   Find the value of

$$\oint_{C} (\frac{-y}{x^2 + y^2})dx + (\frac{x}{x^2 + y^2})dy$$

where C bounds the region interior to the circle $x^2 + y^2 = 4$ and exterior to the
circle $x^2 + y^2 = 1$.

Solution.   Since

$$P(x,y) = \frac{-y}{x^2 + y^2}$$

and

$$Q(x,y) = \frac{x}{x^2 + y^2} ,$$

we find that

$$\frac{\partial Q}{\partial x}(x,y) = \frac{y^2 - x^2}{(x^2 + y^2)^2} = \frac{\partial P}{\partial y}(x,y).$$

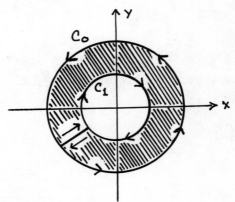

As before the integral in the counterclockwise direction over C is equal to the
integral in the counterclockwise direction over C_0 minus the integral in the
counterclockwise direction over C_1.   By Green's Theorem we have

$$\oint_{C} (\frac{-y}{x^2 + y^2})dx + (\frac{x}{x^2 + y^2})dy = 0.$$

Notice that we did not need to calculate the integral over C_0 or C_1, since Green's
Theorem gave us the result directly.

To find the value of these integrals we would parametrize using polars to
obtain

$$x = 2 \cos \theta \qquad y = 2 \sin \theta \qquad 0 \le \theta \le 2\pi \qquad \text{(on } C_0)$$

and

$$x = \cos \theta \qquad y = \sin \theta \qquad 0 \le \theta \le 2\pi \qquad (\text{on } C_1).$$

Thus,

$$\oint_{C_0} (\frac{-y}{x^2 + y^2})dx + (\frac{x}{x^2 + y^2})dy = \int_0^{2\pi} \frac{4 \sin^2\theta + 4 \cos^2\theta}{4 \sin^2\theta + 4 \cos^2\theta} \, d\theta = 2\pi,$$

and

$$\oint_{C_1} (\frac{-y}{x^2 + y^2})dx + (\frac{x}{x^2 + y^2})dy = \int_0^{2\pi} \frac{\sin^2\theta + \cos^2\theta}{\sin^2\theta + \cos^2\theta} \, d\theta = 2\pi.$$

This verifies our previous answer.

Why isn't the integral around C_1 zero? Certainly C_1 is a simple closed curve and certainly $\frac{\partial P}{\partial y}(x,y) = \frac{\partial Q}{\partial x}(x,y)$. The answer is a simple one. Neither $P(x,y)$ nor $Q(x,y)$ is continuous at $(0,0)$ so the hypothesis of the Fundamental Theorem of Line Integrals is not satisfied. From equation (2), however, we notice an intriguing fact. The integral of this function taken in the counterclockwise direction around any circle of radius $r > 0$ centered at the origin will have value 2π. The theory of complex variables deals with this question and many others in a beautifully elegant fashion.

SECTIONS 19.4-5

The calculation of the surface area for a surface S given by a continuously differentiable function $z = f(x,y)$ is straightforward if you remember the formula

$$S = \iint_\Omega \sqrt{(\frac{\partial f}{\partial x}(x,y))^2 + (\frac{\partial f}{\partial y}(x,y))^2 + 1} \, dxdy. \tag{1}$$

On the other hand, if you don't remember the formula, **difficulties will continually surface.**

We shall example the application of this formula by solving Exercise 8 on page 843. Then, solutions for Exercises 10 and 11 are offered to abet your understanding of Section 19.5.

Solution for Exercise 8 (p. 843). Whenever possible it is advisable to sketch (or at least try to visualize) the surface in question. In this problem we are asked to find the surface area S of the cap which is cut off from a sphere of radius a centered at (0,0,a) by a paraboloid of revolution. The surface area we seek is sketched in Diagram 1. To apply equation (1) we need to find $\frac{\partial z}{\partial x}$ and $\frac{\partial z}{\partial y}$. This is easily done using implicit differentiation. From

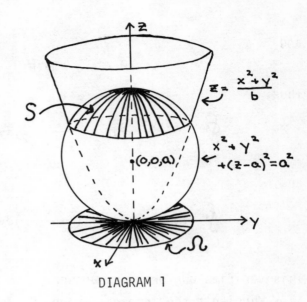

DIAGRAM 1

$$x^2 + y^2 + z^2 - 2az = 0 \tag{2}$$

we obtain

$$2x + 2z \frac{\partial z}{\partial x} - 2a \frac{\partial z}{\partial x} = 0$$

or

$$\frac{\partial z}{\partial x} = \frac{x}{a - z} \, .$$

Similarly,

$$\frac{\partial z}{\partial y} = \frac{y}{a - z} \, .$$

Therefore,

$$S = \iint_{\Omega} \sqrt{\left(\frac{x}{a - z}\right)^2 + \left(\frac{y}{a - z}\right)^2 + 1} \, dxdy.$$

From equation (2) it is easy to see that

$$(a - z)^2 = a^2 - x^2 - y^2$$

so that

$$S = \iint_{\Omega} \frac{a}{\sqrt{a^2 - x^2 - y^2}} \, dxdy.$$

We need to find the radius of the circular region Ω in order to determine limits of integration. We can do this by looking at the intersection of the two surfaces in question. Substitution of $x^2 + y^2 = bz$ into equation (2) gives $z = 2a - b$. Now it should be clear to you why the problem specifies that $a > b > 0$. If this were not the case then the cap cut off by the paraboloid might have turned back under itself requiring us to break the cap into two separate pieces to solve the problem. However, since $a > b$ the intersection must occur in the upper hemisphere. (Why?) The radius of the circular region Ω is therefore $2ab - b^2$ (Verify this). We use polar coordinates to complete the integration.

$$S = \int_0^{2\pi} \int_0^{\sqrt{2ab - b^2}} \frac{a}{\sqrt{a^2 - r^2}} \, r \, dr \, d\theta = 2\pi ab.$$

Solution for Exercise 10 (p. 843). Assume that $f(x,y)$ is continuously differentiable on Ω. To show that the surface S is smooth, we must show that the unit normal $\overline{n}$ to S is continuous throughout S. From Section 17.4 we know that the equation for S can be written as

$$F(x,y,z) = z - f(x,y) = 0 \qquad (x,y) \in \Omega. \tag{3}$$

Consequently, the unit normal to S is given by

$$\overline{n} = \frac{\nabla F}{||\nabla F||} = \frac{\frac{\partial F}{\partial x}\overline{i} + \frac{\partial F}{\partial y}\overline{j} + \frac{\partial F}{\partial z}\overline{k}}{\sqrt{\left(\frac{\partial F}{\partial x}\right)^2 + \left(\frac{\partial F}{\partial y}\right)^2 + \left(\frac{\partial F}{\partial z}\right)^2}} = \frac{-\frac{\partial f}{\partial x}\overline{i} - \frac{\partial f}{\partial y}\overline{j} + \overline{k}}{\sqrt{\left(\frac{\partial f}{\partial x}\right)^2 + \left(\frac{\partial f}{\partial y}\right)^2 + 1}}. \tag{4}$$

Since $f(x,y)$ is assumed to be continuously differentiable on Ω, both $\frac{\partial f}{\partial x}$ and $\frac{\partial f}{\partial y}$ are continuous on Ω and $\overline{n}$ must be continuous throughout S.

Now, assume that S is smooth. This implies that $\overline{n}$ is continuous throughout S so by equation (3) both $\frac{\partial f}{\partial x}$ and $\frac{\partial f}{\partial y}$ exist and are continuous on Ω. Hence $f(x,y)$ is continuously differentiable on Ω.

Solution for Exercise 11 (p. 843). The most direct way to get a handle on the the angle $\gamma(x,y)$ between $\overline{n}$ and $\overline{k}$ is to appeal to the dot product. We have

$$\overline{n} \cdot \overline{k} = ||\overline{n}|| \cdot ||\overline{k}|| \cdot \cos[\gamma(x,y)].$$

Using equation (4) we find that

$$\overline{n} \cdot \overline{k} = \frac{1}{\sqrt{(\frac{\partial f}{\partial x})^2 + (\frac{\partial f}{\partial y})^2 + 1}} ,$$

and since $||\overline{n}|| = ||\overline{k}|| = 1$, Equation 19.4.4

$$\sqrt{(\frac{\partial f}{\partial x})^2 + (\frac{\partial f}{\partial y})^2 + 1} = \sec[\gamma(x,y)],$$

follows. Equation 19.4.5 follows by substitution.

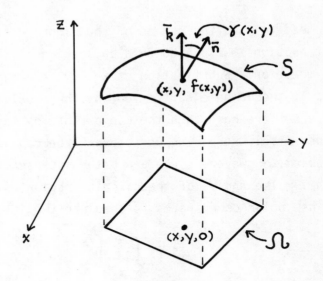

The determination of a general surface integral is a simple consequence of coupling Definition 19.5.2 with your previous knowledge of double integrals. The flux of a vector field V across a piecewise smooth surface S is simply the determination of a particular type of surface integral. Theorem 19.5.3 simplifies the evaluation of the flux for a smooth surface which may be projected onto the xy-plane. If S can be projected onto the xz-plane, we need to find a suitable form for $\sec[\beta(x,z)]$ to obtain a result comparable to that of Theorem 19.5.3. As

$$F(x,y,z) = y - g(x,z) = 0,$$

the result we obtain is

$$\sec[\beta(x,z)] = \sqrt{(\frac{\partial g}{\partial x})^2 + (\frac{\partial g}{\partial z})^2 + 1} .$$

A corresponding equation:

$$\sec[\alpha(y,z)] = \sqrt{(\frac{\partial h}{\partial y})^2 + (\frac{\partial h}{\partial z})^2 + 1} ,$$

applies to surfaces which may be projected onto the yz-plane.

SECTIONS 19.6-7

These final two Sections deal with two famous theorems: The Divergence (Gauss's) Theorem and Stokes' Theorem. The statement of each of these theorems is straightforward. A solution to Exercise 12 (p. 855) and two examples (one for each theorem) are offered below.

Solution for Exercise 12 (p. 855). To verify the Divergence Theorem, we must evaluate (separately) each side of the equation

$$\iint_S (\overline{v}\cdot\overline{n})d\sigma = \iiint_T \text{div } \overline{v} \; dx \; dy \; dz. \tag{1}$$

Easily,

$$\text{div } \overline{v} = \frac{\partial v_1}{\partial x} + \frac{\partial v_2}{\partial y} + \frac{\partial v_3}{\partial z} = -3$$

so that

$$\iiint_T \text{div } \overline{v} \; dx \; dy \; dz = -3 \iiint_T dx \; dy \; dz = -3(4/3)\pi = -4\pi$$

as T is the unit sphere. To evaluate the left hand side of equation (1) we need to find $\overline{n}$, the outer unit normal to the unit ball. The equation for the unit ball is

$$F(x,y,z) = x^2 + y^2 + z^2 - 1 = 0.$$

Thus,

$$\overline{n} = \frac{\nabla F}{||\nabla F||} = \frac{2(x\overline{i} + y\overline{j} + z\overline{k})}{\sqrt{4(x^2 + y^2 + z^2)}} = x\overline{i} + y\overline{j} + z\overline{k} \tag{2}$$

since $x^2 + y^2 + z^2 = 1$. Consequently,

$$\iint_S (\overline{v}\cdot\overline{n})d\sigma = \iint_S [x(1 - x) + y(2 - y) + z(3 - z)]d\sigma = \iint_S (x + 2y + 3z - 1) \; d\sigma.$$

This integral is easy to evaluate if we make use of the symmetry of the unit sphere. First,

$$\iint_S (x + 2y + 3z - 1)d\sigma = \iint_S x \; d\sigma + 2 \iint_S y \; d\sigma + 3 \iint_S z \; d\sigma - \iint_S d\sigma.$$

To evaluate $\iint_S x\, d\sigma$ we project the unit sphere onto the xy-plane as described in

Definition 19.5.2.   Therefore

$$\iint_S x\, d\sigma = \iint_S x\, \sec[\gamma(x,y)]\, dx\, dy = \int_0^1 \int_0^{\sqrt{1-y^2}} \frac{x}{\sqrt{1-x^2-y^2}}\, dx\, dy$$

because

$$\sec[\gamma(x,y)] = \frac{1}{z}\ .$$

But,

$$\int_0^1 \int_0^{\sqrt{1-y^2}} \frac{x}{\sqrt{1-x^2-y^2}}\, dx\, dy = 0$$

so

$$\iint_S x\, d\sigma = 0.$$

In like fashion it may be shown that

$$\iint_S y\, d\sigma = \iint_S z\, d\sigma = 0$$

by projecting the unit sphere onto the yz- and xz-planes respectively. This should
be intuitively clear since the unit sphere is symmetric with respect to the origin.
This leaves us with

$$\iint_S (x + 2y + 3z - 1)\, d\sigma = -\iint_S d\sigma = -4\pi$$

and the Divergence Theorem is verified.

    It is not always the case that the right hand side of equation (1) is the
simplest integral to evaluate.

Example 1.   Verify the Divergence Theorem for the vector field

$$\bar{v} = (\frac{x}{r})\bar{i} + (\frac{y}{r})\bar{j} + (\frac{z}{r})\bar{k},$$

where $r = \sqrt{x^2 + y^2 + z^2}$ and T is the closed ball $x^2 + y^2 + z^2 \leq 4$.

<u>Solution</u>. We quickly find that

$$\frac{\partial v_1}{\partial x} = \frac{r^2 - x^2}{r^3}, \qquad \frac{\partial v_2}{\partial y} = \frac{r^2 - y^2}{r^3}, \qquad \text{and} \qquad \frac{\partial v_3}{\partial z} = \frac{r^2 - z^2}{r^3},$$

so that

$$\text{div } \overline{v} = \frac{3r^2 - x^2 - y^2 - z^2}{r^3} = \frac{2}{r} \, .$$

The integral

$$\iiint_T \text{div } \overline{v} \; dxdydz = \iiint_T \frac{2}{r} \, dxdydz$$

is easier to evaluate if we change to spherical coordinates (Section 18.10). Such is natural since we are integrating within a sphere. We obtain

$$\iiint_T \frac{2}{r} \, dxdydz = 8 \int_0^2 \int_0^{\pi/2} \int_0^{\pi/2} \frac{2}{r} \cdot r^2 \sin\theta \; d\theta d\phi dr = 16\pi.$$

(Why is the result evaluated in the first octant representative?) On the other hand,

$$\iint_S \overline{v} \cdot \overline{n} \; d\sigma = \iint_S d\sigma = 16\pi$$

Since $\overline{n} = \frac{x}{2}\overline{i} + \frac{y}{2}\overline{j} + \frac{z}{2}\overline{k}$, $r = 2$, and $\overline{v} \cdot \overline{n} = 1$.

We conclude with an example verifying Stokes' Theorem.

<u>Example 2</u>. Evaluate $\iint_S [(\text{curl } \overline{v}) \cdot \overline{n}] d\sigma$ over the surface $z = \sqrt{a^2 - x^2 - y^2}$ if

$\overline{v} = 2y\overline{i} - x\overline{j} + z\overline{k}$.

<u>Solution</u>. The surface in this example is the upper hemisphere of a sphere of radius a centered at the origin. We know from equation (2) that

$$\overline{n} = \frac{x}{a}\overline{i} + \frac{y}{a}\overline{j} + \frac{z}{a}\overline{k}$$

on this surface. From the definition (see Formula 19.7.1) we can readily check that

$$\text{curl } \overline{v} = \begin{vmatrix} \overline{i} & \overline{j} & \overline{k} \\ \dfrac{\partial}{\partial x} & \dfrac{\partial}{\partial y} & \dfrac{\partial}{\partial z} \\ 2y & -x & z \end{vmatrix} = -3\overline{k}.$$

Therefore,

$$\iint_S [(\text{curl } \overline{v})\cdot\overline{n}]d\sigma = -3 \iint_S \frac{z}{a} \, d\sigma.$$

If we use the definition of a surface integral (Definition 19.5.2) and project the hemisphere onto the xy-plane, we find

$$\iint_S [(\text{curl } \overline{v})\cdot\overline{n}]d\sigma = -3 \iint_S \frac{z}{a} \, d\sigma = -3 \iint_\Omega dxdy = -3\pi a^2$$

since Ω is a circle of radius a.

To obtain this result using Stokes' Theorem we need to evaluate $\displaystyle\int_C \overline{v}(\overline{r})\cdot d\overline{r}$ where C is the circle $x^2 + y^2 = a^2$. Since

$$d\overline{r} = dx\overline{i} + dy\overline{j} + dz\overline{k}$$

we have

$$\int_C \overline{v}(\overline{r})\cdot d\overline{r} = \int_C (2ydx - xdy + zdz).$$

Since dz = 0 along C we can parametrize the line integral using polar coordinates. Let

$$x = a \cos t \qquad\qquad y = a \sin t,$$

then

$$\int_C \overline{v}(\overline{r})\cdot d\overline{r} = \int_C [2(a \sin t)(-a \sin t) - (a \cos t)(a \cos t)]dt$$

or

$$\int_C \overline{v}(\overline{r})\cdot d\overline{r} = - \int_0^{2\pi} a^2(2 \sin^2 t + \cos^2 t)dt = -3\pi a^2.$$

UNIT STUDY GUIDES

1. OBJECTIVES. At the completion of this Unit you should be able to

   1. Define and use the absolute value symbol.
   2. Use interval notation and the symbols of set theory.
   3. Find the equation of a straight line.
   4. State and use the Factor and Remainder Theorems.
   5. Solve inequalities.
   6. Define a function, a one-to-one function, the inverse of a function, and the composition of two functions, as well as finding the latter two.
   7. Define and find the domain and range of a function.

2. PROCEDURES. (*)

   1. Read Sections 1.1-2 in the text and the Supplement. Work the Exercises in the Supplement.
   2. SSP Section 1.3
   3. SSP Sections 1.4-5
   4. SSP Sections 1.6-8

3. TEXT READING. Sections 1.1-8

4. SUGGESTED EXERCISES.

   p. 14: 1c, 2g, 6
   p. 19: 7, 13, 19, 25
   p. 23: 3, 7, 12, 19, 27
   p. 26: 5, 13, 16, 26, 29, 33, 45, 47
   p. 31: 3, 9, 17c, 20
   p. 37: 5, 11, 21-26

5. SAMPLE QUIZ.

   1. Give a precise definition of a function f.
   2. Find the equation of the straight line which passes through the point $(2,-5)$ and is perpendicular to the line $x + 3y - 4 = 0$.
   3. Solve for real x: $x(x + 2) > 3$.
   4. Find the remainder when $x^5 - 4x^3 + 3$ is divided by $x + 2$.
   5. Let $f(x) = x^2 + 1$ for x in $[-3,2]$ and $g(x) = \dfrac{2}{3 - x}$.

      (a) Find the range of f.   (b) Find g^{-1}, if it exists.

      (c) Evaluate $(g \circ f)(1)$.

(*) The Standard Study Procedure (SSP) is to read the text, read the corresponding Section(s) in the Supplement, and then to work the suggested exercises.

UNIT 2

1. OBJECTIVES. Your primary objective in this Unit is to be able to evaluate limits. While in pursuit of this objective you will call into play the various limit theorems given in the text. You should memorize and understand the definition of a limit. You will <u>not</u> be asked to find a delta for a given epsilon. At the completion of this Unit you should be able to

1. Give a precise definition of $\lim\limits_{x \to c} f(x)$.

2. Evaluate limits for the following six types of expressions as explained in the Supplement.

   1. Polynomials
   2. Rational functions
   3. Fractional exponents
   4. Absolute values
   5. Piecewise defined functions
   6. Conglomerates

2. PROCEDURES.

   1. Read the Preliminary Comments on pp. 10-15 of the Supplement.
   2. Read Section 2.1 of the text and work some exercises on p. 45.
   3. Read Section 2.2 of the text and work some exercises on p. 54.
   4. Read the Supplement corresponding to Sections 2.1 and 2.2 of the text and try more of the exercises on pp. 45 and 54.
   5. SSP Sections 2.3-4

3. TEXT READING. Sections 2.1-4

4. SUGGESTED EXERCISES.

   p. 45: 4, 6, 10, 14, 17, 18, 19
   p. 54: 2, 3, 4, 8, 9, 10, 14, 16
   p. 62: 1(b, c, e, h, s, u, v), 2(c, g), 3, 4a, 6, 7
   p. 67: 1a, 2(f, j, k, o), 4

5. SAMPLE QUIZ.

   1. Give a precise statement of the definition of $\lim\limits_{x \to 2} f(x) = -5$.

   2. Evaluate each of the following limits.

   (a) $\lim\limits_{x \to 2} 4$

   (b) $\lim\limits_{x \to -2} \sqrt{x + 2}$

   (c) $\lim\limits_{x \to -2} \dfrac{x + 2}{x^3 + 8}$

   (d) $\lim\limits_{x \to -5} (1 + \dfrac{x}{5})(\dfrac{-3}{x + 5})$

   (f) $\lim\limits_{x \to 1} \dfrac{|x - 1|}{x - 1}$

   (e) $\lim\limits_{x \to 0} \dfrac{1 - \dfrac{1}{x}}{1 - \dfrac{1}{x^2}}$

   (g) $\lim\limits_{x \to 0} f(x)$

   where $f(x) = \begin{cases} 1/x & x < 0 \\ x & 0 \le x < 3 \\ x^2 + 1 & 3 \le x \end{cases}$

1. OBJECTIVES. Your primary objective in this Unit is to be able to determine whether or not a given function is continuous and to support your answer. You again will be required to evaluate limits. You are not responsible for the greatest integer function. At the completion of this Unit you should be able to

   1. State and use the Intermediate-Value Theorem and the Maximum-Minimum Theorem.
   2. Give a precise statement of the definition of a function continuous at a point.
   3. Determine where a given function is discontinuous.
   4. Determine if a given function is continuous from the left, continuous from the right, or continuous at a given point.

2. PROCEDURES.

   1. SSP Sections 2.5-6
   2. Reread the Preliminary Comments on pp. 10-15 of the Supplement.
   3. Work the Additional Exercises in Section 2.7 of the text.

3. TEXT READING. Sections 2.5-7

4. SUGGESTED EXERCISES.

   p. 72: 2, 3(c, e, g, k, m), 4, 8a
   p. 81: 6, 7, 11, 16, 22, 23, 29, 30, 31, 32
   p. 83: 1, 7, 16, 18, 19, 32, 36

5. SAMPLE QUIZ.

   1. Give a precise statement of the Intermediate-Value Theorem.
   2. Using ε and δ, give a precise statement of the definition of a function continuous at $x = 0$.
   3. Give complete reasons for your answers to the following questions about the function f.

      (a) Is f continuous at $x = -1$?
      (b) Is f continuous from the right at $x = 0$?
      (c) Is f continuous at $x = 1$?
      (d) Is f continuous from the left at $x = 3$?
      (e) Is f continuous at $x = 5$?

$$f(x) = \begin{cases} \dfrac{|2x|}{x} & x \le -1 \\[2mm] \dfrac{x^2 - 1}{x + 1} & -1 < x \le 0 \\[2mm] -3\sqrt{x} & 0 < x < 1 \\[2mm] x^2 - 4 & 1 < x < 3 \\[2mm] 7x - 16 & 3 \le x < 4 \\[2mm] x + 14 & 5 \le x \end{cases}$$

1. OBJECTIVES. Your primary objective in this Unit is to become a whiz at differentiation, one of the two mechanical processes introduced in this course that is extensively applied throughout the calculus. (The other process, integration, will be introduced in Unit 11.) It is imperative that you become skilled and efficient at this process. If you conscientiously work as many of the Exercises as possible, you will discover that the few rules involved are very simple and will quickly become second-nature to you. At the completion of this Unit you should be able to

   1. State and apply the limit definition of a derivative.
   2. Describe the relationship between differentiability and continuity.
   3. Differentiate functions using the sum, difference, product, and quotient rules.
   4. Determine the value of a derivative at a given point.
   5. Use the $\frac{d}{dx}$ and $f'(x)$ notations.
   6. Solve simple "rate of change" problems.

2. PROCEDURES.

   1. Follow the Standard Study Procedure for each of the four Sections of the text covered by this Unit.
   2. Do more of the problems on pp. 101-2, 106-7. Remember to check your algebra, especially when your answers are not in the form given by the text.

3. TEXT READING. Sections 3.1-4

4. SUGGESTED EXERCISES.

   p. 94: 1, 5, 12, 14, 15, 17, 22, 23, 27, 28
   p. 101: 2, 6, 8, 10, 12, 17, 18, 19, 24, 27, 30, 32
   p. 106: 2, 8, 16, 22, 30
   p. 109: 1, 3, 8, 10, 12, 14

5. SAMPLE QUIZ.

   1. Use the definition to calculate $f'(x)$ where $f(x) = \sqrt{x - 3}$.
   2. The dimensions of a rectangle are changing in such a way that the area of the rectangle is always 14 sq. units. Find the rate of change of the base b with respect to the height h where b = 2 units.

   3. (a) Find $\frac{dy}{dx}$, where $y = \frac{x^2 + 2}{x^2 - 2}$.

      (b) Find $\frac{dy}{dt}$ at t = 1, where $y = t^2 + 2t^{-4} - 3t^{-1} + 4(7)^2$.

      (c) Find $f'(t)$, where $f(t) = -3(t^3 - 7t)(8 - 3t^2)$.

      (d) Find $f'(x)$, where $f(x) = x + 4f(x)$.

1. OBJECTIVES. Your primary objective in this Unit is to extend your differentiating wizardry to encompass the all-important chain rule. It gives you the ability to handle much more difficult (messy) differentiations. At the completion of this Unit you should be able to

    1. Differentiate using the chain rule.
    2. Calculate higher order derivatives.
    3. Differentiate x^n for n a rational number.
    4. Determine the domain for the derivative of a given function.

2. PROCEDURES.

    1. Read Section 3.5 of the text, try some of the suggested exercises, read the Supplement and work more of the remaining exercises.
    2. Follow this same procedure for each of the Sections 3.6 and 3.7.
    3. Work some of the Additional Exercises in Section 3.8 of the text.

3. TEXT READING. Sections 3.5-8

4. SUGGESTED EXERCISES.

    p. 115: 2, 4, 13, 14, 20, 22, 26, 28, 31, 37, 39, 41, 52
    p. 119: 7, 11, 15, 17, 26
    p. 123: 3, 6, 8, 13, 20, 21, 22, 23, 26
    p. 125: 1, 5, 6, 20, 21, 28, 33, 38, 43

5. SAMPLE QUIZ.

    1. Determine the domain for the derivative of the function f:

    $$f(x) = \begin{cases} |x + 1| & x \le 0 \\ 3(x - 1)^{1/3} & 0 < x < 9 \\ (x + 15)/4 & 9 \le x \le 13 \\ x - 6 & 13 < x \end{cases}$$

    2. Find the indicated derivatives. It is not necessary to algebraically simplify your answers.
    (a) $\frac{d}{dx}\{1 + \frac{d}{dx}(x^2 + x)\}$.   (b) $\frac{dy}{dt}$ when t = 2, given that $y = x^3$, $x = 2u - u^3$, and $u = 2 - \sqrt{t - 1}$.
    (c) $\frac{dy}{d(x^3)}$ where $y = x^2$.   (d) $\frac{dy}{dx}$ where $y = [(7x^2 + 1)^{2/3} + x^6]^{-1/5}$.
    (e) $(f \circ g)'(x)$, where $f(x) = \sqrt{4 - x^2}$ and $g(x) = (x^3 + 1)^5$.

1. OBJECTIVES. A good bit of the text for this Unit is devoted to proofs of numerous theorems. The proofs are relatively short and straightforward. As an aid to your mathematical training and understanding of the ideas discussed, you should make an effort to follow the proofs in detail. To apply one of these theorems to a problem, you must verify whether or not the hypothesis is true. You must then state explicitly how the theorem implies any results that relate directly to the problem at hand. At the completion of this Unit you should be able to

   1. Solve problems relating the derivative to the slope of tangent and normal lines to a curve.
   2. Differentiate implicitly.
   3. State and apply Rolles Theorem.
   4. State and apply The Mean Value Theorem.
   5. State and apply Theorem 4.2.4.
   6. Define what it means for a function to be increasing or decreasing on an interval.
   7. Determine where a given function is increasing and where it is decreasing.

2. PROCEDURES. Follow the Standard Study Procedure for each of the four Sections covered by this Unit.

3. TEXT READING. Sections 3.9-10; 4.1-2

4. SUGGESTED EXERCISES.

   p. 128: 1d, 3, 5, 6, 11
   p. 133: 7, 15, 19, 22, 25, 26
   p. 138: 3, 7, 8, 9, 10, 13
   p. 146: 1, 3, 8, 9, 11, 19, 25, 26

5. SAMPLE QUIZ.

   1. Give a precise statement of the definition of an increasing function.
   2. Find the equation of a line which passes through the point $(4,0)$ and is normal to the curve $y = \sqrt{x}$.
   3. Given that $xy^2 \frac{dy}{dx} - x^3 = 4y$ find $\frac{d^2y}{dx^2}$ in terms of x, y, and $\frac{dy}{dx}$.
   4. Find all points, c, which satisfy the conclusion of The Mean Value Theorem for
   $$f(x) = \begin{cases} 2x + 2 & -2 \le x < -1 \\ x^3 - x & -1 \le x \le 2 \end{cases} \quad \text{on } [a,b] = [-2,2].$$
   5. Determine the interval(s) where
   $$f(x) = \begin{cases} (x + 2)^2 & -3 \le x \le 0 \\ x + 2 & 0 < x < 1 \\ 7 - x & 1 < x < 2 \\ x^2 & 3 < x \end{cases}$$
   is increasing.

1. OBJECTIVES. This Unit focuses on certain characteristics of functions: local, absolute, and endpoint extrema. At the completion of this Unit you should be able to

   1. Locate all critical points of a function.
   2. Classify local, absolute and endpoint extrema, stating justification for your classification.

   For example, if the function f has a local maximum when x = 2, you should be able to ascertain this fact and support your assertion. The Supplement amply illustrates just how brief, but complete, these supportive reasons should be.

2. PROCEDURES. SSP Sections 4.3-4

3. TEXT READING. Sections 4.3-4

4. SUGGESTED EXERCISES.

   p. 152: 2, 9, 12, 17, 23
   p. 157: 8, 10, 14, 18, 19, 20

5. SAMPLE QUIZ.

   1. List the critical points of the following function. Classify the local, absolute, and endpoint extrema. Completely justify your answers.

   $$f(x) = x^{4/3} - 3x^{1/3} \qquad \text{on} \qquad [-1,8].$$

UNIT 8

1. OBJECTIVES. There is but one objective for this Unit. You should be able to solve extremum word problems. To meet this objective, you should be able to

   1. Give a functional representation of the word problem.
   2. Specify the domain of this function.
   3. Determine with justification the absolute maximum and/or minimum of the function.
   4. State your answer in a complete sentence.

2. PROCEDURES.

   1. SSP Section 4.5
   2. Read the Additional Comments on pp. 77-8 of the Supplement.

3. TEXT READING. Section 4.5

4. SUGGESTED EXERCISES.

   p. 161: 4, 7, 10, 11, 18, 26, 32
   p. 183: 3, 12, 19, 36, 41, 42

5. SAMPLE QUIZ.

   1. Find the point on the arc $y = -1 - 2x$ for x in $[0,3]$ which is closest to the point $(1,2)$.

   2. For this problem, simply specify the function (and its domain) which you would analyze to resolve the following question.

      Suppose that the point $P(a,b)$ lies in the first quadrant $(a > 0, b > 0)$. Further, let P lie on the curve $y = 1/x^2$. The tangent to the curve at P will intersect the positive coordinate axes at points R and S. For which value(s) of _a_ is the length of line segment RS minimal?

1. OBJECTIVES. In addition to those characteristics of functions discussed in Unit 7, this Unit focuses on concavity and points of inflection. At the completion of this Unit you should be able to

   1. Determine the intervals on which a function is concave up or down.
   2. Identify (with justification) points of inflection.
   3. Sketch the graph for a function after having found its extrema, concavity, and inflection points.

2. PROCEDURES. SSP Sections 4.6-7

3. TEXT READING. Sections 4.6-7

4. SUGGESTED EXERCISES.

   p. 167: 6, 10, 15
   p. 171: 13, 18, 19, 21, 29

5. SAMPLE QUIZ.

   1. For the function $f(x) = x\sqrt{4 - x^2}$ on $[-2,1]$:

      (a) List the critical points
      (b) Determine the concavity
      (c) Classify (with full justification) the extrema
      (d) Identify (with full justification) the inflection points
      (e) Sketch the graph.

UNIT 10

1. OBJECTIVES. At the completion of this Unit you should be able to

   1. Solve related rates problems.
   2. Make approximations using differentials and calculate the error in your approximation.
   3. Calculate and use the derivatives of implicitly defined functions.

   You will not be held responsible for the "little-oh" notation which appears in Section 4.9.

2. PROCEDURES.

   1. SSP Section 4.8
   2. SSP Section 4.9

3. TEXT READING. Sections 4.8-9

4. SUGGESTED EXERCISES.

   p. 175: 2, 3, 6, 8, 15, 19, 20
   p. 182: 1, 6, 7, 12, 15, 19
   p. 183: 18, 37, 40

5. SAMPLE QUIZ.

   1. Superman is in level flight at 8 miles above the ground. His flight path passes directly over a rocket installation. How fast is he flying when the distance between him and the rocket installation is 10 miles and this distance is decreasing at a rate of 4 miles per minute?
   2. A stone is projected from ground level vertically upward with an initial speed of 64 feet per second. What is the velocity of the stone after it has traveled 96 feet?
   3. Using differentials, approximate $(-10)^{-1/3}$.

1. OBJECTIVES. A good bit of the text reading for this Unit is devoted to motivating and providing the fundamentals behind the integral. It is not possible to express these ideas with a simpler notation than that used by the text. You should make every effort to follow the text in detail. At the completion of this Unit you should be able to

   1. Calculate the upper and lower sums for a given function and a given partition.
   2. State the Fundamental Theorem of Calculus.
   3. Evaluate integrals using the Fundamental Theorem of Calculus.
   4. Use the integral to find the area of a given region.
   5. Use the integral to solve motion problems.

   Note: The fourth objective is repeated in the next Unit.

2. PROCEDURES.

   1. SSP Sections 5.1-3
   2. SSP Section 5.4
   3. SSP Section 5.5

3. TEXT READING. Sections 5.1-5

4. SUGGESTED EXERCISES.

   p. 196: 7, 9, 11, 13
   p. 203: 1, 2, 3, 4, 8, 9
   p. 207: 1, 4, 9, 10, 12, 15, 17, 22
   p. 210: 1, 5, 9, 10, 12, 14, 17, 18, 19

5. SAMPLE QUIZ.

   1. An object moves along a coordinate line with velocity $v(t) = (t - 2)(t - 1)$. Its initial position at time $t = 0$ is 2 units to the right of the origin. Where is the particle positioned when $t = 6$ and what is the total distance traversed by the particle from $t = 0$ to $t = 6$?
   2. Give a precise statement of the Fundamental Theorem of Calculus.
   3. Evaluate $\displaystyle\int_1^6 \frac{dx}{\sqrt{3x - 2}}$.
   4. Find $L_f(P)$ for $f(x) = 4x - x^2$, x in [0,5], and P = {0,2,3,5}.
   5. Sketch the region bounded by the curves $x = \sqrt{y}$, $y = -x$, and $y = 4 - 2x$. Find the area of the region.

UNIT 12

1. OBJECTIVES. At the completion of this Unit you should be able to

   1. Differentiate definite integrals having variable limits of integration.
   2. Use the Substitution Principle to evaluate definite and indefinite integrals.
   3. Determine the area of a region in the plane by integrating with respect to either "x" or "y".

   The general properties of the integral discussed in Section 5.6 should be viewed as a prerequisite to one's ability to deal with most integration problems.

   You are not responsible for the material in Section 5.12. However, you may find it interesting to read about an alternate definition of the integral.

2. PROCEDURES.

   1. Read Sections 5.6 and 5.7 of the text and work some of the exercises on p. 214.
   2. Read Section 5.8 of the text and try some of the exercises on p. 217.
   3. Read Section 5.8 of the Supplement and work more of the exercises on p. 217.
   4. SSP Section 5.9
   5. SSP Section 5.10
   6. Work as many of the exercises in Section 5.11 of the text as seem necessary.

3. TEXT READING. Sections 5.6-11

4. SUGGESTED EXERCISES.

   p. 214:   2, 5, 8, 11, 12, 13, 14b
   p. 217:   1, 3, 6, 10, 14, 18, 21, 23, 25, 29, 30
   p. 220:   1-19, 21
   p. 227:   2, 4, 10, 12, 14
   p. 227-9: 3, 5, 6, 15, 24-29, 35, 36

5. SAMPLE QUIZ.

   1. Find $H'(x)$, when $H(x) = \int_{x^2}^{2x} x \cdot f(t)dt$.

   2. Evaluate: (a) $\int \frac{x}{\sqrt{x+1}}\,dx$ (b) $\int 5x^2(1 + x^3)^4 dx$.

   3. Sketch the region bounded by the curves $y = |x^2 - 2|$ and $y = 3x + 2$. Set up, but do not evaluate, one or more integrals of the form $\int \cdots dy$ to represent the area of the region.

1. OBJECTIVES. Your first objective is to conscientiously wade through all the detail present in the lengthy text readings for this Unit. These readings focus on a more general view of logarithms and exponentials (anti-logarithms) than you may have seen before. The Supplement begins by highlighting the critical formulas and results that may pull the details together. The balance of the Supplement consists of complete solutions for some of the Suggested Exercises. At the completion of this Unit you should be able to

    1. Evaluate derivatives and integrals involving logarithms.
    2. Evaluate derivatives and integrals involving exponentials.

Note that implicit in each of these objectives is the requirement that you be aware of the myriad of properties surrounding logarithms and exponentials. For example, $\log e^2 = 2$, $\log_2 8 = 3$, $e^{\log x} = x$, $f(x) = \log x$ is increasing on $(0, \infty)$, $\log 1 = 0$, and so on, and $e^{\text{so on}}$. Such "small" points should and will come to you easily after diligent attention to the exercises.

2. PROCEDURES.

    1. Read Sections 6.1 and 6.2 of the text, work some of the Suggested Exercises.
    2. Do the same for Section 6.3.
    3. Do the same for Section 6.4.
    4. Read the commentary part of the Supplement.
    5. Work the rest of the Suggested Exercises.

3. TEXT READING. Sections 6.1-4

4. SUGGESTED EXERCISES. (Asterisk indicates solution given in Supplement.)

p. 241: 1, 2, 5, 6, 9, 10*, 13b, 15, 17*, 18*
p. 248: 1, 5*, 8*, 9, 11, 13, 17, 18, 19, 25*, 26, 28, 29, 31*, 32*, 34, 36, 37, 39*, 41b, 43b*, 45e, 45f*
p. 256: 5, 8, 11*, 18, 20, 25, 27, 28*, 32, 39, 43, 44*, 45, 48, 49*, 50, 57, 59, 65*, 66

5. SAMPLE QUIZ.

    1. Evaluate $f'(x)$ for each of the following:
        (a) $f(x) = \log(e^x)$ (b) $f(x) = \log(x + \log x)$ (c) $x^2 = e^{f(x)}$.

    2. For what real values of x is $\log(x + 2) = \int_2^x \frac{dt}{t}$?

    3. Evaluate: (a) $\int_{e^2}^{e} \frac{\log x}{x} dx$ (b) $\int \frac{e^{2x-1}}{e^{3-x}} dx$.

UNIT 14

1. OBJECTIVES. At the completion of this Unit you should be able to

   1. Solve problems involving exponentials with arbitrary bases to arbitrary powers.
   2. Differentiate logarithmically.
   3. Solve exponential growth and decline problems.
   4. Apply the technique of integration by parts.

2. PROCEDURES.

   1. SSP Section 6.5
   2. SSP Section 6.6
   3. SSP Section 6.7

3. TEXT READING. Sections 6.5-7

4. SUGGESTED EXERCISES. (Asterisk indicates solution given in Supplement.)

   p. 266: 1e, 1i, 1j*, 1ℓ*, 2a*, 2c, 3c, 3e*, 4*, 6e*, 6h*, 7d*, 9c*, 9d*, 9f, 10a*, 10d, 10f*, 10g*, 11c*, 12(e, f, g, i), 15*
   p. 273: 4, 7, 12*, 14, 15
   p. 278: 2, 3, 4, 5, 9, 10, 12, 15, 16, 17, 21, 24

5. SAMPLE QUIZ.

   1. The rate of decay of a radioactive substance is proportional to the amount of substance present. Last year there were 20 lbs. of the substance in your closet. If there are at most 16 lbs. in the closet today, for how many more years might there be at least 3 lbs. present?

   2. Evaluate $\frac{d}{dx}[(2^x + \log_3 x)^x]$.

   3. Evaluate $\int x\, 3^{-2x} dx$.
   4. Sketch the graphs of the following functions labeling two points on each graph.

$$y = 1 + \log_4 |x - 2| \qquad \text{and} \qquad y = \frac{1}{3^{1-x}} .$$

1. OBJECTIVES. At the completion of this Unit you should be able to

    1. Differentiate expressions involving the circular functions.
    2. Integrate expressions involving the circular functions.
    3. Evaluate expressions (including limits) involving the circular functions.

2. PROCEDURES.

    1. Review the material on trigonometry in Section 1.2 of the text and do the
        review exercise on trigonometry in Section 7.1-2 of the Supplement.
    2. Read Section 7.1 of the text and work some of the Suggested Exercises.
    3. Read Section 7.2 of the text and work some of the Suggested Exercises.
    4. Read the balance of the Supplement.
    5. Work more of the Exercises without looking up the relevant formulas in
        the text.

3. TEXT READING. Sections 7.1-2

4. SUGGESTED EXERCISES.

    p. 298: 1, 2, 5, 10, 19, 23, 25, 27, 30, 34, 39, 43, 45, 46a, 50, 56, 57,
             59, 63, 65
    p. 303: 1, 5, 7, 8, 10, 15, 17, 18, 19, 24, 25, 30, 31, 32

5. SAMPLE QUIZ.

    1. Evaluate $\int e^{3x} \sin x \cos x \, dx$.   2. Evaluate $\frac{dy}{dx}$ for $y = \int\limits_{\csc x}^{2} \frac{dt}{t^2 - 1}$.

    3. Evaluate $\lim\limits_{x \to 0} x^2 \csc^2(x/2)$.

    4. Find the equation of the tangent line to the curve $y = \tan^2 x$ where $x = \pi/6$.

UNIT 16

1. OBJECTIVES. At the completion of this Unit you should be able to

    1. Differentiate, integrate, and evaluate expressions involving the inverse
        trigonometric functions.
    2. Solve related rate and extremum problems involving trigonometry.
    3. Solve simple harmonic motion problems.

    Note: Chapter 6 of the text concludes with two optional Sections on the
    hyperbolic functions. You are not responsible for any of this material.
    However, since the hyperbolic functions are intimate companions of the
    trigonometric functions and are of great practical value, an optional
    commentary supporting these very claims is included for your pleasure.

2. PROCEDURES.

    1. SSP Section 7.3
    2. Work the Additional Exercises in Section 7.3 of the Supplement.
    3. Work the Suggested Exercises in Section 7.4.
    4. Read Section 7.5 of the text and work the Suggested Exercises.

3. TEXT READING. Sections 7.3-5

4. SUGGESTED EXERCISES. (Asterisk indicates solution given in Supplement.)

    p. 309: 1-12, 15, 18, 19, 21, 24, 31, 35-38, 44, 47, 48, 49, 53
    p. 311: 6, 7*, 15, 20, 21*, 22*, 23*, 24*, 28*
    p. 317: 1, 5, 6, 9, 10

5. SAMPLE QUIZ.

    1. Evaluate $\sec(\arcsin(-\frac{1}{3}))$.
    2. The sum of the base and altitude of an isosceles triangle is a constant
        20 inches. If the area of the triangle is 48 sq. in. and increasing at
        the rate of 4 in./min., determine the rate of change in one of the base
        angles of the triangle.
    3. Evaluate $\frac{d}{dx}[e^{\arctan 4x}]$.
    4. An object in simple harmonic motion passes through the central point x = 0
        at time t = 0 and every 3 seconds thereafter. Find an equation for the
        motion if v(0) = -4.

1. OBJECTIVES. At the completion of this Unit you should be able to

   1. Use the method of partial fractions to evaluate integrals.

   Naturally, you may use and are still liable for the technique of integration by parts and the Substitution Principle.

2. PROCEDURES.

   1. SSP Section 8.1
   2. Review the Substitution Principle (Unit 12) and integration by parts (Unit 14) as seems necessary. Work more of the exercises on pp. 332-3.
   3. SSP Section 8.2

3. TEXT READING. Sections 8.1-2

4. SUGGESTED EXERCISES.

   p. 332: 3, 4, 6, 7, 9, 10, 15, 18, 19, 21, 22, 24, 29, 31, 32
   p. 341: 1, 3, 4, 6, 11, 12, 15, 18

5. SAMPLE QUIZ.

   1. Evaluate $\int x \arctan x \, dx$.

   2. Evaluate $\int \dfrac{dx}{x^4 - 1}$.

1. OBJECTIVES. At the completion of this Unit you should be able to

    1. Evaluate integrals involving the trigonometric functions.
    2. Evaluate certain types of integrals using a trigonometric substitution.

    You should be aware that quizzes on this Unit may well require you to apply techniques of integration studied earlier: the Substitution Principle, integration by parts, the method of partial fractions.

2. PROCEDURES.

    1. Read Sections 8.3-4 of the text, try a few of the Suggested Exercises, read the Supplement, and then work the remaining Suggested Exercises.
    2. Read Section 8.5 of the Supplement, try the first few Suggested Exercises, read Section 8.5 of the text, work the remaining Suggested Exercises and work the Additional Exercises in the Supplement.
    3. Read Section 8.9 of the Supplement and then turn yourself loose on the problems on pp. 361-3.

3. TEXT READING. Sections 8.3-5, 9

4. SUGGESTED EXERCISES.

    p. 345:   4, 5, 8, 9, 10, 12, 15, 19, 21, 24
    p. 348:   1, 2, 5, 6, 8, 9, 12, 14
    p. 350:   2, 5, 6, 8, 11, 13, 15, 17, 20, 22, 23
    p. 361:   4, 6, 8, 12, 20, 22, 24, 32, 35, 40, 41, 46, 52, 56-60

5. SAMPLE QUIZ.

    Evaluate the following integrals.

    1. $\displaystyle\int \sqrt{x^2 + 6x + 13}\ dx$ 2. $\displaystyle\int x\ \arcsin x\ dx$ 3. $\displaystyle\int \frac{dx}{\csc^4 x}$

1. OBJECTIVES. At the completion of this Unit you should be able to

   1. Integrate rational expressions in sin x and cos x.
   2. Approximate definite integrals using the **Trapezoidal** Rule and Simpson's Rule.
   3. Calculate bounds for the error in your approximate integrations in (2).

   Quizzes on this Unit may require you to apply techniques of integration studied earlier. You can best test your mastery of all the techniques of integration by working on the mixed bag of problems in Section 8.9 without consulting the text or the Supplement.

2. PROCEDURES.

   1. SSP Sections 8.6-7
   2. SSP Section 8.8

3. TEXT READING. Sections 8.6-8

4. SUGGESTED EXERCISES.

   p. 353: 1, 5, 8, 9, 11
   p. 355: 2, 4, 5, 13, 16, 18
   p. 360: 1, 2

5. SAMPLE QUIZ.

   1. Use Simpson's Rule with n = 3 to approximate $\int_0^\pi \cos x\, dx$ to the **nearest** hundredth.
   2. Determine a value of 'n' (number of subdivisions of [a,b]) so the error in approximating $\int_1^3 e^x dx$ by the Trapezoidal Rule is less than .001.

   3. Evaluate $\int \dfrac{dx}{3 \cos x + 5}$.

1. OBJECTIVES. At the completion of this Unit you should be able to

   1. State and use the formula for the distance between two points.
   2. Determine the center and radius of a circle.
   3. State and use the formula for the distance from a point to a line.
   4. Execute and identify a translation of axes.
   5. Given the focus, vertex, axis of symmetry, and directrix find the associated parabola, and given a parabola find these quantities.
   6. Given the foci, center, and vertices find the associated ellipse, and given an ellipse find these quantities.
   7. Given the foci, vertices, center, and asymptotes find the associated hyperbola, and given a hyperbola find these quantities.
   8. Determine whether a given second degree equation is an ellipse, a circle, a parabola, or a hyperbola.
   9. Graphically represent a conic and the 'vital parts' noted in objectives (2), (5), (6), (7).
  10. Write an equation to fit a given locus description.

You are not responsible for the "eccentricity" of a conic. (Pappus is!)

2. PROCEDURES.

Chapter 9 of the Supplement consists of solutions for some of the Suggested Exercises.

   1. Read Sections 9.1-2 of the text and work the Suggested Exercises.
   2. Do the same for each of the Sections 9.3-5.

3. TEXT READING. Sections 9.1-6. (In addition, we recommend that you read about rotation of axes in Section 9.7.)

4. SUGGESTED EXERCISES. (Asterisk indicates solution given in Supplement.)

   p. 369: 1b, 3a*, 3b, 4*, 5*
   p. 376: 1(d,g), 1h*, 2e*, 2(d,h), 3a*, 3c, 9-14
   p. 383: 1, 2, 4*, 6*, 12*, 13, 14*, 15
   p. 388: 1, 4, 6*, 7*, 10, 11, 13, 14*, 17*, 22
   p. 389: 10*, 13*

5. SAMPLE QUIZ.

   1. Identify the conic $9x^2 - y^2 - 18x - 4y + 4 = 0$ and its 'vital parts'. Sketch the graph.
   2. Find the points on the parabola $y^2 - 8x + 2y = 15$ which are 5 units from the directrix for the parabola.
   3. Find and identify the equation of the locus of all points whose distances from (2,1) are equal to one-half their distances from the line $y = -2$.

1. OBJECTIVES. At the completion of this Unit you should be able to

   1. Compute the average value of a function on an interval.
   2. Compute the volume of a solid of revolution.
   3. Compute the volume of solids by parallel cross sections.

2. PROCEDURES.

   1. Read Section 10.1 of the text and work the **Suggested Exercises.**
   2. SSP Sections 10.2-3
   3. Work the Additional Exercises in Sections 10.2-3 of the Supplement.

3. TEXT READING. Sections 10.1-3

4. SUGGESTED EXERCISES.

   p. 398: 6, 7, 9, 12, 13
   p. 407: 4, 8, 10, 11, 15, 19, 20, 21
   p. 414: 4, 10, 12, 13, 15
   p. 426: 5, 9, 14

5. SAMPLE QUIZ.

   1. Determine the average value for the areas of the regions bounded by the
      curves $y = 0$ and $y = tx - x^3$ where $1 \le t \le 4$.
   2. The base of a certain solid is the region bounded by the curves $y = e^x$,
      $y = 2$ and $x = 0$. Cross sections of the solid perpendicular to the x-axis
      are semicircles. Determine the volume of the solid.
   3. Set up, but do <u>not</u> evaluate, the integral(s) needed to represent the volume
      of the solid obtained by revolving about the line $x = 3$ the region
      bounded by the curves $y = \sin x$, $y = 1/2$ and $x = 0$.

      (a) Sketch the region R
      (b) Set up the integral(s) using the shell method
      (c) Set up the integral(s) using the disc method

1. OBJECTIVES. At the completion of this Unit you should be able to

   1. Solve word problems involving the notion of work.
   2. Solve fluid pressure problems.
   3. Solve revenue stream problems.

   Each of these objectives centers on a simple formula. In addition, the first objective requires that you understand some elementary physics. You will have to do some interpretive thinking. This is known as Crooke's Law. Thus, you should be able to solve all of the work problems either by Hooke or by Crooke.

2. PROCEDURES.

   1. Read Section 10.4 of the text and work the Suggested Exercises.
   2. Do the same for Section 10.5.
   3. Do the same for Section 10.6.

3. TEXT READING. Sections 10.4-6

4. SUGGESTED EXERCISES. (Asterisk indicates solution given in Supplement.)

   p. 418: 2c*, 3, 5, 7b*, 12*, 13, 14*
   p. 422: 2*, 4, 7b*, 8
   p. 425: 1, 6a*

5. SAMPLE QUIZ.

   Set up, but do <u>not</u> evaluate, the integrals needed to solve the following problems. Provide a carefully labelled diagram where appropriate.

   1. The trapezoidal gate of a dam is 8 feet high, 20 feet across at the top and 14 feet across at the bottom. Determine the depth of the water which will exert one-half the force on the gate as when the dam is full.
   2. Let S be a continuous revenue stream at the rate of $(1000 + 60t) per year. What is the present value of the second and third years of revenue? Assume continuous compounding at 5%.
   3. A steam shovel is excavating sand. Each load weighs 500 lb. when excavated. The shovel lifts each load to a height of 15 feet in 1/2 minute, then dumps it. A leak in the shovel lets sand drop out while the shovel is being raised. The rate at which sand leaks out is 160 lb/min. Calculate the work done by the shovel in raising one load.

1. OBJECTIVES. At the completion of this Unit you should be able to

    1. Convert a polar coordinate equation into rectangular coordinates, and conversely.
    2. Sketch the graph of a polar curve.
    3. Determine the points of intersection and collision for two polar curves.

    You are not explicitly responsible for symmetry tests in polars, but it would help your graphing technique if you were capable of making these tests. It would also be helpful for you to be able to recognize the more general polar families of cardioids, conics, lines, circles, petals, and lemniscates.

2. PROCEDURES.

    1. SSP Section 11.1
    2. SSP Section 11.2
    3. Work <u>all</u> of the Additional Exercises in Section 11.2 of the Supplement.
    4. SSP Section 11.3

3. TEXT READING. Sections 11.1-3

4. SUGGESTED EXERCISES.

    p. 436: 1-8, 11, 25, 34, 35, 39, 43
    p. 440: 2, 5, 12, 14
    p. 442: 5, 6, 7, 9

5. SAMPLE QUIZ.

    1. Sketch the graphs and determine the collision points for $r^2 = \cos 2\theta$ and $r = -2 + 3\cos\theta$.
    2. On separate axes sketch the following polar curves:

        (a) $r^2 = 1$, $0 \le \theta < \pi/4$

        (b) $r\sec\theta = -2$

        (c) $r = \dfrac{3}{1 - 2\sin\theta}$

        (d) $r^2(\cos^2\theta + 4\sin^2\theta) = 4$

UNIT 24

1. OBJECTIVES. At the completion of this Unit you should be able to

    1. Compute the area of a region whose boundaries are expressed in polar coordinates.
    2. Sketch the graph and write the equation of the line tangent to a curve given parametrically.
    3. Find the equation (in rectangular and polar form) of the tangent line to a polar curve.
    4. Compute the length of a curve (arc) given parametrically. This includes the special cases of polar curves and curves given as $y = f(x)$.
    5. Compute the area of a surface of revolution.

You are not responsible for the material in Section 11.6. A solution for Exercise 1 (p. 454) is given in the Supplement and you should read it.

2. PROCEDURES.

    1. SSP Section 11.4
    2. Work all of the Additional Exercises in Section 11.4 of the Supplement.
    3. SSP Section 11.5
    4. SSP Section 11.7
    5. SSP Section 11.8
    6. Read the Further Comments on Integration on pp. 247-252 of the Supplement.

3. TEXT READING. Sections 11.4-5, 11.7-8

4. SUGGESTED EXERCISES.

    p. 446: 5, 9, 15
    p. 453: 4, 6, 11, 15, 17, 22-26
    p. 463: 1-11, 14, 15, 23, 26, 30
    p. 470: 3, 10, 15, 17

5. SAMPLE QUIZ.

    1. Find the area of the polar region which is bounded by the curves $r = \dfrac{1}{1 - \cos \theta}$ and $r \cos \theta = -1$.
    2. Set up, but do not evaluate, the integral(s) needed to find the surface area when the curve $x = y^2 + 1$, $1 < y < 2$, is revolved about the x-axis.
    3. Find the equation of the tangent line to the curve $r = \sec \theta + \csc \theta$ when $\theta = \pi/6$.
    4. Set up, but do not evaluate, the integral(s) needed to find the length of the polar curve $r = \cos 3\theta$.

1. OBJECTIVES. At the completion of this Unit you should be able to

    1. Define the terms: sequence, limit of a sequence (using ϵ and N), bounded sequence, monotone (increasing et al.) sequence.
    2. Determine the boundedness, monotonicity, and convergence (or divergence) of a sequence.
    3. Apply L'Hospital's Rule.
    4. Evaluate improper integrals.

2. PROCEDURES.

    1. Read the Preliminary Comments on pp. 253-7 of the Supplement.
    2. SSP Sections 12.1-2
    3. SSP Section 12.3
    4. SSP Sections 12.4-6
    5. SSP Section 12.7

3. TEXT READING. Sections 12.1-7

4. SUGGESTED EXERCISES. (Asterisk indicates solution given in Supplement.)

    p. 478: 6, 9, 14, 15, 25
    p. 488: 5, 9, 26, 27, 29, 30
    p. 491: 3*, 8*, 10*, 12*, 18*, 20*, 23*, 24*, 27, 33, 36, 38
    p. 495: 4, 7, 11, 17
    p. 500: 10, 13, 20, 21, 29
    p. 506: 4, 9, 10, 14, 17, 18, 19, 28, 42
    p. 512: 6, 18, 20, 31

5. SAMPLE QUIZ.

    1. Give a precise definition of a bounded sequence.
    2. Give an example of a sequence which converges to π but is not monotonic.
    3. Show that the sequence $\{\frac{n}{2^n}\}$ is monotonic.
    4. Evaluate:

      (a) $\displaystyle\lim_{n \to \infty} (1 - \frac{3}{n})^n$ (b) $\displaystyle\lim_{x \to 0} \frac{1 - \sec x}{x \sin x}$ (c) $\displaystyle\int_0^3 \frac{dx}{(x - 1)^3}$

UNIT 26

1. OBJECTIVES. At the completion of this Unit you should be able to

    1. Define an infinite series.
    2. Use the sigma notation.
    3. Determine the sum of geometric series and certain series which telescope.
    4. Determine the absolute convergence, conditional convergence or divergence
        of a given series by applying the various tests (e.g., Integral Test) and
        theorems (e.g., Theorem on Alternating Series).

2. PROCEDURES.

    1. SSP Section 13.1
    2. SSP Section 13.2
    3. SSP Sections 13.3-4
    4. SSP Section 13.5
    5. Work the Additional Exercises in Section 3.12 of the text.

3. TEXT READING. Sections 13.1-5

4. SUGGESTED EXERCISES. (Asterisk indicates solution given in Supplement.)

    p. 517: 2d, 2i*, 6b*, 6e, 6f, 8b*
    p. 524: 1, 10, 14, 21, 29
    p. 530: 6, 7, 11, 12, 18, 21
    p. 534: 4, 7, 8, 10, 16, 18, 21, 22, 26
    p. 539: 3, 5, 13, 14, 20, 21
    p. 572: 9, 12, 15, 24, 27, 32, 36

5. SAMPLE QUIZ.

    1. Evaluate:

    (a) $\displaystyle\sum_{j=2}^{5} (-1)^j \, j^2$ (b) $\displaystyle\sum_{k=1}^{\infty} \frac{1}{(k + 1)(k + 3)}$ (c) $\displaystyle\sum_{k=1}^{\infty} \frac{(2)^{2k-1}}{5^{k+2}}$

    2. Determine the absolute convergence, conditional convergence, or
        divergence of each of the following series. Give full justification for
        your conclusions.

        (a) $\displaystyle\sum_{k=2}^{\infty} (-1)^k \frac{k + 1}{\sqrt{k^7 - 2k}}$ (b) $\displaystyle\sum_{k=1}^{\infty} (-1)^{k+1} (1 - \frac{1}{k})^k$

        (c) $\displaystyle\sum_{k=4}^{\infty} (-1)^k \frac{1}{2 + \sqrt{k}}$ (d) $\displaystyle\sum_{k=1}^{\infty} \frac{k! \, 3^k}{(2k + 1)!}$

1. OBJECTIVES. At the completion of this Unit you should be able to

    1. Determine the Taylor series expansion for a given function about a given point.
    2. Use Taylor polynomials with remainder term to approximate integrals and values of functions.
    3. Determine the region of convergence for a power series.

    On occasion you will find it useful to be able to integrate and differentiate power series. Further, there are certain functions whose Taylor series expansions you should memorize (see p. 289 of the Supplement). You are not responsible for Section 13.11 on the Binomial Series but you should read this brief Section anyway.

2. PROCEDURES.

    1. SSP Sections 13.6-7
    2. Read Section 13.8 of the text and work the Suggested Exercises.
    3. SSP Sections 13.9-10
    4. Work the Additional Exercises in Section 13.12 of the text.

3. TEXT READING. Sections 13.6-10

4. SUGGESTED EXERCISES.

    p. 546: 5, 11, 17, 19
    p. 552: 1, 5, 7, 9
    p. 556: 1, 5, 13
    p. 562: 7, 9, 13, 16, 20, 21
    p. 568: 1b, 1f, 4, 11
    p. 572: 19, 25, 38, 42, 46, 47, 51

5. SAMPLE QUIZ.

    1. Estimate $\displaystyle\int_1^4 x^2 e^{-x} dx$ to an accuracy of 0.01.

    2. Determine the Taylor Series expansion about $a = 3$ for the function
       $f(x) = \dfrac{x}{1 + x}$.

    3. Determine the region of convergence for the Taylor Series $\displaystyle\sum_{n=0}^{\infty} \dfrac{(x - 2)^n}{3^n \cdot n}$.

1. OBJECTIVES. At the completion of this Unit you should be able to

    1. Use the distance formula in three dimensions.
    2. Compute the norm of a vector.
    3. Compute the dot product of two vectors.
    4. Compute the component of one vector in the direction of another.
    5. Determine and use the symmetric equations for and the vector parametrization of a line.
    6. Determine and use the scalar and vector forms for the equation of a plane.
    7. Compute the cross product of two vectors.

You are not responsible for the material in Section 14.8 but you should read and enjoy this brief Section anyway.

2. PROCEDURES. Chapter 14 of the Supplement consists exclusively of solutions for some of the Suggested Exercises.

    1. Read Section 14.1 of the text and work the Suggested Exercises.
    2. Repeat this procedure for Sections 14.2-3 and each of the Sections 14.4-7.
    3. Work the Additional Exercises in Section 14.9 of the text.

3. TEXT READING. Sections 14.1-7, 14.9

4. SUGGESTED EXERCISES. (Asterisk indicates solution given in Supplement.)

    p. 577: 2c, 4, 7*, 8
    p. 588: 9, 11, 12, 13, 14b*, 15, 19
    p. 595: 2, 9*, 10*, 14, 18*, 20*
    p. 601: 4, 5, 6, 9d*, 12*, 14b*
    p. 608: 3, 5, 10, 16*, 20*, 21*
    p. 614: 12*, 16*, 20, 21a, 22a, 23*, 24b*, 25a
    p. 618: 1, 3*, 4b*, 4d*, 6*, 7*, 13*, 14

5. SAMPLE QUIZ.

    1. Find the coordinates of a point on the line
$$\frac{x-1}{2} = \frac{y+2}{1} = \frac{z-3}{4}$$
       which is 2 units from the point $(5,0,11)$ (on the line).
    2. Find an equation for the plane which passes through the point $(2,1,-3)$ and the line determined by the planes $3x + y - z = 2$ and $2x + y + 4z = 1$.
    3. Compute the component of $\overline{a} = (1,3,-2)$ in the direction of $\overline{b} = (2,-4,1)$.
    4. Compute $\overline{a} \cdot [\overline{b} \times \overline{c}]$ for $\overline{a} = (2,1,3)$, $\overline{b} = (1,-1,0)$, $\overline{c} = (-3,2,4)$.

1. OBJECTIVES. At the completion of this Unit you should be able to

    1. Differentiate and integrate vector functions.
    2. Find the tangent vector and the tangent line to a given curve at a given point.
    3. Determine the angle of intersection of two curves.
    4. Determine the velocity, speed and acceleration of an object moving along a path $\overline{r}(t)$.
    5. Compute the arc length of a curve C.
    6. Compute the curvature of a curve C.

2. PROCEDURES. Chapter 15 of the Supplement consists primarily of solutions for some of the Suggested Exercises.

    1. Read Section 15.1 of the text and work the Suggested Exercises.
    2. Repeat this procedure for each of the Sections 15.2-6.
    3. Read the Concluding Comments on Vectors on pp. 326-8 of the Supplement.

3. TEXT READING. Sections 15.1-6

4. SUGGESTED EXERCISES. (Asterisk indicates solution given in Supplement.)

    p. 625: 6, 9, 15*, 17*, 18*
    p. 630: 4, 6, 8, 12, 13, 23*, 25*
    p. 635: 1e, 5, 8*, 9*, 10*
    p. 643: 2*, 5*, 6*, 9*
    p. 649: 1*, 9*, 10*
    p. 657: 5, 10*, 11*, 14, 16

5. SAMPLE QUIZ.

    1. Compute $\frac{d}{dt}[(t\overline{i} - 3t\overline{j}) \times (-t^2\overline{j} + 5t\overline{k})]$.

    2. Find the tangent line to the curve $\overline{r}(t) = e^t\overline{i} - 2t\overline{j} + t^2\overline{k}$ at $t = 2$.

    3. The path of a plane motion is the curve
$$\overline{r}(t) = (t^4 - 6t^2)\overline{i} + (2t + 3t^2 + t^3)\overline{j}.$$
Compute the speed at those points where the acceleration is zero.

    4. Represent the length of the curve $\overline{r}(t) = t\overline{i} + t^2\overline{j} + t^3\overline{k}$, $t \in [0,2]$, as an integral.

    5. Compute the curvature of the spiral of Archimedes $r = a\theta$.

UNIT 30

1. OBJECTIVES. At the completion of this Unit you should be able to

   1. Determine the domain and range for a function of several variables.
   2. Identify quadric surfaces.
   3. Use level curves to represent a function of two variables.
   4. State and use the definition for the partial derivative of a function of several variables.
   5. Compute the first and second order partials for a function of several variables.
   6. Define an open set, a closed set, the neighborhood of a point, and the interior and boundary of a set.
   7. Define the limit of a function of several variables.

2. PROCEDURES.

   1. SSP Sections 16.1-3
   2. SSP Sections 16.4-6

3. TEXT READING. Sections 16.1-6

4. SUGGESTED EXERCISES.

   p. 662: 7, 8, 13, 15
   p. 668: 1, 11, 16, 17, 23, 26, 29, 30
   p. 675: 4, 10, 11, 15
   p. 681: 1, 8, 17
   p. 685: 3, 5, 7, 10
   p. 691: 1, 6

5. SAMPLE QUIZ.

   1. Determine the domain and range for the function $f(x,y) = \dfrac{1}{|3 - \sqrt{25-x^2-y^2}|}$.
   2. Draw the level curves for the function $f(x,y) = 2x - y^2$.
   3. Give an example of a set in the xy-plane which is neither open nor closed.
   4. Using the definition compute $\dfrac{\partial f}{\partial x}$ for $f(x,y) = x^2/y$.
   5. Compute $\dfrac{\partial^2 f}{\partial y \partial x}$ for $f(x,y,z) = xy^2 + x^3z + yzx$.
   6. Identify the surface $x^2 - 3y^2 + z = 0$ and find its traces.

1. OBJECTIVES. At the completion of this Unit you should be able to

   1. Find the gradient for a function of several variables.
   2. Find the directional derivative of a function at a given point in an indicated direction.
   3. Compute derivatives using the chain rule.
   4. Find the equations for tangent and normal lines to a curve $f(x,y) = c$ at a given point.
   5. Find the equation for a tangent plane to a surface $z = f(x,y)$ (or $g(x,y,z) = 0$) at a given point.
   6. Find and classify the stationary points of a function of two variables.

2. PROCEDURES.

   1. SSP Sections 17.1-2
   2. SSP Section 17.3
   3. SSP Section 17.4
   4. SSP Section 17.5

3. TEXT READING. Sections 17.1-5

4. SUGGESTED EXERCISES. (Asterisk indicates solution given in Supplement.)

   p. 698: 5, 20
   p. 707: 3, 12, 15, 19
   p. 716: 3, 11, 13, 15, 21, 22, 24
   p. 725: 7, 11, 16, 18, 23, 28, 29
   p. 731: 2*, 3, 9, 13*

5. SAMPLE QUIZ.

   1. Find the directional derivative of $f(x,y) = x^2 e^y - z$ at (2,1,1) toward the point (1,3,-1).
   2. Given that $u = x^2 - 3y^2 x$, $x = 2s - 3t$ and $y = 4st^2$, find $\frac{\partial u}{\partial s}$ when $s = t = 1$.
   3. Find the equation for the tangent plane to the surface $z^2 + 9x^2 + 4x^3 y = 26$ at the point (1,2,3).
   4. Find and classify the stationary points for $f(x,y) = x^3 - 3x + 4y - y^2$.

UNIT 32

1. OBJECTIVES. At the completion of this Unit you should be able to

   1. Use the Second-Partials Test to classify the stationary points of a function of two variables.
   2. Use the method of Lagrange multipliers to find extrema of a function subject to a side condition.
   3. Compute the differential of a function and use it to make approximations.
   4. Determine a function given its gradient.

2. PROCEDURES.

   1. SSP Section 17.6
   2. Read Section 17.7 of the text and work the Suggested Exercises.
   3. Read Section 17.8 of the text and work the Suggested Exercises.
   4. SSP Section 17.9

3. TEXT READING. Sections 17.6-9

4. SUGGESTED EXERCISES. (Asterisk indicates solution given in Supplement.)

   p. 736: 3, 5, 13, 18, 19*, 22*, 23*
   p. 743: 5, 11*, 15*, 18*, 19*
   p. 748: 7, 9, 16a
   p. 755: 5, 11

5. SAMPLE QUIZ.

   1. Find and classify the stationary points of $f(x,y) = x^3 - 3x + 4y - y^2$.
   2. Show that the cube has the largest volume of all rectangular solids with a given surface area.
   3. Use differentials to approximate $\sin(\pi/5)\cos(7\pi/8)$.
   4. Find a function $H(x,y)$ such that $x^2 y^3 \overline{i} + H(x,y)\overline{j}$ is the gradient for some function.

1. OBJECTIVES. At the completion of this Unit you should be able to

    1. Use multiple-sigma notation.
    2. Calculate the upper and lower sums for a given function of two variables and a given partition.
    3. Use repeated (iterated) integrals to evaluate double integrals.
    4. Use repeated integrals to represent the area of a planar region.
    5. Use repeated integrals to represent the volume of a solid.

2. PROCEDURES.

    1. SSP Section 18.1
    2. SSP Section 18.2
    3. SSP Sections 18.3-4

3. TEXT READING. Sections 18.1-4

4. SUGGESTED EXERCISES. (Asterisk indicates solution given in Supplement.)

    p. 759: 3, 5, 9
    p. 769: 5*, 6*, 8*
    p. 781: 9*, 11*, 12*, 17, 19*, 21, 22, 25*, 26*

5. SAMPLE QUIZ.

    1. Evaluate $L_f(P)$ for $f(x) = x^2 + (y - 1)^2$ on
        $R = \{(x,y): -2 \leq x \leq 1, 0 \leq y \leq 3\}$ where the partition $P = P_1 \times P_2$ is given by $P_1 = \{-2,1\}$ and $P_2 = \{0,2,3\}$.

    2. Evaluate $\iint_\Omega x \, dy \, dx$ with Ω as the triangle with vertices (0,0), (3,1), (1,5).

    3. Set up, but do not evaluate, the repeated integrals needed to represent the volume of the solid bounded by $2x + 3y + 4z = 12$ and the coordinate planes.

1. OBJECTIVES. At the completion of this Unit you should be able to

    1. Evaluate double integrals in polar coordinates.
    2. Represent areas and volumes using double integrals in polar coordinates.
    3. Sketch the region of integration for a given polar integral.
    4. Convert double integrals from rectangular to polar coordinates, and conversely.
    5. Evaluate triple integrals.
    6. Represent volumes and masses using triple integrals.

    You are not responsible for Sections 18.8-10 but physical scientists, in particular, would be well-advised to read over the material in these Sections.

2. PROCEDURES.

    1. SSP Section 18.5
    2. SSP Sections 18.6-7

3. TEXT READING. Sections 18.5-7

4. SUGGESTED EXERCISES. (Asterisk indicates solution given in Supplement.)

    p. 790: 1*, 7*, 12*, 14*
    p. 800: 3, 5-10, 15*, 18

5. SAMPLE QUIZ.

    1. Sketch the region of integration for $\int_0^{\pi/3} \int_{\cos\theta}^1 r\, dr\, d\theta$.

    2. Convert $\int_0^2 \int_0^y x\, dx\, dy$ to polar coordinates.

    3. Set up, but do not evaluate, the triple integral(s) needed to represent the mass of the solid bounded by $z = \sqrt{x^2 + y^2}$ and $z = \sqrt{9 - x^2 - y^2}$ if the density at each point is proportional to its distance from the y-axis.

    4. Rewrite $\int_0^1 \int_0^x \int_0^x dz\, dy\, dx$ as one or more triple integrals of the form

        $$\iiint \cdots dy\, dx\, dz.$$

1. OBJECTIVES. At the completion of this Unit you should be able to

    1. Evaluate line integrals over piecewise smooth curves.
    2. State and apply the Fundamental Theorem for Line Integrals.
    3. State and apply Green's Theorem.

2. PROCEDURES.

    1. SSP Sections 19.1-2
    2. SSP Section 19.3

3. TEXT READING. Sections 19.1-3

4. SUGGESTED EXERCISES.

    p. 827: 4, 7, 11, 16
    p. 832: 2, 5, 6, 11
    p. 836: 2, 6, 9, 12, 13, 16

5. SAMPLE QUIZ.

    1. Compute the line integral of $\overline{F}(x,y) = (x - y)\overline{i} + xy^2\overline{j}$ on the line segment from (1,2) to (2,6).
    2. Does the Fundamental Theorem for Line Integrals apply to the line integral problem 1? Explain your answer.

    3. Evaluate $\int_C xydx + (\frac{x^2}{2} + y + 3x)dy$ by Green's Theorem integrating over C counterclockwise, where C is the triangle with vertices (0,0), (1,0), (1,2).

UNIT 36

1. OBJECTIVES. At the completion of this Unit you should be able to

   1. Compute surface area.
   2. Evaluate surface integrals.
   3. Compute the flux of a vector field $\bar{v}$ across a smooth surface S.
   4. Define and find the divergence and curl of a vector field.
   5. State and apply the Divergence Theorem.
   6. State and apply Stokes' Theorem.

2. PROCEDURES.

   1. SSP Sections 19.4-5
   2. SSP Sections 19.6-7

3. TEXT READING. Sections 19.4-7

4. SUGGESTED EXERCISES. (Asterisk indicates solution given in Supplement.)

   p. 843: 2, 8*, 10*, 11*
   p. 850: 1f, 2c, 4a, 5b, 6d
   p. 854: 4, 8, 12*, 17, 19
   p. 860: 6, 16, 18, 21

5. SAMPLE QUIZ.

   1. Evaluate the flux for the vector field $\bar{v}(x,y,z) = x\bar{i} + 2y\bar{j} - 3z\bar{k}$ over the surface $z = x^2 + y^2$, $z \leq 1$. Take $\bar{n}$ as the upper unit normal.
   2. Compute the divergence for $\bar{v}(x,y,z) = x^2y\bar{i} + z\bar{j} + xyz^3\bar{k}$.
   3. Verify Stokes' Theorem on the upper half of the unit sphere $x^2 + y^2 + z^2 = 1$ with $\bar{v}(x,y,z) = x\bar{i} + 2y\bar{j} + z\bar{k}$.